VII International Symposium and Young Scientists School on Modern Problems of Laser Physics 2016

Journal of Physics: Conference Series Volume 793

Novosibirsk, Russia
22 - 28 August 2016

ISBN: 978-1-5108-3679-2
ISSN: 1742-6588

Printed from e-media with permission by:

Curran Associates, Inc.
57 Morehouse Lane
Red Hook, NY 12571

Some format issues inherent in the e-media version may also appear in this print version.

Printed by Curran Associates, Inc. (2017)

For permission requests, please contact the Institute of Physics at the address below.

Institute of Physics
Dirac House, Temple Back
Bristol BS1 6BE UK

Phone: 44 1 17 929 7481
Fax: 44 1 17 920 0979

techtracking@iop.org

Additional copies of this publication are available from:

Curran Associates, Inc.
57 Morehouse Lane
Red Hook, NY 12571 USA
Phone: 845-758-0400
Fax: 845-758-2633
Email: curran@proceedings.com
Web: www.proceedings.com

Table of contents

Volume 793

VII International Symposium and Young Scientists School "Modern Problems of Laser Physics"

22–28 August 2016, Novosibirsk, Russian Federation

Accepted papers received: 14 December 2016
Published online: 16 February 2017

Preface

Papers

Preface

General Information

This volume of the Journal is devoted to the VII International Symposium and Young Scientists School "Modern Problems of Laser Physics" (MPLP-2016)[1]. I was held in Novosibirsk Akademgorodok, Russia, 22–28 August 2016. Akademgorodok is the well-known Siberian Scientific Centre of Russian Academy of Sciences (RAS). It was founded in 1957. Since that time it is the place of many international scientific meetings, because it reflects the unique and fruitful symbiosis of many research institutions and Novosibirsk State University at one territory.

Since the first MPLP meeting in 1995 the Symposium usually gathers scientists from many countries, carrying out their investigations at the forefront of laser physics, quantum metrology and high-resolution spectroscopy, physics of ultracold atoms, molecules and ions, atom optics, ultrafast phenomena and attoscience, quantum optics and information, nonlinear optics and applications of laser radiation from THz to UV radiation ranges in medicine, geophysics, chemistry and microbiology. Traditionally the Symposium is the place where scientists can discuss new trends in modern laser physics, generate new ideas as well as initiate further collaborations.

Organizers

The Symposium was traditionally organized by the *Institute of Laser Physics of the Siberian Branch of Russian Academy of Sciences* [2] as a host organization in close collaboration with

- *Novosibirsk State University* [3]
- *Institute of Spectroscopy of the Russian Academy of Sciences* [4]
- *International Laser Center of M.V. Lomonosov Moscow State University* [5]
- *Federal State Unitary Enterprise "VNIIFTRI"* [6]

Sponsors

The organization of the Meeting would have been impossible without the financial support from our sponsors:

- *Federal Agency for Scientific Organizations* [7]
- *Russian Foundation for Basic Research* [8]
- *Novosibirsk Regional Fund of Support of Science and Innovative Activities* [9]
- *"Nauchnoe Oborudovanie" Group of Companies* [10]

[1] mplp2016.laser.nsc.ru
[2] www.laser.nsc.ru
[3] www.nsu.ru
[4] isan.troitsk.ru
[5] www.ilc.edu.ru
[6] www.vniiftri.ru
[7] fano.gov.ru
[8] www.rfbr.ru
[9] fondnid.ru
[10] www.spegroup.ru

Committees

1. International Advisory Committee

The Organizers thank all members of the International Advisory Committee for their valuable contributions to the Program of the meeting.

Sergei Arakelian	Vladimir State University, Vladimir, Russia
Viktor Balykin	Institute of Spectroscopy RAS, Troitsk, Russia
Nicolò Beverini	Dipartimento di Fisica, Universita' di Pisa, Italy
Razvan Dabu	National Institute for Nuclear Physics and Engineering, Bucharest-Magurele, Romania
Evgueni Dianov	Fiber Optics Research Center at GPI RAS, Russia
Martial Ducloy	Laboratoire de Physique des Lasers, Universite Paris-Nord, France
Elisabeth Giacobino	Laboratoire Kastler Brossel, Université Paris 6, France
Hidetoshi Katori	University of Tokyo, Science and Technology Agency, Japan
Sergei Kilin	B.I. Stepanov Institute of Physics NASB
John Kitching	National Institute of Standards and Technology (NIST), Boulder, USA
Vasiliy Klimov	P.N. Lebedev Physics Institute RAS, Moscow, Russia
Nikolai Kolachevsky	P.N. Lebedev Physics Institute RAS, Moscow, Russia
Leong Chuan Kwek	Centre for Quantum Technologies, National University of Singapore
Gerhard Leuchs	Universität Erlangen-Nürnberg, Institut für Optik, Germany
Filippo Levi	INRIM, Optic Division, Torino, Italy
Helen Margolis	National Physical Laboratory, Teddington, Middlesex, UK
Vitaly Pal'chikov	Scientific Research Institute "VNIIFTRI", Moscow region, Mendeleevo, Russia
Kyung Hyun Park	THz Photonics Creative Research Center, Electronics and Telecommunications Research Institute, Korea
Ekkehard Peik	Physikalisch-Technische Bundesanstalt (PTB), Braunschweig, Germany
Hélène Perrin	Université Paris 13, Villetaneuse, France
Ernst Rasel	Institute of Quantum Optics, Hanover, Germany
Philip Russell	Max Planck Institute for the Science of Light (MPL), Erlangen, Germany
Aleksandr Sergeev	Institute of Applied Physics RAS, Nizhny Novgorod, Russia
Anatoly Shalagin	Institute of Automation and Electrometry SB RAS, Novosibirsk, Russia
Feng Song	Nankai University, Tianjin, China
Aleksey Taichenachev	Institute of Laser Physics SB RAS, Novosibirsk, Russia
Sergei Turitsyn	Aston University, Birmingham, UK
Antoine Weis	University of Fribourg, Fribourg, Switzerland
Jun Ye	JILA, NIST and University of Colorado, Boulder, USA
Victor Zadkov	Institute of Spectroscopy RAS, Troitsk, Russia

2. Local Organizers

Chairman of the Symposium:

Prof. Sergei N. Bagayev

Academician

Scientific Director of the Institute of Laser Physics SB RAS

Organizing Committee:

Dr. Vladimir I. Denisov

Chairman of the Local Committee

Deputy Director of the Institute of Laser Physics SB RAS

Prof. Alexey V. Taichenachev

Chairman of the Local Program Committee

Corresponding Member of the Russian Academy of Sciences, Director of the Institute of Laser Physics SB RAS

Dr. Pavel V. Pokasov

Vice-Chairman of the Local Committee

Scientific secretary of the Institute of Laser Physics SB RAS

Dr. Denis V. Brazhnikov

Scientific Secretary of the Symposium

Senior Researcher of the Institute of Laser Physics SB RAS

Alexander M. Shilov

Head of the School for Young Scientists of MPLP-2016

Researcher of the Institute of Laser Physics SB RAS

Elena S. Razum

Chairman of the Financial Committee

Deputy director of the Institute of Laser Physics SB RAS

Yury T. Shestakov

Chairman of the Technical Committee

Deputy director of the Institute of Laser Physics SB RAS

Invited Speakers

Name	Organization	Title of a talk
Anton Afanasiev	Institute of Spectroscopy RAS, Troitsk, Russia	Single photon transport by a moving atom
Alexey Akimov	Texas A&M University, USA	Collisions in ultra-cold thulium atoms
Alexander Apolonskiy	Ludwig-Maximilians-Universität München, Garching, Germany	21 century mid-infrared biomedical absorption spectroscopy: conventional FTIR vs Laser Time Domain
Sergei Arakelian	Vladimir State University, Vladimir, Russia	Laser-induced semiconductor fractal structures with topological quantum effects
Sergei Babin	Institute of Automation and Electrometry SB RAS, Novosibirsk, Russia	New schemes and regimes of Raman fiber lasers
Jacopo Belfi	Physics Department, University of Pisa, Italy	Laser gyroscopes and their applications in fundamental physics, in metrology and in seismology
Silvia Bergamini	The Open University, Milton Keynes, UK	Rydberg interactions for quantum-enhanced protocols with mixed states
Dmitry Churkin	Novosibirsk State University, Novosibirsk, Russia	Spatio-temporal dynamics in fibre lasers
Boris Dubetsky	Hallandale, USA	Atom interferometers' phases at the presence of heavy masses. Their use to measure Newtonian gravitational constant: Optimization, Error model, Perspectives.
Bess Fang	LNE-SYRTE, Observatoire de Paris, Paris, France	Continuous cold-atom inertial sensor with 1 nrad/s rotation stability
Guoying Feng	Institute of Laser & Micro/Nano Engineering, Sichuan University, Sichuan, P.R.China	Random lasing based on doped nanocrystals
Andrey Goncharov	Institute of Laser Physics SB RAS, Novosibirsk, Russia	An optical frequency standard based on ultra cold magnesium atoms
Vladimir Haisler	A.V. Rzhanov Institute of Semiconductor Physics SB RAS, Novosibirsk, Russia	Vertical-cavity surface-emitting lasers for chip-scale atomic clocks
John Harvey	University of Auckland, Auckland, New Zealand	Recent developments in femtosecond fibre lasers
David Hutchinson	Department of Physics, University of Otago, New Zealand	Ultracold atoms for simulation of many body quantum systems
Stepan Ignatovich	Institute of Laser Physics SB RAS, Novosibirsk, Russia	Yb:YAG/I_2 optical frequency standard at 515 nm with instability at the level 10^{-15}
Boris Knyazev	Novosibirsk State University and Budker Institute of Nuclear Physics SB RAS, Novosibirsk, Russia	Recent advances in the terahertz photonics and spectroscopy at Novosibirsk free electron laser
Hans-Christian Koch	Paul Scherrer Institut, Villigen, Switzerland	Atomic magnetometry for the neutron electric dipole moment experiment at Paul Scherrer Inst.
Nikolay Kolachevsky	P.N. Lebedev Physical Institute RAS,	Cryogenic optical cavities for laser frequency

	Moscow, Russia	stabilization
Dmitry Kolker	Institute of Laser Physics SB RAS & Novosibirsk State University, Novosibirsk, Russia	Wide tunable OPO at MID-IR spectral region pumped by Q-switch Nd:YAG laser
Andrey Kolovsky	L.V. Kirensky Institute of Physics SB RAS, Krasnoyarsk, Russia	Wave-packet dynamics of ultra-cold atoms in 2D optical lattices subject to synthetic electric and magnetic fields
Leong Chuan Kwek	Centre for Quantum Technologies, National University of Singapore, Singapore	Hybrid quantum system: Superconducting resonator-Rydberg system
Thomas Legero	Physikalisch-Technische Bundesanstalt (PTB), Braunschweig, Germany	Ultrastable, 10 mHz linewidth lasers based on cryogenic silicon resonators
Gerhard Leuchs	Max Planck Institute for the Science of Light, Erlangen-Nuremberg, Germany	The quantum vacuum as a dielectric
Kun Liu	Tianjin University, Tianjin, China	Fiber-optic pressure and temperature sensor
Evgueni Martynovich	Irkutsk Branch of Institute of Laser Physics SB RAS, Irkutsk, Russia	Investigation of single defects created in crystals by laser emission and hard radiation
Gennadii Matvienko	V.E. Zuev Institute of Atmospheric Optics SB RAS, Tomsk, Russia	The interaction of intensive femtosecond radiation with atmospheric media
Alexander Mayorov	Institute of Laser Physics SB RAS, Novosibirsk, Russia	Application of laser technologies in manufacture of elements of cardiovascular prosthesis
Gaetano Mileti	University of Neuchâtel, Neuchâtel, Switzerland	Double-resonance spectroscopy in Rubidium vapour-cells for high performance and miniature atomic clocks
Alexander Nevsky	Heinrich-Heine-University of Dusseldorf, Germany	An ultra-stable silicon cryogenic optical resonator
Boris Nyushkov	Institute of Laser Physics SB RAS, Novosibirsk, Russia	Fiber-based femtosecond optical frequency comb stabilized to iodine frequency standard
Maxim Okhapkin	Physikalisch-Technische Bundesanstalt (PTB), Braunschweig, Germany	Search for the low-energy isomer in ^{229}Th
Vitaly Ovsiannikov	Voronezh State University, Voronezh, Russia	Higher-order effects on uncertainties of clocks of Mg atoms in an optical lattice
Kyung Hyun Park	THz Photonics Creative Research Center, Electronics and Telecommunications Research Institute, Seoul, Korea	Role of photonics in terahertz technologies for industrial applications
Hélène Perrin	Université Paris 13, Villetaneuse, France	Probing superfluidity in a quasi two-dimensional Bose gas through its local dynamics
Alexander Plekhanov	Institute of Automation and Electrometry SB RAS, Novosibirsk, Russia	Spaser as novel versatile biomedical tool
Alexander Razhev	Institute of Laser Physics SB RAS, Novosibirsk, Russia	Pulsed UV laser technologies for ophthalmic surgery
Ernst Rasel	Institute of Quantum Optics, Hanover, Germany	Optical spectroscopy of atomic Bloch bands
Valentin Rudenko	M.V. Lomonosov Moscow State University, Sternberg Astronomical Institute	Cryogenic opto-acoustical gravitational wave antenna (Cryo-OGRAN)

Igor Ryabtsev	A.V. Rzhanov Institute of Semiconductor Physics SB RAS, Novosibirsk, Russia	Controlling the interactions between cold Rydberg atoms by a time-varying electric field
Alexander Shkurinov	M.V. Lomonosov Moscow State University, Moscow, Russia	Gas, plasma and nano-cluster medium for generation of intense THz radiation
Sergei Slyusarev	VNIIFTRI, Mendeleevo, Moscow Region, Russsia	Optical frequency standard based on cold strontium atoms
Igor Sokolov	Peter the Great St. Petersburg Polytechnic University, St. Petersburg, Russia	Light localization in cold and dense atomic ensemble
Feng Song	Nankai University, Tianjin, P.R. China	Influence of energy transfer upconversion on the thermal deposition in Nd doped lasers
Aleksey Taichenachev	Institute of Laser Physics SB RAS, Novosibirsk, Russia	Recent advances in precision spectroscopy of ultracold atoms and ions
Vladimir Trunov	Institute of Laser Physics SB RAS, Novosibirsk, Russia	New trends in ultrahigh intensity coherent beam combining
Andrey Turlapov	Institute of Applied Physics RAS, Nizhniy Novgorod, Russia	Near-field interference in a chain of fluctuating Bose condensates
Ken-ichi Ueda	Institute for Laser Science, University of Electro-Communications, Tokyo, Japan	Challenge for thermal-lens-free ceramics lasers
Ichiro Ushijima	Center for Advanced Photonics, RIKEN, Saitama, Japan	Cryogenic optical lattice clocks towards an uncertainty of sub-10^{-18} level
Sergey Vyatchanin	M.V. Lomonosov Moscow State University, Moscow, Russia	Speed meter based on dissipative coupling
Antoine Weis	University of Fribourg, Fribourg, Switzerland	Characterizing and imaging magnetic nanoparticles by optical magnetometry
Martin Zeppenfeld	Max Planck Institute of Quantum Optics, Garching, Germany	Controlled molecular ensembles at cold and ultracold temperatures
Alexey Zheltikov	M.V. Lomonosov Moscow State University, Moscow, Russia	Nonlinear optics in the mid-infrared; a new morning
Jean-Jacques Zondy	School of Sciences & Technology, Nazarbaev University, Astana, Kazakhstan	6 microns generation of mid-infrared light using orientation-patterned gallium phosphide

School for Young Scientists

Also the School for Young Scientists was organized as a part of the Symposium. Several top-ranking scientists delivered their lectures on “hot” topics of modern laser physics and applications:

Prof. Sergey Babin [1]

Institute of Automation and Electrometry SB RAS, Novosibirsk, Russia

Prof. Valentin Rudenko [2]

P.K. Sternberg Astronomical Institute, M.V. Lomonosov Moscow State University, Moscow, Russia

Prof. Nicolò Beverini [3]

Department of Physics, University of Pisa, Pisa, Italy

Prof. Boris Vainer [4]

Rzhanov Institute of Semiconductor Physics SB RAS & Novosibirsk State University, Novosibirsk, Russia

Prof. David Hutchinson[5]

University of Otago, National Centre of Research Excellence (Dodd Walls Centre), New Zealand

Prof. Jean-Jacques Zondy [6,7]

Nazarbaev University, School of Science and Technology, Astana, Kazakhstan

Prof. Boris Knyazev [8]

Novosibirsk State University &

Budker Institute of Nuclear Physics SB RAS, Novosibirsk, Russia

List of lectures:

[1] *New technologies in fibre optics*

[2] *First registration of gravitational waves by using the large laser interferometers with suspended mirrors*

[3] *Sagnac effect and gyroscopes*

[4] *Lasers and infrared thermography: harmony, mutual assistance and reciprocal gain*

[5] *A brief history of Bose-Einstein condensation of ultracold gases*

[6] *Continuous-wave Optical Parametric Oscillators: Theory and Applications*

[7] *Optimally-coupled cw Intracavity Second-Harmonic Generation in solid-state lasers ring lasers: Observation of cascaded Kerr-lens modelocking*

[8] *Beams with angular orbital momentum: a step into the terahertz range*

Results

About 160 participants took part at the MPLP-2016 from 13 countries: China, France, Germany, India, Italy, Japan, Kazakhstan, New Zealand, Russia, Singapore, South Korea, Switzerland and the United Kingdom. The Symposium Program was consisted of invited (30 min), oral (15 min) and poster presentations.

Besides the scientific sessions and lectures, all the participants had opportunity to participate in wide cultural events during the Meeting. Also some of them took the interesting journey to the Altai Republic, where they were able to prolong communication in informal atmosphere among beautiful Altai nature.

Figure 1. Group photo of the participants.

Figure 2. Prof. Mileti gives a talk about the high-performance rubidium atomic clock.

Figure 3. Cultural trip to the lake "Teletskoe" (Altai Republic).

Meeting of the International Advisory Committee was held on Wednesday, August 24. One of the results of that meeting was to continue the tradition of MPLP symposia and to organize the next one in 2018 or 2019.

The Organizers would like to thank participants of the MPLP-2016 for their valuable contributions to success of the Symposium and their interest. We also appreciate the help of all persons, who were involved into the organizing process.

We are looking forward to future fruitful and interesting MPLP meetings!

The Editors

Peer review statement

All papers published in this volume of *Journal of Physics: Conference Series* have been peer reviewed through processes administered by the proceedings Editors. Reviews were conducted by expert referees to the professional and scientific standards expected of a proceedings journal published by IOP Publishing.

IOP Conf. Series: Journal of Physics: Conf. Series **793** (2017) 012001 doi:10.1088/1742-6596/793/1/012001

Single photon transport by a moving atom

A E Afanasiev[1], P N Melentiev[1], A A Kuzin[1,2], A Yu Kalatskiy[1,2] and V I Balykin[1]

[1] Institute for Spectroscopy, Russian Academy of Sciences, 5 Phizicheskaya str., Troitsk 108840, Moscow, Russia

[2] Moscow Institute of Physics and Technology, 9 Institutskiy per., Dolgoprudniy 141700, Moscow region, Russia

E-mail: afanasiev@isan.troitsk.ru

Abstract. The results of investigation of photon transport through the subwavelength hole in the opaque screen by using single neutral atom are represented. The basis of the proposed and implemented method is the absorption of a photon by a neutral atom immediately before the subwavelength aperture, traveling of the atoms through the hole and emission of a photon on the other side of the screen. Realized method is the alternative approach to existing for photon transport through a subwavelength aperture: 1) self-sustained transmittance of a photon through the aperture according to the Bethe's model; 2) extra ordinary transmission because of surface-plasmon excitation.

It follows from basic principles of wave physics that the passage of a wave through a hole that is considerably smaller than the wavelength can be neglected. In the classical work by Bethe [1] on transmission of light by a nanohole in an infinitely thin and perfectly conducting screen, a simple expression for the transmission efficiency of light has been obtained, which is scaled in relation to the hole size as $(r/\lambda)^4$, where r is the radius of the hole, and λ is the wavelength. Under assumptions made on the screen, the transmission efficiency falls rapidly when the wavelength becomes greater than the hole radius. Ebbesen and collaborators [2] discovered the Extraordinary Optical Transmission (EOT) through sub-periodic hole array. The EOT is mediated by the aid of electromagnetic surface modes supported by the holey surfaces [3]. The majority of researchers agree that the central role in this phenomenon is played by surface waves, such as *surface plasmons*.

Here we demonstrate a new physical approach for an effective light transmission through nanohole. It is based on the photon transport that involves the participation of a particle other than a plasmon, namely, a *neutral atom*. In this scheme, a *single atom* transfers a *single photon* through a nanohole. The proposed scheme is an another mechanism of photon's transport through the nanohole which supplement existing [1, 2]. Besides the using of a new particle for photon transport it opens up a new possibilities for surface science. It is possible to use such scheme for investigation of van der Waals interaction because of atom-surface interaction [4]. Another application is for atom-plasmon interaction investigation. Indeed the subwavelength hole is a highly nonlinear plasmonic element. So, the interaction of excited atom with such structure opens a new way for tailoring the spectral properties of materials [5].

The basic idea of the photon transport by a moving atom is presented in figure 1 [6]. An atom moving toward a metal screen with a hole absorbs a photon of laser radiation immediately in front of the hole. If the lifetime of the excited atom is substantially larger than the time of

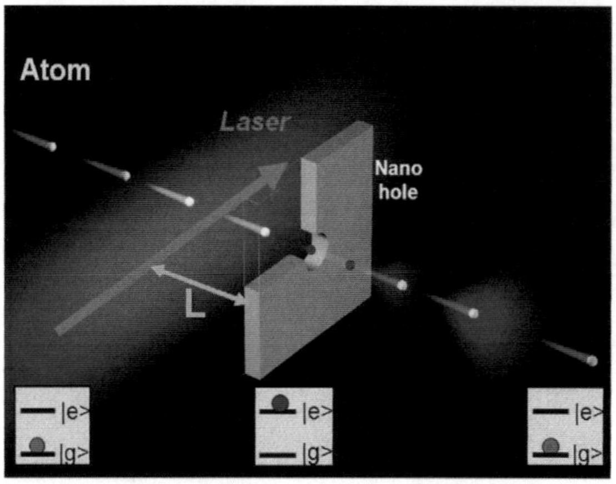

Figure 1. Scheme of a photon transport by atom through nanohole.

flight of the atom through the nanohole (in a real experiment, the nanochannel), the transition of the atom from the excited state to the ground state with emission of a photon can occur on the other side of the screen, which means the transfer of the energy of the photon through the nanohole.

At the wavelength $\lambda = 800$ nm and the nanohole radius $r = 50$ nm, the ratio of the probability of passage of a photon involving the participation of an atom to the probability of passage of a photon alone is $\eta \sim 2 \times 10^4$. The physical reason for such high photon transfer efficiency is the reduction of the "single photon wave packet"due to its absorption by the atom and, as a result, its localization in a volume less than the wavelength and the nanohole size. In principle this scheme allows to transform "single-photon in single-mode wave packet"of the laser light into a "single-photon but multimode wave packet"in free space.

The experiment was performed with *Rb* atoms. A beam of *Rb* atoms is directed to the sample with nanoholes. Atoms were excited into a long-lived excite state $5D_{5/2}$ at the transitions $5S_{1/2} \rightarrow 5P_{3/2}$ and $5P_{3/2} \rightarrow 5D_{5/2}$. The decay of the $5D_{5/2}$ state via the channel $5D_{5/2} \rightarrow 6P_{3/2} \rightarrow 5S_{1/2}$ with the emission of a photon at a wavelength of 420 nm takes place with a characteristic lifetime of 500 ns. Such a long lifetime makes it possible for the atom to transfer the photon energy over a distance of about 150 μm. This value suffices to ensure the flight of the atom in the excited state through the nanohole. The probability of emission of a blue photon by the atom is about 2%. The detection of blue photon gives an evidence of atom passage through the hole in the excited 5D state. As a consequence, this fact gives an evidence of photons transfer at 776 nm and 780 nm wavelengths (through the $5D_{5/2} \rightarrow 6P_{3/2} \rightarrow 5S_{1/2}$ decay channel). Our calculations show that for one detected blue photon there are about 20 photons at 776 nm and 780 nm.

The photon transfer efficiency depends on the nanohole size, the material of the screen, and the velocity and the scheme of energy levels of the atom. At small sizes of the nanohole, the photon transfer efficiency decreases substantially because of the interaction of the excited atom with the surface. As a result the surface of the nanohole in the screen causes deexcitation of the atomic state. The described scheme of the atom interaction with the surface offers opportunities to study quantum friction [7] and strength of atom-surface interaction [8].

This work was supported by the Russian Science Foundation (project no. 14-12-00729).

IOP Conf. Series: Journal of Physics: Conf. Series **793** (2017) 012001 doi:10.1088/1742-6596/793/1/012001

References

[1] Bethe H A 1944 Theory of diffraction by small holes *Phys. Rev.* **66** 163

[2] Ebbesen T W, Lezec H J, Ghaemi H F, Thio T and Wolff P A 1998 Extraordinary optical transmission through sub-wavelength hole arrays *Nature* **391** 667

[3] GarcaVidal F J, Martn-Moreno L, Ebbesen T W and Kuipers L 2010 Light passing through subwavelength apertures *Rev. Mod. Phys.* **82** 729

[4] Bloch D and Ducloy M 2005 Atom-wall interaction *Advances In Atomic, Molecular, and Optical Physics* **50** 91–154

[5] Aljunid S A, Chan E A, Adamo G, Ducloy M, Wilkowski D and Zheludev N I 2016 Atomic response in the near-field of nanostructured plasmonic metamaterial *Nano Lett.* **16** 3137

[6] Afanasiev A E, Melentiev P N, Kuzin A A, Kalatskiy A Yu and Balykin V I 2016 Photon transport through a nanohole by a moving atom *New J. Phys.* **18** 053015

[7] Scheel S and Buhmann S Y 2009 Casimir-Polder forces on moving atoms *Phys. Rev.* A **80** 042902

[8] Persson B N J and Lang N D 1982 Electron-hole-pair quenching of excited states near a metal *Phys. Rev.* B **26** 5409–15

Jump electroconductivity in the laser deposited nanoclustered structures

A Antipov, A Shagurina, A Osipov, A Istratov, I Skryabin and S Arakelian

Stoletovs Vladimir State University, 87 Gorky str., Vladimir 600000, Russia

E-mail: arak@vlsu.ru

Abstract. The quantum states verification in cluster semiconductor/metallic structures by jump/tunneling electroconductivity and possible mechanisms for their implementation are considered in experiment and theory. By our laser ablation technique we have nanostructurized the films for which the ability to control the change in their electrical properties does exist by variation of the topology for the system. The granular conductivity specificity has been under study. The current-voltage characteristics behavior has been measured for a nanocluster bimetallic film (Au+Ag), and the experiments for multilayer bimetal thin films of the different composition have been carried out. Two associated mechanisms for electroconductivity occur in the case, i.e. tunnel transition for electrons and electron activation in the frames of the shell model for a cluster system, in dependence on the nanostructure topology.

1. Introduction

The physical properties of nanocluster systems are very sensitive to the form, size and distance between their composing elements. The fact is well known for any material in general, but to change these fixed lattice parameters and to carry out the stable conditions for the ordinary solid state object/crystal structure we need both to put the object under extremely high pressure ($\gtrsim 10^6$ atm) and to work in a low (liquid He) temperature range ($\lesssim 30$ K) [1].

In contrast, nanocluster structures can be easily modified both in the necessary direction and by the controlled way in femto- nanophotonics experiments. The variation of the above enumerated topology parameters can result in a new type of correlation/quantum states for charge particles [2]. Moreover, electronic energetic bands of the materials can vary dramatically in the case, resulting in a new physical behavior of the system in general.

In the problem of superconductivity the question is, how to fabricate the coupling states (around the forbidden band) at high (nitrogen) temperature ($\gtrsim 140$ K) for charge particles being responsible for electroconductivity. In fact, the superconductivity enhancement temperature may be the result of the mechanism reducing to a temporary change of atom positions, e.g. for $YBa_2Cu_3O_6$/YBCO – under high intensity IR-laser radiation, and more preferable – by X-ray [3].

Thus, the topology mechanisms may play a principal role in the physical electrical property of nanostructures. The fact is true for nanoclusters size ~10 nm and their density ~10^{12} cm^{-2} [4].

In the present paper we give the results of our experiments for the quantum states verification in nanocluster structures due to jump/tunneling electroconductivity. The possible mechanisms of its development are discussed by us in the frame of both tunnel coupled electron system and shell-like model for electron energetic levels in nanoclusters, for which Bohr radius in comparable to the cluster size.

The initial colloidal systems were obtained by continuous laser irradiation of silver and gold targets. The targets were placed in deionized water [5]. The average size of the obtained nanoparticles was about 50 nm and 10 nm, while their concentration was 1 μg/ml . The particle size was controlled by the dynamic light scattering.

For the deposition of the bimetallic structures, the colloidal systems of both metals were mixed in equal proportions. The glass substrate was placed in the mixed solution. The deposition of thin film was made by laser irradiation as described in [6–8]. As a laser source we used Yb-fiber laser (λ=1.06 μm) with pulse duration of 100 ns and with laser repetition rate of 20 kHz. The average power was from 1.5 W up to 4.5 W, the diameter of the laser beam was around 5 μm. The array of nanoparticles was formed on the substrate surface by repeated (up to 10 times) laser beam scanning along one and the same direction. The scanning speed was varied from 0.6 mm/s up to 1.5 mm/s. The obtained in this way array of nanoparticles has the length of about 100 μm, the average transverse dimension was about 5 μm and the average height was about 55 nm.

2. Electroconductivity in nanostructures: experiments and brief interpretation

We will consider below the quantum mechanism of electroconductivity for two models: tunneling electron transport and/or shell-like structure with transition from coupled electron state to free one, being associated with valence/conductivity bands.

Experimental results for voltage (U) and ampere (I) behavior and electroresistance (R) for bimetallic system (Au+Ag) are shown in figure 1. They demonstrate the data for different thickness of the bimetal (Au+Ag) film (figure 1a) and concentration of Ag-substance in (Ag+Ni) system (figure 1c), consequently. We can see the jumps on the obtained I(U) dependences for different thickness. For three cases we show AFM-images for the nanostructures under the measurements (figure 1b). A more detailed behavior of the dependences for electroresistance R vs nanoparticle size near the initial part of the displayed dependences is presented in insert sections in the right part of figure 1a.

It is interesting that the dependence in figure 1a for R may be presented as the analogue of Kondo dependence on the temperature for R(T) in contrast with our case for the dependence R(h) on h. The theoretical description for jump conductivity (figure 1b) will be presented in the next division.

Computer simulation results are shown in figure 2. We carried out the computer simulation when a single electron conductivity was modelled for the metallic granular film electroconductivity due to tunneling effect.

We used the algorithm procedures taking into account: granular film growth; neighborhood definition; scaling/transition from computer data to physical parameters; tunneling processes for electron transport; current magnitude/volt-ampere characteristics. The results are shown in figure 2. The control parameters for calculation: F – deposition rate (by speed) being equivalent to T – arbitrary unit (a.u.) time, R – particle radius, S – distance between neighboring particles and/or the different order distance for the particle location/subsequent localization of the particles over the spatial distribution, H – average thickness of the surface layer/film, W – standard deviation for the thickness value.

To have the reasonable results in comparison with our experiments we can manipulate few adjustable parameters (see figure 2a): monolayer F=0.4; R=2.5 nm; S=10 nm; T=2 a.u., H=12 nm and fixed W=7.5556; T=4 a.u., H=24 nm and fixed W=11.9458. The variation of dependences vs different W, can also be shown in figure 2b: H=24 nm, R=2.5 nm, S=10 nm; T=2 a.u., monolayer F=0.8, W=6.9702 nm; T=4 a.u., monolayer F=0.4, W=11.8588 nm; T=8 a.u., monolayer F=0.2, W=13.6342 nm.

Thus, there is a qualitative agreement between the computer simulation approach and experimental data (cf. figures 2, 3). Because of flexible experimental conditions, the adjustable parameters can be introduced both in experiment and computer simulation consistently; so, the coincidence of the results (at least qualitative) is not artificial.

Tunnel coupled electroconductivity. According to the formulas of elementary quantum mechanics textbook (see e.g. Abrikosov, 2009) the quantum effects in nanoparticles are determined by the band gap value (E_g) vs dimension (size/distance R for the particles):

$$E_g(R) = E_g(\infty) + \frac{\hbar^2\pi^2}{2R^2}\left(\frac{1}{m_e} + \frac{1}{m_h}\right) - \frac{1.8e^2}{\varepsilon R}$$

where m_e and m_h – effective mass of electron and hole, ε – optical dielectric constant of solid state, $E_g(\infty)$ – band gap of monolith sample, second term – dimension contribution, last one – Coulomb forces.

Then, the current density j for electron transport by tunneling mechanism in cluster system can be presented as a formula [9]:

$$j \propto U\sqrt{\frac{E}{k_BT + 2eUd/L}exp\left(\frac{eUd/L - E}{k_BT}\right)}$$

where U – potential difference between macrocontacts, E – energy for particles transfer from infinity/ionization energy, e – electron charge, d – average distance between neighboring clusters, L – length of conductive layer between macrocontacts, k_B – Boltzmann constant, T – temperature.

According to the approach we counted the resistivity (R_{cal}), and compared it with our experimental results (R_{meas}) in a large range of values for Au as an example.

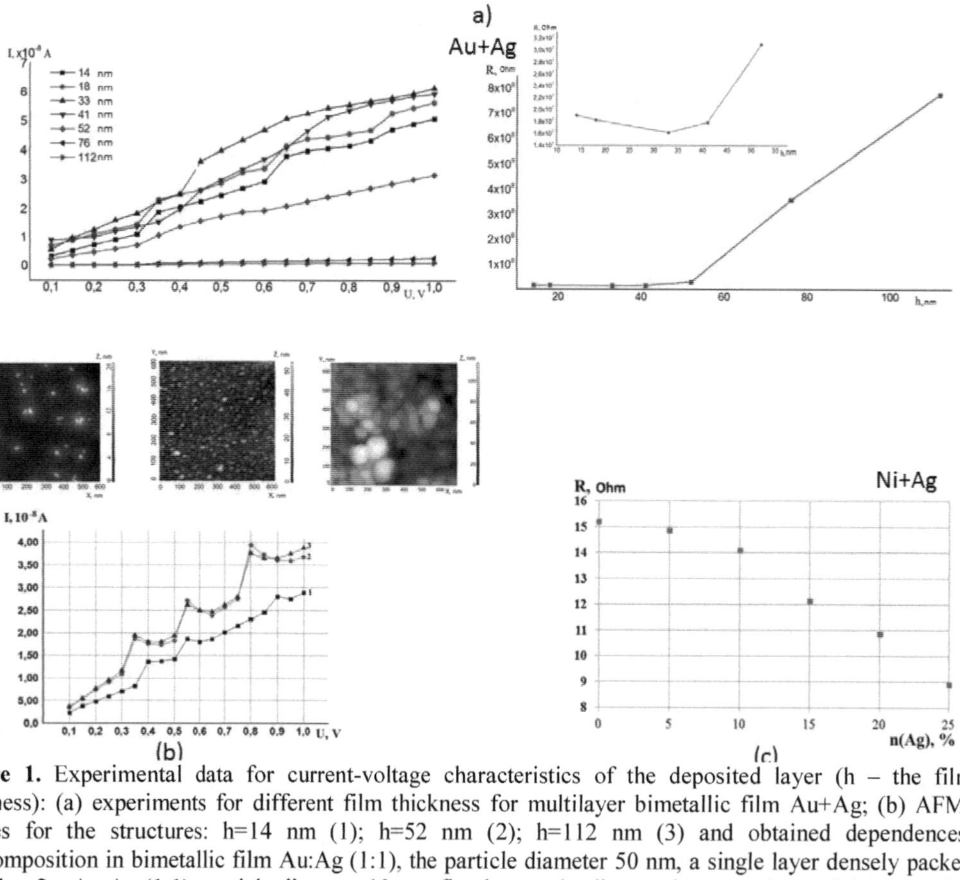

Figure 1. Experimental data for current-voltage characteristics of the deposited layer (h – the film thickness): (a) experiments for different film thickness for multilayer bimetallic film Au+Ag; (b) AFM-images for the structures: h=14 nm (1); h=52 nm (2); h=112 nm (3) and obtained dependences; 1 – composition in bimetallic film Au:Ag (1:1), the particle diameter 50 nm, a single layer densely packed particles; 2 – Au:Ag (1:1), particle diameter 10 nm, five layers, the distance between the particles 4 nm; 3 – Au:Ag (1:1), particle diameter 10 nm, five layers, the distance between the particles 2 nm; (c) electroresistivity R of the array of nanoparticles (Ni+Ag) vs mass concentration of Ag nanoparticles.

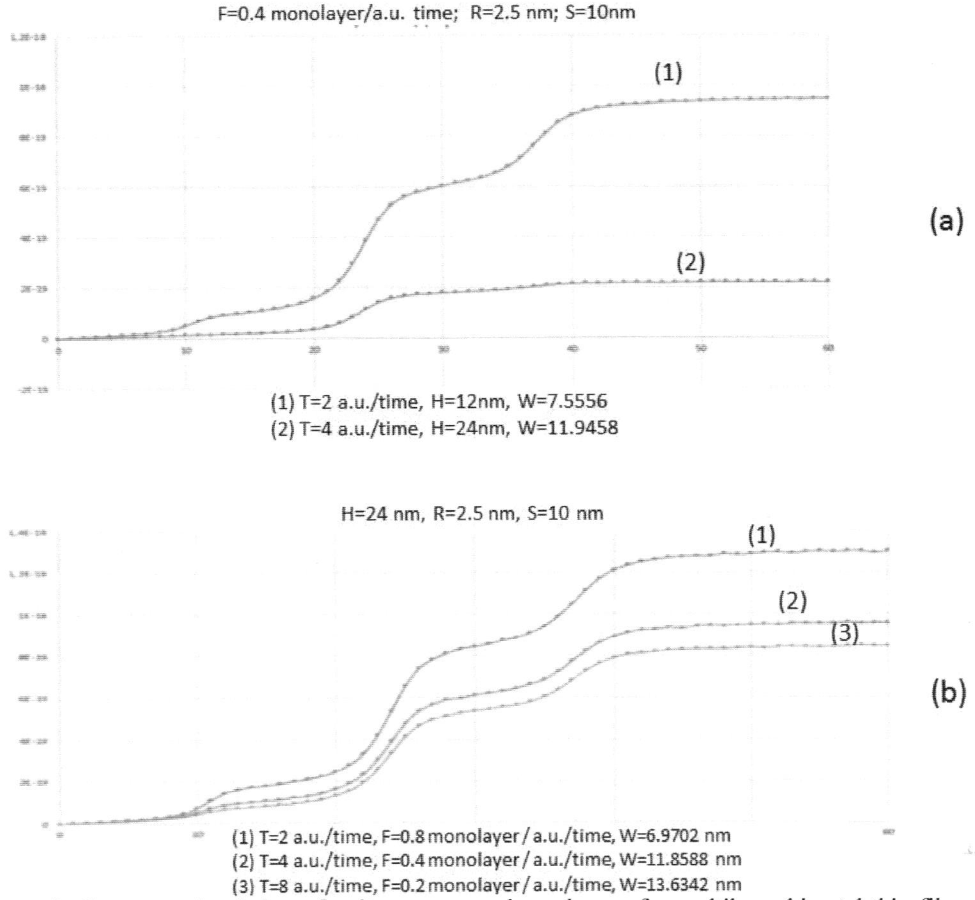

Figure 2. Computer simulation of voltage-ampere dependences for multilayer bimetal thin films of different parameters.

The results are shown in figure 3 for different topology of nanostructures. In general there is a good coincidence meanwhile we take into account both different types of nanostructures and a large range of values. Moreover, the best coincidence is the for the last case: $R_{calc}=5.5\cdot10^{7}$ Ohm and $R_{meas}=3.6\cdot10^{7}$ Ohm.

Thus, we can nanostructurize the film for which the ability to control the change of its electrical resistance is realized by the topology variation of the system [10]. The problem is to find the optimal relationship between some spatial parameters.

In fact, earlier we obtained in experiment that the tunnel transition for electrons is effective for $a_{cluster}$ =120 nm and Λ=8 nm for the semiconductor PbTe surface nanostructurized sample [11]. On the other hand, the manifestation of some rather unusual size effects in granular thin films prepared by the pulsed laser deposition technique (when an average grain size becomes comparable with the thermal de Broglie wavelength) was discussed in the paper [10]. This gives us, in fact, the opportunities to make functionally variable electronic devices being modified in the right direction. The experiments carried out under different conditions (figure 1) by using a few experimental adjustable parameters in correspondence with the computer simulation procedure (figure 2), are necessary for that as it was mentioned above.

For the cluster structures considered by us the tendency to superconductivity can be obtained due to the boundary topology of the ordered clusters. In our previous experiments we observed the decrease of the resistance by several times (at the level of several units of Ohms) for Ni-nanostructure by the same reason [11] when nanocluster size decreased from 100 nm to 10 nm. Some explanations were given in the frame of surface fractal structures with dominant contribution of topological factors.

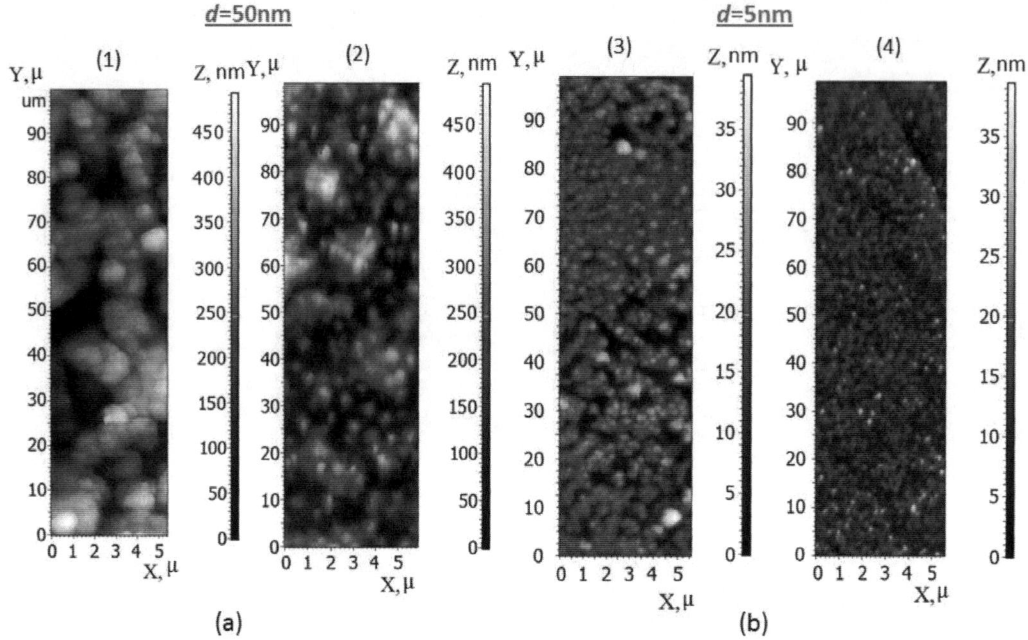

Figure 3. Measurement electroresistance vs nanoparticle size (*d*) and topology (in a large range of values):

(a) d=50 nm: (1) $R_{meas}=6.1\cdot10^{11}$ Ohm, $R_{calc}=2.8\cdot10^{11}$ Ohm,
(2) $R_{meas} = 8.3\cdot10^{10}$ Ohm, $R_{calc}=1.9\cdot10^{10}$ Ohm;
(b) d=5nm: (3) $R_{meas} = 5.4\cdot10^{8}$ Ohm, $R_{calc}=9.3\cdot10^{8}$ Ohm,
(4) $R_{meas} = 3.6\cdot10^{7}$ Ohm, $R_{calc}=5.5\cdot10^{7}$ Ohm.

3. Quantum activation mechanism for electroconductivity

In experiment, with different topology structures, we observed the competition between: (1) the increase of conductivity while opening new channels and (2) the increase of resistance by expanding the area for the grain/cluster. In addition, (3) the thermal activation and/or variation of the potential barrier value also works due to intrinsic electric field in the structure. The last process can result in the current value enhancement in the formed island films by jump effect. We are going now to discuss the effect of electron conductivity in a shell-like structure of the cluster.

The reasonable model for jump conductivity (see figure 1b) in such granular film structure may be associated with the mixture of metal and insulator [12]. This is equivalent to our case due to the existence of the distance between clusters which for a large value results in high electroresistance. In fact, we have a localization of charges in space ~ cluster size *a*, and electrons locked inside a nanoscale cluster (with multi-electron states in the shell model of energetic levels).

Now we mention an activated contribution to the jump conductivity associated with the electron transition mechanism, i.e. activation of electrons in up (delocalized) Habbard band from both down Habbard band in «insulator» and metallic band (energy interval E_a). Such electron jumps occur due to different transitions of electrons between discrete energetic states within each localized zone (cf. [12, 13]).

Thus, the modification of a potential well due to the topology of nanostructures can result in the effect with additional contribution to jump conductivity for discrete energetic levels for electrons.

4. Conclusion

In the paper the electroconductivity for the laser-induced surface and cluster/thin films nanostructures with controlled topology have been under study.

The quantum state verification in cluster structures obtained in experiment, i.e. jump/tunneling electroconductivity, and some possible mechanisms of explanation for nanocluster systems were introduced. Our experiments and theoretical discussions may be reasonable for the design the variable photonic devices in hybrid schemes with light and electrons [14].

Thus, relying on the fundamental physical effects which demonstrate the dimension like behavior in nanostructures we can hope to create (under laser induced modification of the material surface and thin films) the next generation of hybrid schemes of photonics and optoelectronics on new physical principles.

Acknowledgments

This study was performed within the design part of the government contract for research work with VIGU (no. 16.440.2014/K) and partly supported in by the Russian Foundation for Basic Research (project 16-42-330461).

References

[1] Drozdov A P, Eremets M, Troyan I A, Ksenofontov V and Shylin S I 2015 *Nature* **525** 73–76
[2] Arakelian S M, Kucherik A O, Prokoshev V G, Rau V G, Sergeev A G 2015 *Introduction to Femtonanophotonics: Fundamental and Laser Methods of Controlled Fabrication and Diagnostics of Nanostructured Materials*, ed S M Arakelyan (Moscow: Logos Publ.) p 744
[3] Cavalleri A *et al.* 2014 Superconductivity without cooling. Research NEWS. Max Plank Institute
[4] Dragunov V P, Neizvestnyi I G, Gridchin V A 2006 *Fundamentals of Nanoelectronics* (Moscow: Logos Publ.) p 496
[5] Arakelyan S M, Veiko V P, Kutrovskaya S V *et al.* 2016 *J. Nanopart. Res.* **18** 155
[6] Antipov A A, Arakelyan S M, Vartanyan T A, Itina T E, Kutrovskaya S V, Kucherik A O, Sapegina I V 2015 *Optics and Spectroscopy* **119** 119
[7] Antipov A A, Arakelyan S M, Kutrovskaya S V *et al.* 2012 *Bulletin of the Russian Academy of Sciences: Physics* **76** 611
[8] Abramov D V, Antipov A A, Arakelian S M *et al.* 2014 *Laser Phys.* **24** 074010
[9] Kozhevin V M, Yavsin D A, Smirnova I P *et al.* 2003 *Physics of the Solid State* **45** 1993
[10] Sergeenkov S, Cichetto L, Diaz J C C A *et al.* 2016 *J of Physics and Chemistry of Solids* **98** 38
[11] Arakelian S, Emel'yanov V, Kutrovskaya S *et al.* 2016 *Opt. Quant. Electron.* **48** 342
[12] Demishev S V, Pronin, A A, Glushkov V V *et al.* 2002 *JETP* **95** 123
[13] Meilikhov E Z 1999 *JETP* **115** 1484
[14] Arakelian S, Kucherik A, Kutrovskaya S 2016 *Journal of Russian Laser Research* **37** 494
[15] Abramov D, Arakelian S, Kochuev D *et al.* 2016 *Nanosystems: Physics, Chemistry, Mathematics* **7** 220
[16] Abramov D V, Arakelian S M, Makov S A *et al.* 2013 *Technical Physics Letters* **39** 14
[17] Abrikosov A A 2009 *Principles of metal theory.* (Moscow: FIZMATLIT) p 600
[18] Arakelian S, Kutrovskaya S, Kucherik A *et al.* 2016 *Proc. of SPIE* **9884** 988425
[19] Emel'yanov V I, Koroteev N I 1981 *Sov. Phys. Usp.* **24** 864
[20] Konov V I 2015 *Qantum Electron.* **45** 1043
[21] Kucherik A O, Arakelian S M, Garnov S V *et al.* 2016 *Quantum Electron.* **46** 627
[22] Kucherik A, Arakelian S, Vartanyan T *et al.* 2016 *Optics and Spectroscopy* **121** 263
[23] Shi L, Rohringer P, Suenaga K *et al.* 2016 *Nature Materials* **15** 634
[24] Ravagnan L, Siviero F, Lenardi C *et al.* 2002 *Phys. Rev. Let.* **89** 285506
[25] Rozhkova N N, Gribanov A V, Khodorkovskii M A 2007 *Dimond and related materials* **16** 2104
[26] Kao T-Y, Reno J L, Hu Q 2016 *Nature Photonics* **10** 541
[27] Hu Y, Wang Z, Weng Z *et al.* 2016 *Applied Optics* **55** 3226

Fiber-based femtosecond optical frequency comb stabilized to iodine frequency standard

S N Bagayev, V I Denisov, A S Dychkov, N A Koliada, B N Nyushkov, V S Pivtsov, S A Farnosov and A A Antropov

Institute of Laser Physics SB RAS, 13/3 Lavrentyev ave., Novosibirsk 630090, Russia

E-mail: clock@laser.nsc.ru

Abstract. A fiber-based femtosecond optical frequency comb spanning wavelengths from 1 to 2 μm was stabilized precisely to an iodine frequency standard by means of heterodyne optical phase-locked loops. It enables transfer of frequency stability across electromagnetic spectrum and implementation of compact optical clocks with $\sim 10^{-15}$ long-term instability.

1. Introduction

Modern techniques of time and frequency metrology take advantage of so-called optical frequency combs (OFCs) generated by femtosecond laser systems which are precisely stabilized to reference frequency standards [1, 2]. These OFCs are used to solve such tasks as precise measurements of optical frequencies, transfer of stability and accuracy from frequency standards to required frequencies (frequency synthesis), implementation of optical clocks, and timing of ultrafast processes [3-5]. Further expansion of metrological applications of highly-stable femtosecond OFCs beyond labs will become possible with more compact and reliable design of the OFC-based devices. Thus, the use of fiber-optic femtosecond OFC generators with stabilization to compact optical frequency standards is now regarded as a promising approach for development of transportable (or even mobile) versions of optical frequency synthesizers and optical clocks [6-8].

2. Experimental

Our experimental study was aimed at precise stabilization of a fiber-based octave-spanning OFC to an iodine optical frequency standard. The comb was generated by means of a home-made femtosecond erbium fiber laser similar to [9] and a hybrid highly-nonlinear fiber [7, 8]. Optical phase-locked loops (OPLLs) were applied to stabilize the comb teeth to the optical standard as depicted in figure 1. The corresponding short-wavelength ($\lambda_1 \sim 1$ μm) and frequency-doubled long-wavelength ($\lambda_2 \sim 2$ μm) teeth were locked to the fundamental laser wavelength ($\lambda_{os} \sim 1$ μm) of the optical standard.

To ensure reliable and tight locking of the relatively noisy optical frequency comb to the optical standard, the used OPLLs must be fast enough and feature broadband noise suppression [10]. On this basis we designed feedback loops which incorporate a miniature intracavity electro-optic phase modulator [11] and an extracavity fiber-coupled acousto-optic frequency modulator (Brimrose). These elements allowed fast frequency control and implementation of OPLLs with bandwidths of ~220 kHz and ~100 kHz for the short-wavelength and long-wavelength sides of the OFC, respectively. In order to extend the effective dynamic range of the OPLLs and provide continuous long-term stabilization of the comb, the fast frequency controls were assisted by the relatively slow controls: PZT-based cavity length control, pump power control, and temperature control.

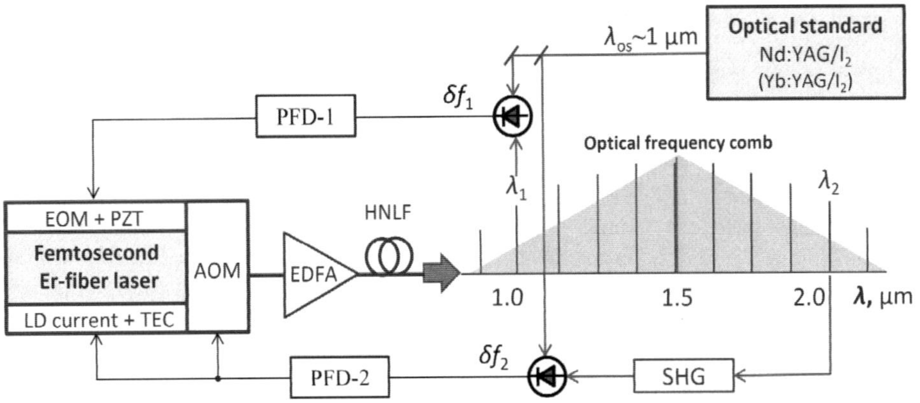

Figure 1. Implemented scheme of precise stabilization of a fiber-based femtosecond OFC to an iodine optical frequency standard: EOM – electro-optic phase modulator, PZT – piezoelectric translator, AOM – acusto-optic frequency modulator, LD – pumping laser diode, TEC – thermoelectric converter, EDFA – erbium-doped fiber amplifier, HNLF – highly-nonlinear fiber, PFD-1,2 – phase-frequency detectors, SHG – second harmonic generation; δf_1 and δf_2 stand for the beat radiofrequencies controlled by the PFDs.

To the best of our knowledge, such combination of controls including a fiber-coupled acousto-optic frequency modulator was not yet used for stabilization in the early experimental and commercial OFCs generated by femtosecond erbium fiber laser systems. Therefore, stabilization performance in those systems was limited by the relatively slow pump power control of erbium fiber lasers. The millisecond life time of the upper laser level limits the bandwidth of the corresponding laser frequency control via population inversion variation to a value less than 10 kHz [12].

The optical standard is based on a home-made frequency-doubled Nd:YAG (optionally Yb:YAG) laser stabilized to a narrow nonlinear saturated absorption resonance in the hyperfine structure of molecular iodine. Its long-term frequency instability was measured to be as low as ~10^{-15} [13].

3. Results and discussion

The implemented design of OPLLs has ensured tight locking of the OFC to the reference optical standard. Simultaneous continuous measurements (with a 1 second rate) of the beat frequencies δf_1 and δf_2 have revealed just millihertz-scale residual random deviations of the corresponding short-wavelength (λ_1~1 µm) and frequency-doubled long-wavelength (λ_2~2 µm) comb teeth against the optical standard frequency (figure 2). The residual root-mean-square phase errors (measured in the same way as in [10]) were sufficiently less than 1 rad^2 in both of the OPLLs.

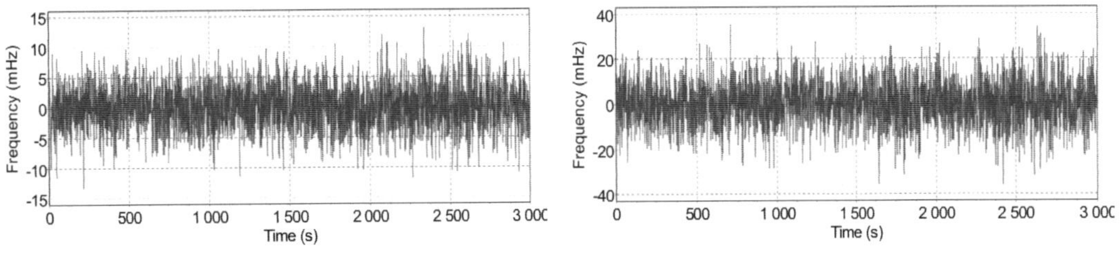

Figure 2. Residual random deviations of the OFC teeth against the optical standard frequency: the left graph – measurement of the beat frequency δf_1 (at 20 MHz) obtained with the short-wavelength tooth (λ_1~1 µm); the right graph – measurement of the beat frequency δf_2 (at 10 MHz) obtained with the frequency-doubled long-wavelength tooth (λ_2~2µm). Gate time of the frequency counters was 1 s.

An original measurement approach was used to evaluate transfer of stability from the optical standard frequency ($f_{os} \approx 2.8\times10^{14}$ Hz) to the octave comb width ($\Delta f_{comb}=f(\lambda_1) - f(\lambda_2) \approx 1.4\times10^{14}$ Hz). We mixed the second harmonic of the beat frequency δf_1 with the beat frequency δf_2 to obtain the corresponding difference radiofrequency f_{rf} that can be defined as:

$$f_{rf} = 2\delta f_1 - \delta f_2 = 2\Delta f_{comb} - f_{os}.$$

Instability of this easily measurable radiofrequency (f_{rf} ~30 MHz) can be attributed to residual random deviations of the doubled comb width ($2\Delta f_{comb}$) against the optical standard frequency (f_{os}). It is important to note that the comb offset contribution to the frequency f_{rf} is reduced to zero. Continuous measurement of the frequency f_{rf} has provided us with necessary data for calculation of the Allan deviation shown in figure 3. It characterizes an added instability (resulting from imperfection of the comb width stabilization by OPLLs) which is additive to the instability of the optical standard. Since the comb width is a multiple of the intermode frequency spacing ($f_{im} \sim 10^8$ Hz), the added instability for the frequency f_{im} is the same as for the comb width. This feature allows evaluation of the stability transfer also to the radio frequency band by the OFC-based optical clock. Since the added instability in our setup is a few orders of magnitude less than the inherent instability of the used iodine frequency standard [13] (at least at averaging times from 1 to 1000 s), we suppose that our setup transfers stability of the optical standard frequency to the OFC with negligible impairment. Continuous locking of the comb teeth to the optical standard frequency was maintained for hours during experiments. It was limited mainly by the finite range of a PZT-based slow frequency control.

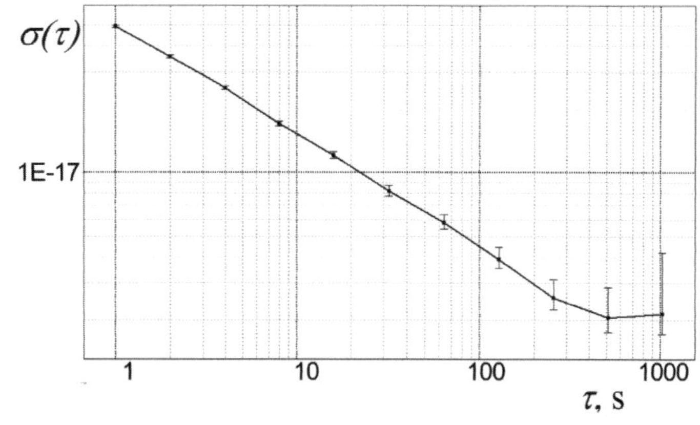

Figure 3. Allan deviation calculated for residual random deviations of the doubled comb width ($2\Delta f_{comb}$) against the optical standard frequency (f_{os}).

4. Conclusion

We have demonstrated precise and reliable stabilization of a compact fiber-based octave-spanning OFC to an advanced iodine optical frequency standard with $\sim10^{-15}$ long-term instability. The original arrangement of high-bandwidth OPLLs employing fast EOM- and AOM-based controls has enabled stability transfer from the optical standard to the OFC with negligible impairment. Unlike conventional stabilization of octave-spanning OFCs, the applied stabilization scheme does not employ any nonlinear interferometer. This feature results in compactness and reduced complexity of the setup. Such stabilized fiber-based OFC allows implementation of a compact transportable optical clock.

References

[1] Ye J and Cundiff S T 2005 *Femtosecond Optical Frequency Comb Technology: Principle, Operation and Application* (New York: Springer)

[2] Udem T, Holzwarth R and Hansch T 2009 *Eur. Phys. J. Spec. Top.* **172** 69
[3] Beverini N, Carelli G, De Michele A, Maccioni E, Nyushkov B, Sorrentino F and Moretti A 2005 *Opt. Lett.* **30** 32
[4] Bagayev S N, Denisov V I, Klementyev V M, Korel I I, Kuznetsov S A, Pivtsov V S and Zakharyash V F 2004 *Laser Phys.* **14** 1367
[5] Bagayev S N, Trunov V I, Pestryakov E V, Frolov S A, Leshchenko V E, Kokh A E and Vasiliev V A 2014 *Laser Phys.* **24** 074016
[6] Gubin M A, Kireev A N, Konyashchenko A V, Kryukov P G, Shelkovnikov A S, Tausenev A V and Tyurikov D A 2009 *Appl. Phys.* B **95** 661
[7] Korel I I, Nyushkov B N, Denisov V I, Pivtsov V S, Koliada N A, Sysolyatin A A, Ignatovich S M, Kvashnin N L, Skvortsov M N and Bagaev S N 2014 *Laser Phys.* **24** 074012
[8] Pivtsov V S, Nyushkov B N, Korel' I I, Koliada N A, Farnosov S A and Denisov V I 2014 *Quantum Electron.* **44** 507
[9] Denisov V I, Ivanenko A V, Nyushkov B N and Pivtsov V S 2008 *Quantum Electron.* **38** 801
[10] Beverini N, Prevedelli M, Sorrentino F, Nyushkov B and Ruffini A 2004 *Quantum Electron.* **34** 559
[11] Nyushkov B N, Pivtsov V S, Koliada N A, Kaplun A B and Meshalkin A B 2015 *Quantum Electron.* **45** 486
[12] Haverkamp N, Hundertmark H, Fallnich C and Telle H R 2004 *Appl. Phys.* B **78** 321
[13] Denisov V I, Ignatovich S M, Kvashnin N L, Skvortsov M N and Farnosov S A 2016 *Quantum Electron.* **46** 464

Optical frequency standard with ytterbium single ion

S V Chepurov[1], A A Lugovoy[1], S N Kuznetsov[1,2], K M Rumynin[1,2], M V Okhapkin[1,3], A V Taichenachev[1,2], V I Yudin[1,2] and S N Bagayev[1]

[1]Institute of Laser Physics SB RAS, 13/3 Acad. Lavrentyev ave., Novosibirsk 630090, Russia
[2]Novosibirsk State University, 2 Pirogova str., Novosibirsk 630090, Russia
[3]Physikalisch Technische Bundesanstalt, Braunschweig 38116, Germany

E-mail: svc04@ngs.ru

Abstract. We report on the progress in development of a highly accurate optical frequency standard based on the single ion of ytterbium-171 at the Institute of Laser Physics, Novosibirsk.

1. Introduction

Frequency standards play very important role in both fundamental research and various applications. This is primarily related to the fact that the accuracy of modern frequency standards, which implement a reference for one of the basic units of measure of the SI system (second) is several orders of magnitude higher than the accuracy of references for other physical values. Development of a new generation of highly accurate optical frequency standards is a fundamental problem of laser spectroscopy. Currently, the standards based on atoms or ions localized in space are considered to be the most stable ones.

A promising candidate for the use in these optical frequency standards is the ^{171}Yb ion. The specific features of the energy level diagram of ytterbium ion make it possible to develop optical frequency standards based on two transitions (figure 1): (i) the $^2S_{1/2}$ (F = 0) - $^2D_{3/2}$ (F = 2) quadrupole transition with a wavelength of 436 nm and a natural linewidth of 3.1 Hz and (ii) the $^2S_{1/2}$ (F = 0) - $^2F_{7/2}$ (F = 3) octupole transition at 467 nm with a natural lifetime of several years.

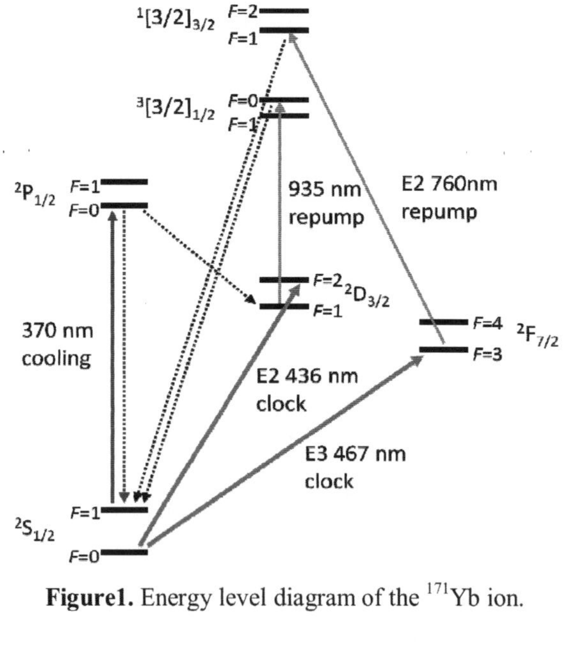

Figure1. Energy level diagram of the ^{171}Yb ion.

The largest frequency shift that contributes to the systematic uncertainty of many atomic frequency standards is the interaction of the thermal blackbody radiation with the atomic eigenstates. Presence of two ultra narrow optical transitions in the same thermodynamic environment makes possible implementation of so called "synthetic" frequency standard with suppressed blackbody radiation (BBR) frequency shift [1]. In ^{171}Yb ion at room temperature the residual BBR shift is estimated to be on the order of 10^{-18} for the "synthetic" frequency which is a combination of the octupole (467 nm) and the quadrupole (436 nm) optical transition frequencies. Thus, the "synthetic" frequency standard based on $^{171}Yb^+$ can be practically immune to the blackbody radiation shift.

Miniature endcap trap is used for capturing and retaining the single ion by means of a quadrupole radio frequency potential (figure 2).

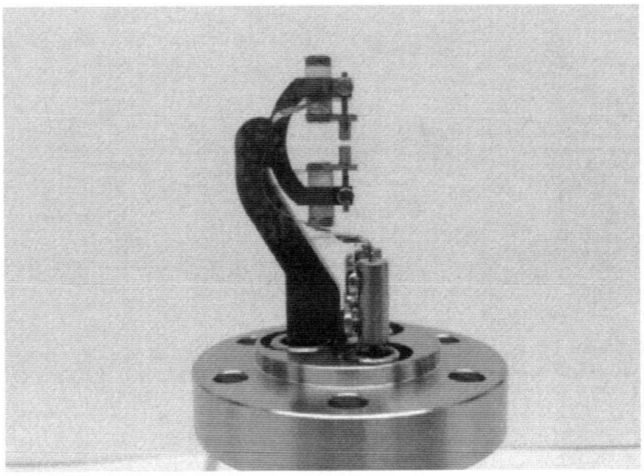

Figure 2. Radiofrequency ion trap.

Open design allows laser beams conveniently access the trapped ion. Outer trap electrodes, ytterbium oven, electron source and two additional electrodes are used to apply constant voltages for stray field compensation.

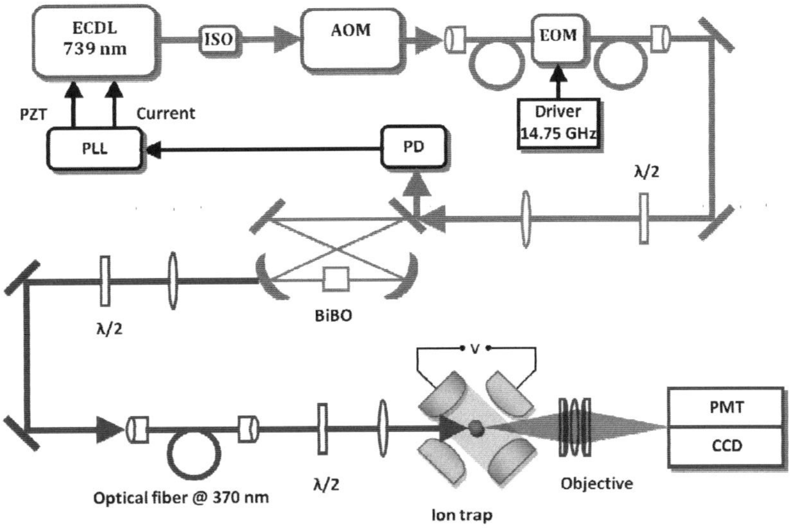

Figure 3. Laser system for Doppler cooling of the ion. ECDL - external cavity diode laser, ISO - optical isolator, AOM - acousto-optical modulator, EOM - electro-optical modulator, PZT - piezo-electric transducer, PLL - phase-locked loop, PD - photodetector, PMT - photo-multiplier tube, CCD - charge-coupled device camera.

The quasicycling $^2S_{1/2}$ (F = 1) - $^2P_{1/2}$ (F = 0) electric dipole transition with natural linewidth of 23 MHz at 370 nm is used for Doppler cooling and detection of the ion (figure 1). Doppler cooling of the ion is performed with the help of a frequency modulated radiation of the diode laser resonantly frequency doubled in a nonlinear crystal [2] (figure 3).

Up to 100 μW of the UV light is available for the ion cooling. A commercial external cavity diode laser with 10 mW at 739 nm is frequency-modulated in a high speed EOM at 14.75 GHz to generate a sideband that excites the $^2S_{1/2}$ (F = 0) - $^2P_{1/2}$ (F = 1) hyperfine component of the cooling transition. ECDL frequency is resonantly doubled in a cavity-enhanced SHG BIBO crystal. The enhancement ring cavity has a finesse of 400 and FSR of ~740 MHz chosen to be resonant with the sidebands of the modulated light. The laser is frequency locked to the cavity for short-term stability.

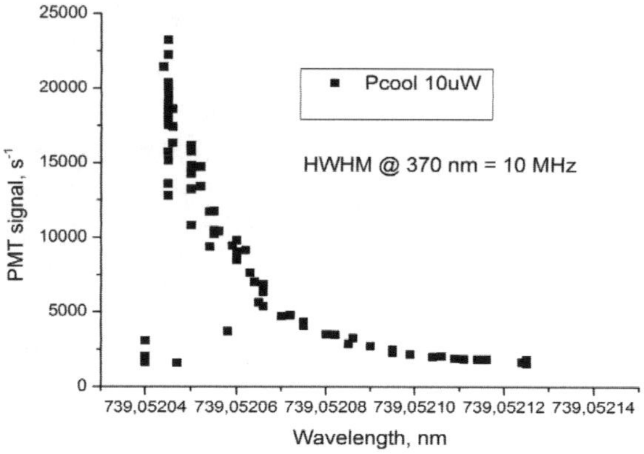

Figure 4. Trapped ^{171}Yb single ion (left) and resonance fluorescence signal (right) from the cooled single ion.

Using the developed laser system ytterbium ions can be captured and cooled in the ion trap. Figure 4 shows a CCD camera image of the trapped ^{171}Yb single ion and a spectral profile of the resonance fluorescence signal on the cooling transition. The signal has a HWHM of ~ 10 MHz which indicates the absence of the Doppler broadening.

Narrow line probe laser is constructed to excite the ion clock transition. The radiation from a frequency doubled extended-cavity diode laser at a wavelength of 871 or 934 nm is used (figure 5).

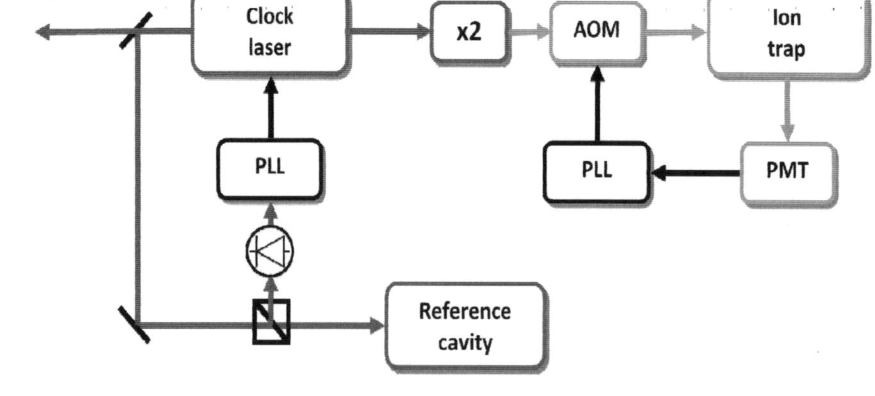

Figure 5. Laser system for excitation of the ^{171}Yb ion clock transition.

The linewidth of the free-running laser is decreased to ~ 1 Hz by the Pound-Drever-Hall frequency stabilization to a high-finesse Fabry-Perot etalon made of ultralow expansion (ULE) glass (figure 6). To suppress fluctuations of the cavity length caused by low-frequency mechanical (acoustic) perturbations a suspension mount design is employed [3]. The geometry of suspension allows to reduce sensitivity of the cavity to horizontal and vertical vibrations by one order of magnitude. In addition a vacuum chamber is mounted on a passive vibration isolation platform.

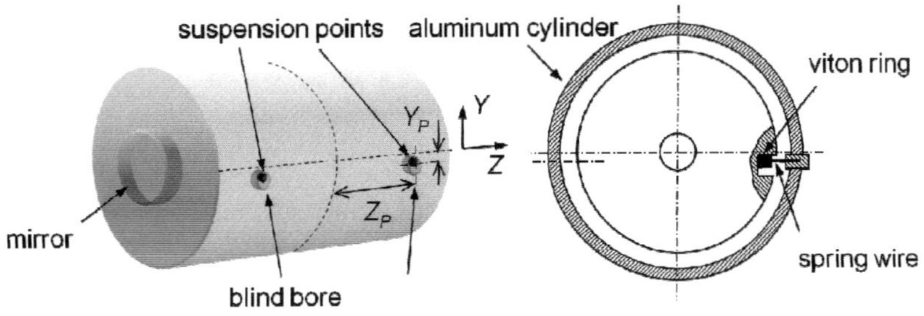

Figure 6. Vibration-insensitive design of the reference cavity.

Cavity length is 102.5 mm, diameter is 50 mm. Suspension point coordinates are Y_P = - 0.9 mm, Z_P = 35 mm, blind bore depth and diameter are 10 mm and 8 mm correspondingly.

The ULE cavity finesse was measured to be ~3×10^5. The coefficient of cavity thermal expansion (CTE) is nearly zero at the temperature of 26° C, the corresponding thermal frequency drift of the cavity is ~ 0.25 Hz/s (figure 7).

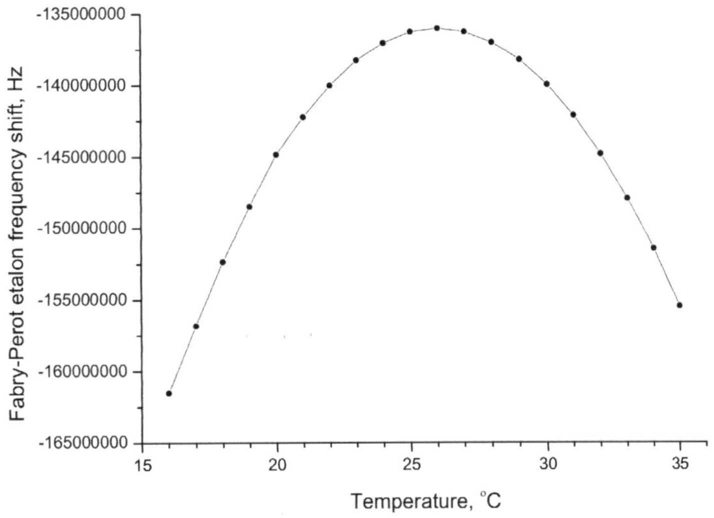

Figure 7. Thermal frequency drift of the reference cavity.

To obtain an excitation spectrum of the ion clock transition a specially tailored laser pulse sequence was employed (figure 8). The cooling and repumping laser pulses were alternating with the clock laser. The ion is laser cooled for typically ~20 ms and the induced fluorescence light is detected by a photomultiplier. For the cooling power of about 5 μW typical fluorescence count rate at the half-maximum of the cooling resonance is 10 kHz, with a background count rate of 500 Hz. Then the HFS sideband of the cooling laser is switched off and within a time of 20 ms the ion is being transferred with high probability to the $^2S_{1/2}(F = 0)$ ground state via non-resonant excitation to the $^2P_{1/2}(F = 1)$ state and the fluorescence signal will disappear. The cooling and repumper lasers are then blocked by mechanical shutters and the magnetic field is switched from the high value of 400 μT needed for cooling to 1 μT for the clock transition spectroscopy. Within about 10 ms when the magnetic field has settled to its new value, a 4-20 ms pulse from the probe laser is applied. As long as probe pulse spectral width is larger than the natural linewidth of the atomic transition, it determines the frequency resolution of the resonance signal via the Fourier transformation, while the laser intensity and the detuning from resonance determine the effective Rabi frequency for the excitation.

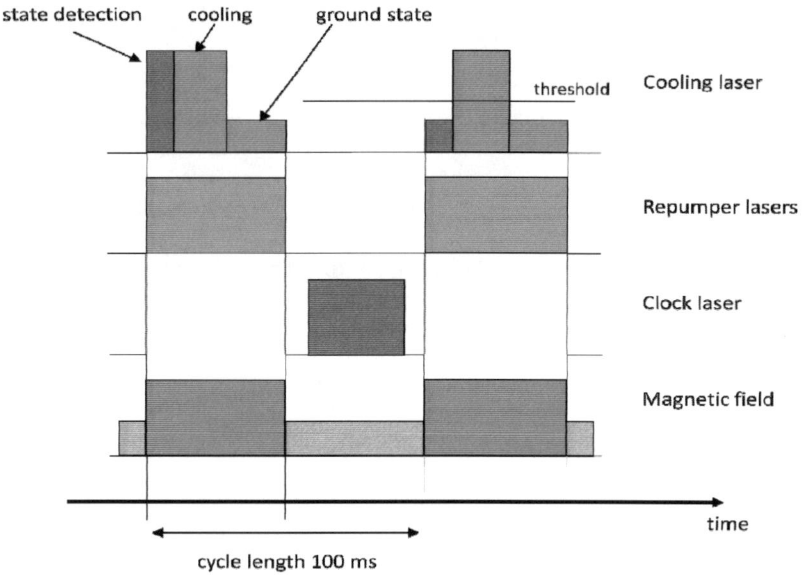

Figure 8. Time sequence for the interrogation of the $^{171}Yb^+$ clock transition.

The cooling and repumper lasers are unblocked in a few milliseconds after the end of the probe pulse to avoid disturbing by stray light. The ion state is detected on the cooling transition by so-called electron shelving method introduced by H. Dehmelt [4]: when the cooling laser is on, the fluorescence signal appears immediately if the ion is in the ground state, while no signal is detected during detection time of 4 ms when the ion was excited to the $^2D_{3/2}(F = 2)$ state. The excitation attempt is valid if the following requirements are met: (1) the fluorescence rate during cooling step exceeds a threshold of 5 kHz which ensures that the ion is in the F = 1 ground state during the cooling step and is cooled to low temperatures, (2) the fluorescence rate during the ground state preparation step (after turning off the cooling sideband) is below the threshold of 5 kHz , this means that the ion is in the F = 0 ground state when the probe laser pulse is applied. The sequence of interrogation cycles is repeated several times at a certain frequency of the probe laser. The excitation spectrum of the ion clock transition is obtained by repeating the procedure described above for different probe laser frequencies.

Acknowledgment
This work received financial support from the Russian Scientific Foundation under the grant 16-12-00052.

References

[1] Yudin V I *et al.* 2011 *Phys. Rev. Lett.* **107** 030801
[2] Chepurov S V, Lugovoy A A, Kuznetsov S N 2014 *Quantum Electronics* **44** 527
[3] Nazarova T *et.al.* 2006 *Appl. Phys.* B **83** 531
[4] Dehmelt H 1982 *IEEE Trans. Instrum. Meas.* **31** 83

Properties of aqueous solutions in THz frequency range

O Cherkasova[1], M Nazarov[2, 3] and A Shkurinov[2, 4]

[1] Institute of Laser Physics of SB RAS, Novosibirsk 630090, Russia
[2] Crystallography and Photonics Federal Research Center, RAS, Moscow 117342, Russia
[3] Kurchatov Institute National Research Center, pl. Akad. Kurchatova 1, Moscow 123182, Russia
[4] Lomonosov Moscow State University, Moscow 119991, Russia

E-mail: o.p.cherkasova@gmail.com

Abstract. Terahertz time-domain spectroscopy has been used for measuring of bovine serum albumin and glucose solutions response. The transmission and the attenuated total internal reflection geometries have been combined for analyzing the dielectric properties of aqueous solutions spectra at 0.07-3.2 THz.

1. Introduction

It is known that the terahertz (THz) signal is sensitive to the glucose concentration in blood in the case of in vivo reflection studies of human or animal skin [1]. The origin of that sensitivity is the properties of water - the main component of tissues and fluids in a body. The state of the water itself, which is prevailing in biological samples, represents a characteristic that can be measured using broadband THz spectroscopy [2, 3]. Unfortunately, THz time-domain spectroscopy (TDS) usually captures not the centre of a low-frequency 'peak' of water absorption (in the region near 0.03 THz), but its high-frequency 'tail', starting from the frequency of 0.2 THz (figure 1) [2]. Nevertheless, THz-TDS has become an established method which complements the existing methods of diagnostics of biological tissues and solutions. However, there is still no consensus how to interpret the THz signal contrast in scanning the skin in vivo and how to predict the changes in the THz response of solutions.

In present work the THz-TDS has been used for measuring of bovine serum albumin (BSA) and glucose solutions in a wide range of concentrations and with additional low frequency part (0.07-0.2) of the THz spectra. To detect reliably small-scale changes in solutions we have performed measurements using both transmission and in attenuated total internal reflection (ATR) configurations. Combining the results of both configurations, the reliable range of the obtained complex dielectric function spectrum can be considerably broadened to the most important low frequency part. We also analyzed the change in the THz absorption during incubation of BSA with glucose. This is a model experiment for the study of the process of protein glycation. Protein glycation is accelerated under hyperglycemic conditions resulting in loss of the structure and biological functions of proteins.

2. Experimental setup and samples

The THz time-domain spectrometer [4, 5] and the calculation methods [6, 7] are described in detail in our previous works. The measurements of solution parameters were performed sequentially using two THz time-domain spectrometers: a 'low-frequency' spectrometer – in transmission configuration using "thick" -500-μm cell, and a 'high-frequency' spectrometer – in attenuated total internal reflection configuration using a Dove prism and also in transmission configuration for high frequencies using "thin" cell (100 μm). In the low-frequency spectrometer [5] a multi-dipole emitting antenna at an average power of 10 μW (in the THz range) has been used and the detection was performed by conventional single-dipole antenna. This efficient emitter allows one to transmit through a 0.5-μm-thick water layer with frequencies up to 1 THz. For a stable measurement of the extremely low for THz-TDS frequencies (~0.07 THz), a long time sampling of 40 ps with respect to the THz pulse centre has been applied.

$$\varepsilon_{water} = \varepsilon_{\infty} + \frac{\Delta\varepsilon_1}{1+i\omega\tau_1} + \frac{\Delta\varepsilon_2}{1+i\omega\tau_2} + \frac{A_1}{\omega_1^2 - \omega^2 + i\gamma_1\omega} +$$

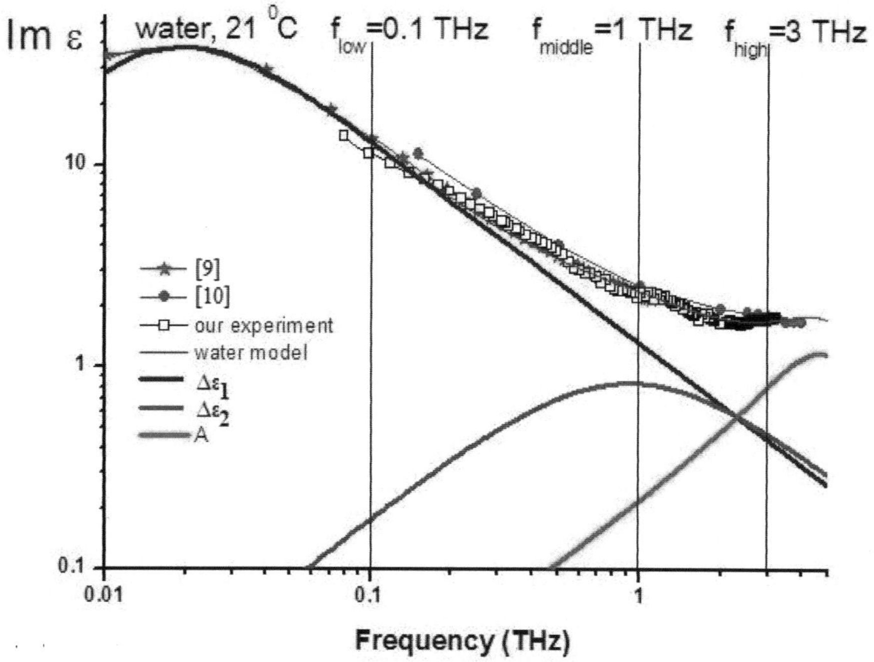

Figure 1. Spectra of the imaginary part of the dielectric permittivity Im{ε(f)} of distilled water in the THz frequency range described by two-component Debye model.

In the high-frequency spectrometer [4], the LT-GaAs surface and a 1-mm-thick ZnTe crystal have been used as an emitter and a receiver, respectively [5]. The average-power THz radiation of 100 nW was sufficient to measure the reflection spectrum in the range of 0.3 – 3.2 THz. As the reference transmitted or reflected THz signal was used with the air or distilled water instead of the solution.

We used solutions of BSA and D-glucose (Sigma, USA) in a double-distilled water. Glucose solutions at a concentration of 270 – 840 mg mL^{-1} and BSA solutions at a concentration of 30 – 500 mg mL^{-1} were measured at the room temperature of 22°C. The solution volume for a single measurement in the ATR configuration constituted 800 μL (that provides a 1 mm thick layer on the prism surface), whilst in the transmission configuration, it was 400 μL (for 500 μm cell). To carry out the glycation reaction, BSA was dissolved in phosphate buffer (50 mM, pH 7.4) at a concentration of 50 mg mL^{-1} and incubated with glucose (0.5 M) for 96 hours at a temperature of $T = 47$ °C.

3. Approaches to the description of the dielectric function of solutions in the THz range

To fit dielectric function spectra ε(f) we apply well-known two-component Debye model [7, 8] (figure 1). To clear out which process is responsible for THz sensitivity of biomolecule solution we divide available frequency range into three parts: "low" –slow relaxation dominates (f≤0.1 THz); "middle"- for fast relaxation (0.1<f<3) and "high" – for additional Lorentz term (f>3 THz) (see figure 1). We choose appropriate experimental method for each spectral range (thick cell transmission (500 μm) for "low" frequencies, thin cell (100 μm) and ATR for "middle" frequencies, ATR for "high" frequencies). Actually all methods have considerable intersection in frequencies, so we combine the results obtained by different methods and this way obtain more precise data and a more broadband frequency range than ever published in THz literature for water solution spectroscopy.

Even for pure water, the multi-parametric model of dielectric function is not exact and applicable in a very wide frequency range. In describing a real solution with a single varied parameter (for example, the concentration of dissolved substances), it does not make any sense to vary plenty of parameters in the expression for dielectric function of the solution, as it is now accepted in most of the publications.

Our approach to the description of the dielectric function of protein or sugar solutions in the THz range is the following:

1. To systematize the known parameters of water model and refine the range of reasonable values for each of them at room temperature.
2. In our view, the reference parameters of each relaxation process should be taken from a corresponding frequency range and appropriate measuring method.
3. The slow Debye γ-relaxation is the only significant process and other terms/processes introduce negligible contribution in the changes of dielectric permittivity of aqueous solutions in THz-TDS frequency range

Thus, for approximating the dielectric function spectrum, certain constraints should be set on each parameter, proceeding from the data taken from the frequency range appropriate to each of the processes. In our case, to analyse the data in the range of 0.07 – 3.2 THz at room temperature (20 – 25 °C), we select the following parameter ranges, which do not contradict the most reliable published data: ε_∞=2.1-2.7, $\Delta\varepsilon_1$=72-75, τ_1=8.3-9.5 ps, $\Delta\varepsilon_2$=1.4-2.1, τ_2=130-360 fs, ε_s=77-80.2. We obtain for a most simple model the relevant set of parameters which accurately enough describe both our experimental results and the data published by other authors [9, 10] for distilled water in the frequency range of 0.07 – 3.2 THz (figure 1).

4. Results

We analyzed the reasons for the THz transmission changes of studied solutions comparing experimental spectra to the model dielectric function of water. The insertion of glucose into water leads only to an increase of relaxation time τ_1 of the slow Debye process or as well to a decrease of relaxation amplitude of this solution (table 1). This simple approach describes observed spectral changes in a broad frequency range and for a number of concentrations from 10 mg/ml to saturated solution. The similar changes are observed for BSA spectra. Here ε_s – is the sum of amplitudes off all terms in the model, this value can by precisely measured by dielectric spectroscopy in low frequencies.

We perform fitting simultaneously for both Re and Im parts of ε(f), which complicates an exact agreement in a wide frequency range, but increase the repeatability of obtained values. To describe the influence of solute concentration, we fit experimental complex spectra by a model only with the first Debye term's amplitude - $\Delta\varepsilon_1$ varied (see table 1). We suggest that, instead of complicating the models used for the analysis of experimental data, it is necessary to improve the accuracy and repeatability of experiments and to expand the spectral range of a measurement.

Table 1. The parameters of dielectric function for glucose and BSA solution.

C $mg \cdot mL^{-1}$	ε_s	ε_∞	$\Delta\varepsilon_1$	τ_1 ps	$\Delta\varepsilon_2$	τ_2 fs	ε 0.1 THz	ε 3.0 THz
0	78.5	2.5±0.1	75±6	9.5±0.8	1.5±0.05	230±30	7.8-i11.4	3.1-i1.7
270	60.0	2.49±0.1	53±2	9.5±0.3	1.6±0.05	230±20	6.9-i9.4	2.9-i1.5
810	34.3	2.5±0.02	29±2	9.5±0.3	1.7±0.03	230±20	6.4-i5.1	2.8-i1.3
100 (BSA)	64	2.5±0.02	65±2	9.5±0.3	1.6±0.03	230±20	7.3-i10.5	3.0-i1.5

As well we have found that increasing of BSA concentration in the solution results in a decrease of the amplitude $\Delta\varepsilon_1$ (figure 2). We have not confirmed anomalous changes observed in papers [11, 12] at low concentrations and at low frequencies. The agreement with published data is only in following: the dependence of $\Delta\varepsilon_1$ from concentration of a solute is nonlinear and has a bend at a particular concentration. We found this concentration for the case of BSA to be 30±5 mg/ml.

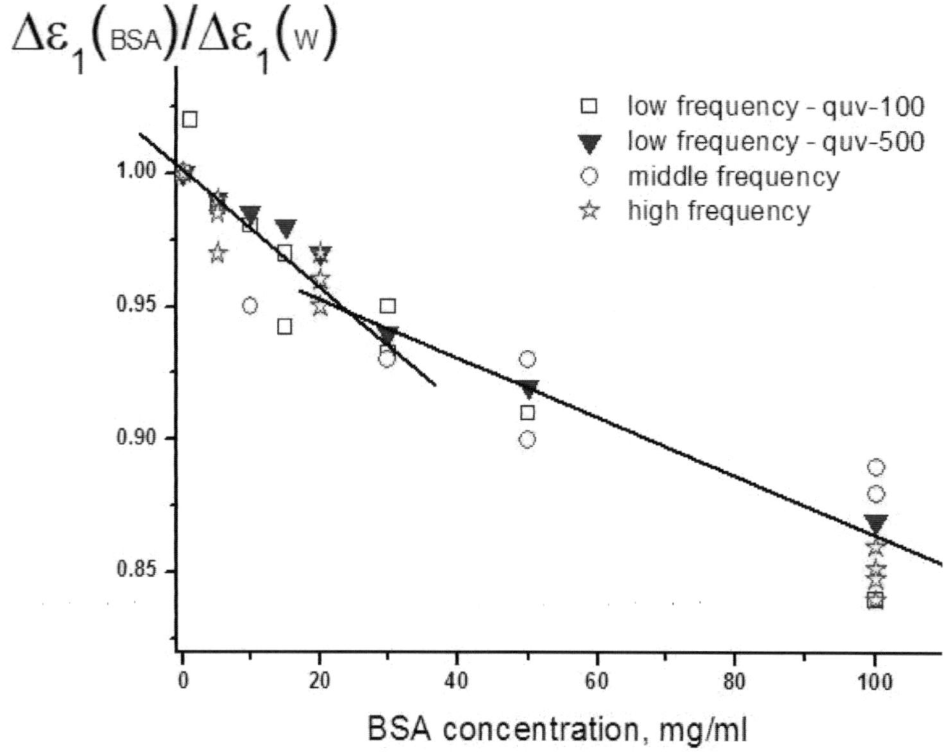

Figure 2. The dependence of slow Debye process amplitude $\Delta\varepsilon_1$ from BSA concentration in water.

Of practical interest is not a study of aqueous solutions of proteins or sugars separately, but a study of their mixtures. It is known that a high glucose level in human blood leads to glycation of proteins. The BSA (50 $mg \cdot mL^{-1}$) was incubated in a phosphate buffer (50 mM, pH 7.4) in the presence of glucose (0.5 M) for 96 hours at $T = 47$ °C.

Figure 3 shows variations of transmission coefficient T_{int} during incubation of BSA with glucose. An averaged transmission coefficient of solution is normalized on an averaged transmission coefficient of pure water. The transmission coefficient of BSA incubated in buffer alone did not change significantly. This indicates that BSA does not spontaneously modify under incubation conditions.

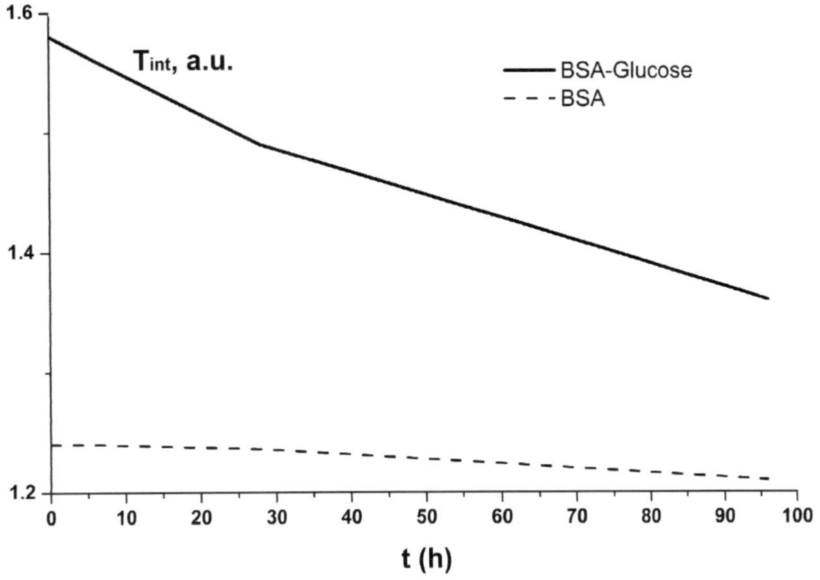

Figure 3. Dependence of an averaged transmission coefficient T_{int} at 0.1-1.0 THz from time t (h) after the start of incubation of BSA with glucose.

It was found that value of THz absorption coefficient of glycated albumin solution varies considerably during incubation. At the early stage of incubation, the amplitude of the first Debye term $\Delta\varepsilon_1$ was 88% relative to the value for pure water. After 96 hours of incubation BSA with sugar the amplitude of the first Debye term $\Delta\varepsilon_1$ was 92% relative to that for pure water [13]. During incubation sugar molecules form covalent bonds with protein molecule. The incubation mixture contains a significantly smaller part of sugar molecules bound with water molecules. In other words, the amount of free water molecules is increased at the final stage of incubation. As a result, the imaginary part of the dielectric constant is increased and this in turn leads to a reduction in the transmission of incubation mixture at 96 hour of incubation.

5. Conclusion

Dielectric properties of BSA and glucose solutions were systematically measured at 0.07-3.2 THz frequency range. It was found that the most significant is the reduction $\Delta\varepsilon_1$ (or increase τ_1) of the slow Debye relaxation process with increasing concentrations of solute. Glycation of BSA results in a change the parameters of slow Debye relaxation. During incubation of BSA with sugars the incubation mixture contains a significantly smaller part of sugar molecules bound with water molecules and the amount of free water molecules is increased at the final stage of incubation.

Acknowledgments
This work has been supported by RFBR (grant № 14-02-00846, № 14-02-00979 and № 16-52-00222).

References

[1] Cherkasova O, Nazarov M and Shkurinov A 2016 *Opt. Quant. Electron.* **48** 217
[2] Nazarov M, Cherkasova O and Shkurinov A 2016 *Quantum Electron.* **46** 488–95

[3] Cherkasova O, Nazarov M, Angeluts A and Shkurinov A 2016 *Optics and Spectroscopy* **120** 50
[4] Nazarov M, Shkurinov A, Kuleshov E and Tuchin V 2008 *Quantum Electron.* **38** 647–54
[5] Nazarov M, Shkurinov A, Angeluts A and Sapozhnikov D 2009 *Radiophysics and Quantum Electronics* **52** 595
[6] Angeluts A, Balakin A, Evdokimov M, Esaulkov M, Nazarov M, Ozheredov I, Sapozhnikov D, Solyankin P, Cherkasova O and Shkurinov A 2014 *Quantum Electron.* **44** 614
[7] Yada H, Nagai M and Tanaka K 2008 *Chem. Phys. Lett.* **464** 166
[8] Fitzgerald A, Pickwell-MacPherson E and Wallace V 2014 *PLOS ONE* **9** e99291
[9] Vinh N, Allen S and Plaxco K 2011 *Am. Chem. Soc.* **133** 8942
[10] Shiraga K, Suzuki T, Kondo N, De Baerdemaeker J and Ogawa Y 2015 *Carbohydr. Res.* **406** 46
[11] Bye J, Meliga S, Ferachou D, Cinque G, Zeitler J and Falconer R 2014 *J. Phys. Chem.* A **118** 883–8
[12] Sushko O, Dubrovka R and Donnan R 2015 *The Journal of Chemical Physic* **142** 055101-1–9
[13] Cherkasova O, Nazarov M and Shkurinov A 2016 *Proc. SPIE* **9917** Saratov Fall Meeting 2015: Third International Symposium on Optics and Biophotonics and Seventh Finnish-Russian Photonics and Laser Symposium (PALS) 991706 doi:10.1117/12.2229741

Atom interferometers' phases at the presence of heavy masses; their use to measure Newtonian gravitational constant; optimization, error model, perspectives

Boris Dubetsky

1849 S Ocean Dr, Apt 207, Hallandale, Florida 33009, USA

E-mail: bdubetsky@gmail.com

Abstract. The contribution to the phase of the atom interferometer caused by the gravity field of a massive test mass is considered. This contribution can be extracted by applying the double difference technique to measure the Newtonian gravitational constant G. Estimates and further calculations showed that after choosing the largest (given the current state of the art) multiphoton wave vector, the time delay between pulses, the mass of the test body and the signal optimization in respect to atomic positions and velocities, one should be able to obtain an accurate G measurement of 200 ppb, which is 2-3 orders of magnitude more accurate than what can currently be obtained. Calculated variations of the phases under the small deviations of atomic variables made it clear that atom clouds' radii and temperatures have to be as small as 100 micron and 100 pK, which has also been achieved already.

1. Introduction

Among the fundamental physical constants c, ħ, G, the Newtonian gravity constant (CODATA, 2015)

$$G=6.67408(31)\times10^{-11}\,m^3kg^{-1}s^{-2} \tag{1}$$

is measured at the lowest accuracy of only 46ppm. In this talk, we consider the possibility of increasing this accuracy by using the atom interferometry technique [1]. This technique was first applied [2] by using a test mass moving vertically around the trajectories of atom clouds. For this talk, we assume [3] that the test mass has a cuboid shape with a small cuboid hole for atoms to go through, and that this cuboid shape consists of 2 parts moving horizontally to and from atom clouds (see figure 1). We calculated the phase double difference

$$\Delta^2\phi=\Delta\phi_a-\Delta\phi_b, \Delta\phi=\phi(z_1,v_{1z})-\phi(z_2,v_{2z}), \tag{2}$$

where $\phi(z,v)$ is a phase of the atom interferometer in which atoms are launched vertically from point z with velocity v. Since both atom clouds are irradiated by the same field and stay on the same platform, the vibration contributes equally to the phases and that contribution is excluded in the 1st order phase difference $\Delta\phi$. When test mass halves are joined or separated (see figure 1*a* and *b*), phase differences are equal to $\Delta\phi_{a,b}$. The part of the phase difference caused by Earth's gravitational field is evidently the same for both differences, and this part is eliminated in the phase double difference. Therefore, $\Delta^2\phi$ depends only on the gravitational field of the test mass, which is linear in G.

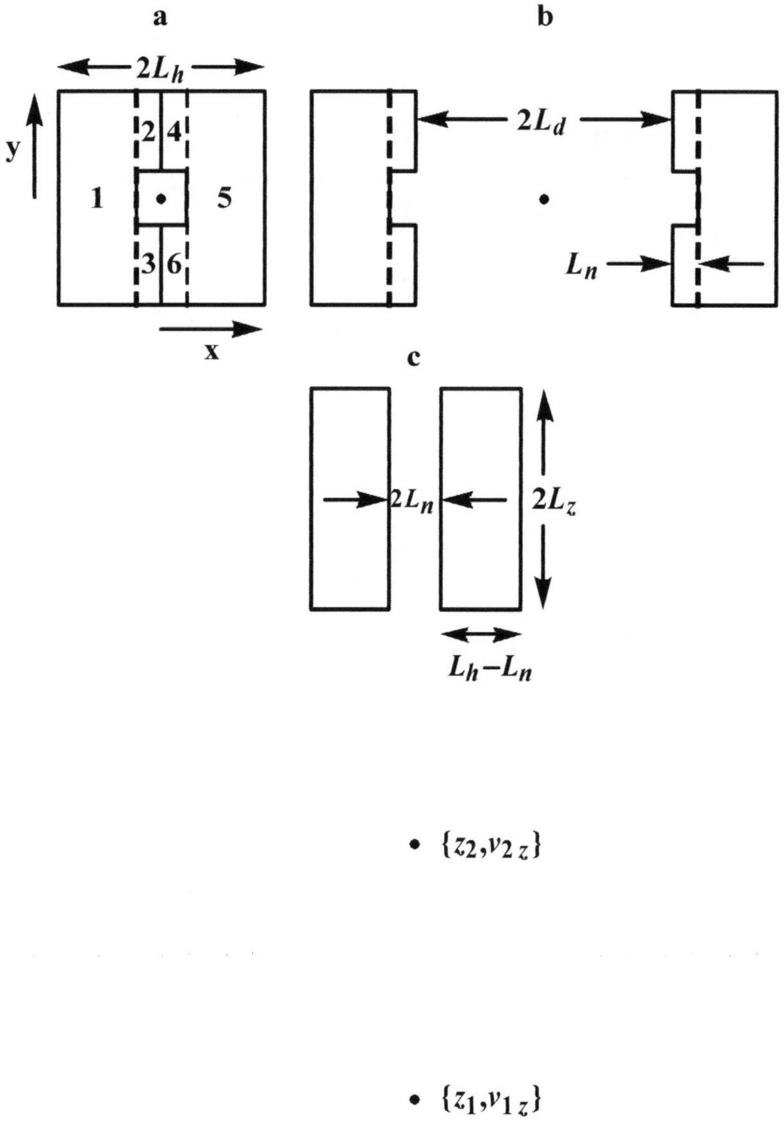

Figure 1. The test mass as a whole is cuboid with a narrow hole for Raman fields and atom trajectories. Atoms are launched from the points z_1 and z_2 with velocities v_{1z} and v_{2z}. Test mass consists of 2 halves. (a) Top view. Joined halves. (b) Top view. Halves separated on the distance $2L_d$. (c) Side view, cross section x=0 for joined halves.

2. Optimization

In contrast to [3] (where the calculation was made for the parameters' modest values), we calculated the phases for ^{87}Rb and the maximal value of the parameters achieved at the current state of the art in atom interferometry, i.e. for the time delay between pulses T=1.15 s [5], the effective wave vector $k=7.248\times10^{8}$ m^{-1}, the test mass M=1080 kg of Pb [6], and the phase noise $\phi_{err}=10^{-4}$ rad. The chosen value of k vector can be obtained by using multiphoton processes, first considered in [1]. This value is 45 times greater than the wave vector associated with 2-quantum Raman process in ^{87}Rb. It was achieved [7] using a sequential technique.

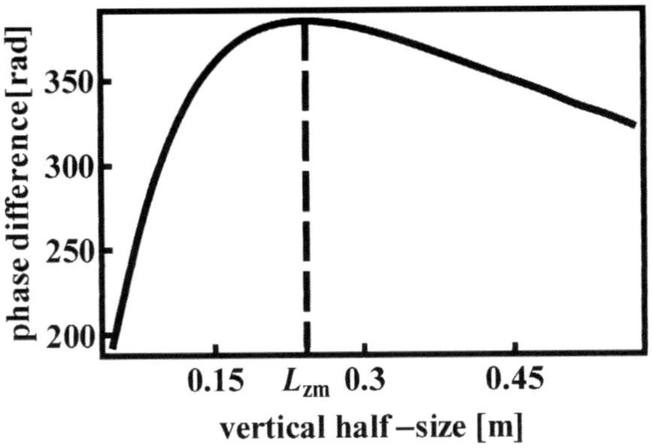

Figure 2. Dependence of the maximum of phase difference on cuboid half-size.

The contribution to the phase caused by the test mass δϕ was studied in detail in review [8]. We obtained a response linear in the test mass gravity **δg** assuming that [9] the magnitude of this field is small,

$$\delta g \ll g, \tag{3}$$

where g is the magnitude of Earth's gravitational field. The ratio δg/g is a small parameter of our theory. Evidently, only this part contributes to the double difference, $\Delta^2\phi=\Delta^2\delta\phi$. Using the Wigner representation of the atomic density matrix (first applied for atom interferometry in [10]), we showed that δϕ consists of 3 parts, the classical part, the recoil term and the Q-term. The recoil term was obtained without expanding over recoil velocity $\hbar\mathbf{k}/M_a$, M_a is the atomic mass, while for the Q-term, we used perturbation over gravity curvature tensor. Calculations performed in [8] showed that for the chosen value of the wave vector, the recoil term would overcome the classical part, while the Q-term is 2-3 orders of magnitude smaller, and we did not include it in the calculations presented here. From equations (62, 66, 73) in [8], the following expression is obtained for the sum of the classical part and the recoil term:

$$\delta\phi(z,v)=\mathbf{k}\cdot\int_0^T dt\left\{(T-t)\delta g\left[\mathbf{X}^{(0)}(\mathbf{x},\mathbf{v},T+t_1+t)+\frac{\hbar\mathbf{k}}{2M_a}(T+t)\right]+t\delta g\left[\mathbf{X}^{(0)}(\mathbf{x},\mathbf{v},t_1+t)+\frac{\hbar\mathbf{k}}{2M_a}t\right]\right\}, \tag{4}$$

where $\mathbf{x}=\{0,0,z\}$ and $\mathbf{v}=\{0,0,v\}$ are the initial atom cloud's position and velocity, $\mathbf{X}^{(0)}(\mathbf{x},\mathbf{v},t)$ is the atom trajectory, t_l is the time delay between the moment of the atoms' launching and 1st Raman pulse. Evidently, $\Delta^2\delta\phi$ achieves the maximum when $\{z_1,v_{1z}\}$ is the point of absolute maximum, and $\{z_2,v_{2z}\}$ is the point of absolute minimum. We found these extrema iteratively using a reasonably wide area and reasonably small steps in z and v. The sizes of the area and steps were restricted by my PC's speed and power. For the given test mass M and density ρ, we found extrema as a function of cuboid vertical halfsize L_z, see figure 2. To get a maximum signal, we recommend choosing L_{zm} shown in figure 2 as vertical halfsize of test mass. Values of the other optimum parameters of the system are presented in table 1.

Table 1. Maximal value of the phase double difference and optimal values of the test mass and atom clouds variables.

phase difference	$\Delta^2\delta\phi = 386.52738$
vertical half-size	$L_{zm} = 0.24219097$m
horizontal half-size	$L_{hm} = 0.22162188$m
1st cloud position	$z_{1\,m} = -6.8517823$m
1st cloud velocity	$v_{1\,m} = 11.158930$m/s
2nd cloud position	$z_{1\,m} = -6.2983410$m
2nd cloud velocity	$v_{1\,m} = 11.172994$m/s

One can estimate the accuracy of measurement as

$$err \equiv \frac{\delta G}{G} = \frac{\phi_{err} = 10^{-4}\,\text{rad}}{387\,\text{rad}} \approx 200\ \text{ppb}. \tag{5}$$

This accuracy is more than 200 times better than that claimed in CODATA 2015.

3. Error model

To achieve highly precise measurements of the interferometers' phases, one has to prepare both the atomic and proof mass system with great accuracy. The most challenging part is precisely positioning the atom clouds [3]. The extrema of the clouds positions and velocities are, obviously, preferable here. That is why the extrema (found above) in {z,v} space allow one not only to maximize the response, but also to make less stringent the requirements for atom clouds' position, velocity, temperature and size because the response becomes quadratic on variations of these variables near the extrema.

Let us now allow small variations of the atom clouds' initial positions, velocities (atomic variables) and small displacements of the test body halves (see figure 3). In this talk we calculated linear and quadratic dependences on atomic variables only. Their relative weights are presented in table 2.

One sees that in spite of using extremum points $\{z_i.v_i\}$ linear terms are not equal to 0. It is because extrema $\{z_i,v_i\}$ have been found in section 2 approximately. One can find that coefficients in the linear dependences small enough so that, for allowed variations of position and velocity (see below table 3), the quadratic dependencies are not overcome by linear contributions.

One can use nonlinear terms to estimate atom clouds' radii and temperatures. Consider for example relative contribution

$$\delta\phi_z = \alpha z_i^2 \tag{6}$$

If Raman fields are sufficiently flat to neglect ac-Stark shift variation across the atom cloud, and if Raman pulses are sufficiently short to neglect the Doppler broadening of the Raman transition, then one needs only to average (6) over atoms' spatial distribution.

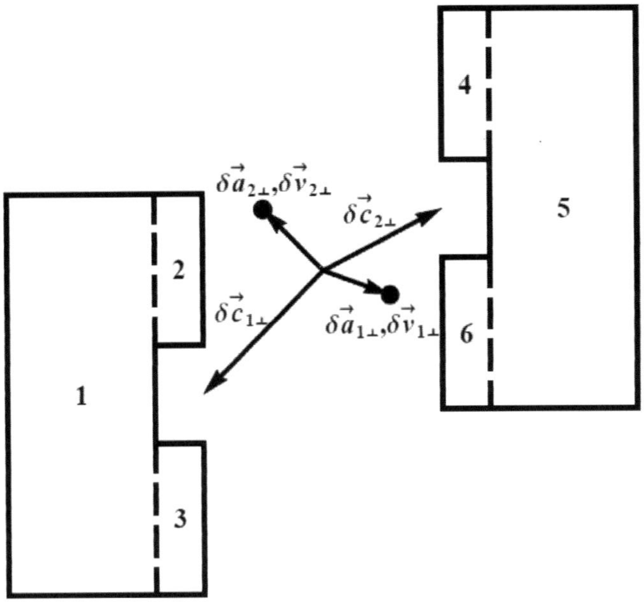

Figure 3. Top view. Small variations of the atomic and proof mass variables.

Table 2. Error model for 1080 kg test mass, atomic variables.

Term	Relative weight
Linear in position	$0.0012117481\ \delta z_1$ $0.0011188391\ \delta z_1$
Linear in velosity	$-2.5456033 \times 10^{-6}\ \delta v_{1\,z}$ $-2.3703001 \times 10^{-6}\ \delta v_{2\,z}$
Nonlinear in position	$6.6840356\ (\delta x_1^2+\delta y_1^2)$ $-13.368071\ \delta z_1^2$ $1.4586375\ (\delta x_2^2+\delta y_2^2)$ $-2.9172749\ \delta z_2^2$
Nonlinear in velocity	$9.0832760\ (\delta v_{1\,x}^2+\delta v_{1\,y}^2)$ $-18.166552\ \delta v_{1\,z}^2$ $1.9881322\ (\delta v_{2\,x}^2+\delta v_{2\,y}^2)$ $-3.9762644\ \delta v_{2\,z}^2$
Position-velocity cross-term	$15.578255\ (\delta x_1 \delta v_{1\,x}+\delta y_1 \delta v_{1\,y})$ $-31.156509\ \delta z_1 \delta v_{1\,z}$ $3.4018485\ (\delta x_2 \delta v_{2\,x}+\delta y_2 \delta v_{2\,y})$ $-6.8036970\ \delta z_2 \delta v_{2\,z}$

For Gaussian distribution,

$$\frac{\exp\left(-\delta z_i^2 / \delta z_{i0}^2\right)}{\sqrt{\pi}\delta z_{i0}},$$

after averaging one gets

$$\langle \delta\phi_z \rangle = \frac{\alpha}{2}\delta z_{i0}^2. \tag{7}$$

Requiring it to be equal to the expected relative error of phase measurement, err; one finds for atom cloud radius

$$\delta z_{i0} = \sqrt{\frac{2err}{\alpha}}. \tag{8}$$

In the same manner, we determine atom clouds velocities' variations and temperatures. These quantities are presented in table 3 for relative error value (5). Even though it is challenging to cool to those temperatures and focus on the radii, the temperatures and radii are higher than those achieved in article [12].

Table 3. Atom interferometers' parameters one should hold to achieve measurements of 200 ppb.

1st cloud vertical radius [m]	0.00017297989
1st cloud vertical velocity [m/s]	0.00014838628
1st cloud vertical temperature [K]	$1.1519682 \times 10^{-10}$
1st cloud horizontal radius [m]	0.00024463051
1st cloud horizontal velocity [m/s]	0.00020984989
1st cloud horizontal temperature [K]	$2.3039364 \times 10^{-10}$
2nd cloud vertical radius [m]	0.00037028943
2nd cloud vertical velocity [m/s]	0.00031717019
2nd cloud vertical temperature [K]	$5.2630531 \times 10^{-10}$
2nd cloud horizontal radius [m]	0.00052366833
2nd cloud horizontal velocity [m/s]	0.00044854639
2nd cloud horizontal temperature [K]	1.0526106×10^{-9}

4. Conclusion

We showed that using the atom interferometry technique, for the chosen geometry of the test body, at the positions and velocities of the atom clouds determined by the optimization, can give the double difference of the atomic interferometers phases as large as 387 rad at the phase noise level 10^{-4} rad. This should allow one to measure the Newtonian gravitational constant G with an accuracy of 200 ppb, which is 2 to 3 orders of magnitude better than that currently achieved using conventional methods. To achieve this result one has to realize **SIMULTANEOUSLY** (a) sequential technique to increase the effective wave vector, (b) small radii and (c) low temperatures of the atom clouds, determined by using the built error model. Each of the parameters is within the current state of the art in atomic interferometry.

References

[1] Dubetsky B Ya, Kazantsev A P, Chebotaev V P, Yakovlev V P 1984 *JETP Lett.* **39** 651
[2] Fixler J B, Foster G T, McGuirk J M, Kasevich M A 2007 *Science* **315** 74
[3] Dubetsky B 2014 Optimization and error model for atom interferometry technique to measure Newtonian gravitational constant *Preprint* Physiscs/1407.7287
[4] Snadden M J, McGuirk J M, Bouyer P, Haritos K G and Kasevich M A 1998 *Phys. Rev. Lett.* **81** 971
[5] Dickerson S M, Hogan J M, Sugarbaker A, Johnson D M S and Kasevich M A 2013 *Phys. Rev. Lett.* **111** 083001
[6] Biedermann G W, Wu X, Deslauriers L, Roy S, Mahadeswaraswamy C and Kasevich M A 2015 *Phys. Rev.* A **91** 033629
[7] Kovachy T, Asenbaum P, Overstreet C, Donnelly C A, Dickerson S M, Sugarbaker A, Hogan J M and Kasevich M A 2015 *Nature* **528** 530
[8] Dubetsky B, Libby S B and Berman P 2016 *Atoms* **4** 14
[9] Dubetsky B 2008 unpublished work.
[10] Dubetsky B, Kasevich M A 2006 *Phys. Rev.* A **74** 023615
[11] Rosi G, Sorrentino F, Cacciapuoti L, Prevedelli M and Tino G M 2014 *Nature* **510** 518
[12] Kovachy T, Hogan J M, Sugarbaker A, Dickerson S M, Donnelly C A, Overstreet C and Kasevich M A 2015 *Phys. Rev. Lett.* **114** 143004

Double-resonance spectroscopy in Rubidium vapour-cells for high performance and miniature atomic clocks

M Gharavipour[1], C Affolderbach[1], S Kang[2], and G Mileti[1]

[1]Laboratoire Temps-Fréquence, Institut de Physique, Université de Neuchâtel, Switzerland

[2]Time and Frequency Division, National Institute of Standards and Technology, Boulder, Colorado, USA

Email: gaetano.mileti@unine.ch

Abstract. We report our studies on using microwave-optical double-resonance (DR) spectroscopy for a high-performance Rb vapour-cell atomic clock in view of future industrial applications. The clock physics package is very compact with a total volume of only 0.8 dm^3. It contains a recently in-house developed magnetron-type cavity and a Rb vapour cell. A homed-made frequency-stabilized laser system with an integrated acousto-optical-modulator (AOM) – for switching and controlling the light output power– is used as an optical source in a laser head (LH). The LH has the overall volume of 2.5 dm^3 including the laser diode, optical elements, AOM and electronics. In our Rb atomic clock two schemes of continuous-wave DR and Ramsey-DR schemes are used, where the latter one strongly reduces the light-shift effect by separation of the interaction of light and microwave. Applications of the DR clock approach to more radically miniaturized atomic clocks are discussed.

1. Introduction

Microwave-optical double-resonance (DR) spectroscopy in alkali vapour cells constitutes simultaneously a field of basic research and of more application-oriented developments of precision instruments such as frequency standards and magnetometers [1]. In the case of atomic frequency standards (atomic clocks), the frequency of an oscillator, for example quartz, is stabilized to the atomic resonance by an electronic servo-loop (see figure 1).

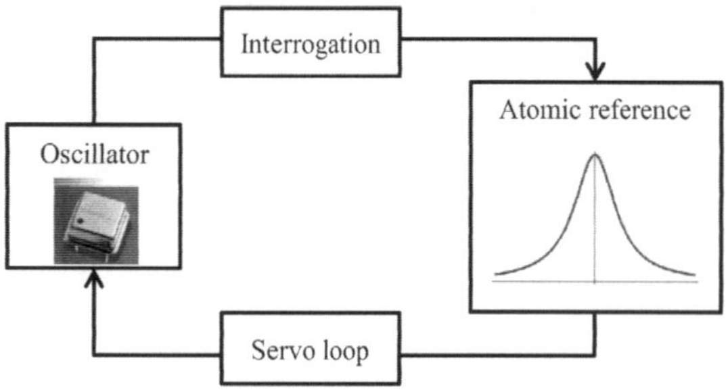

Figure 1. Basic principle in atomic clocks.

Historically, the first investigations were realized with alkali plasma discharge lamps as optical source while the radio-frequency (rf) or microwave field was applied using different techniques such as simple metallic loops, antennas, horns or resonant cavities. The various types of optical pumping processes [2] and the AC Stark shift induced on the rf or microwave atomic transitions by the resonant optical field were investigated in great details from both the experimental and the theoretical points of view [3, 4]. Different techniques inspired by the Nuclear Magnetic Resonance (NMR) methods were also developed in order to study the sources of relaxations (in particular the collisions) [5]. In the case of Rubidium atomic vapour cells, the presence of two isotopes with partially overlapping optical transitions allowed to improve the efficiency of the hyperfine optical pumping, thanks to the so-called "isotopic filtering", and resulted in the realization of commercial atomic frequency standards that were required in many applications such as network synchronization, secure telecommunication and satellite positioning and navigation systems [6, 7, 8]. Recently portable standards such as the space passive hydrogen maser (SPHM) [9] or laser-beam CS beam clocks (LPCs) [10] show the short-term stability for one second averaging time 7×10^{-13} – 1.5×10^{-12} and the long-term stability for one day 1×10^{-14} – 3×10^{-14} with the volume range typically from 13 to 28 dm^3. In compact cell-type clocks of interest here, the accuracy and the medium- to long-term frequency stability are ultimately limited by the fact that the atoms are stored in a thermal vapour cell with unavoidable collisions which effects are very sensitive to the environmental changes and other aging processes.

In this present investigation we report on our highly-compact Rb atomic clock based on microwave-optical DR interrogation in both continuous-wave (CW) mode [11] and pulsed [12] Ramsey-DR scheme [13, 14]. We show that with both CW-DR and Ramsey-DR schemes, the achieved short- and the long-term frequency stability of our Rb atomic clock are comparable with the ones from SPHM or cold atoms portable clocks [15].

2. Double-Resonance schemes

In a DR atomic clock two resonant electromagnetic fields of light and microwave are required for both CW-DR and Ramsey-DR schemes. The optical field by using a laser, in our case acting on Rb D2 line at 780 nm, prepares –with optical pumping– and probes –with optical detection– the atomic states of the reference –in our case Rb. The microwave field samples the so-called clock frequency which is the frequency of the clock transition of the atomic reference by using a microwave generator. In our ^{87}Rb atomic clock the clock transition is the $5^2S_{1/2}$ $|F_g\!=\!1, m_F\!=\!0\rangle \leftrightarrow |F_g\!=\!2, m_F\!=\!0\rangle$ with the frequency of ν_{Rb} = 6 834 682 610.900312 Hz [16]. The three steps of optical pumping, microwave interrogation and optical detection are applied simultaneously in the case of CW-DR, while they are separated in time in the case of Ramsey-DR scheme (see figures 2(b) and 2(c)). This separation can suppress the light shifts (LS) effect in Ramsey-DR scheme, but it requires an additional component like an acousto-optical-modulator (AOM) as an optical switch. It makes this scheme slightly more complex and increases the volume and the power consumption compared to CW-DR scheme.

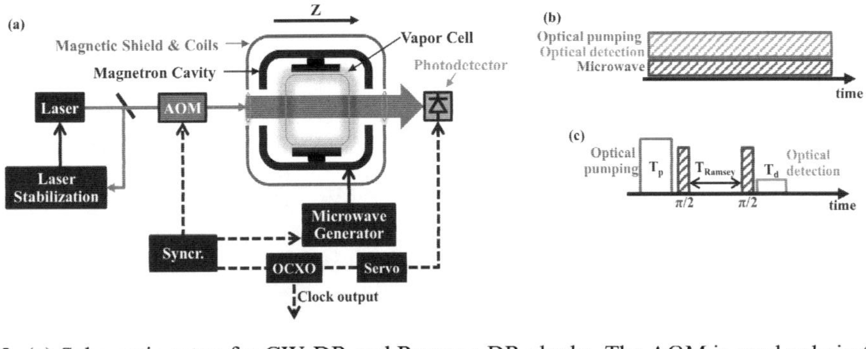

Figure 2. (a) Schematic setup for CW-DR and Ramsey-DR clocks. The AOM is used only in the case of Ramsey-DR scheme. (b) CW-DR sequences. (c) Ramsey-DR sequences.

3. High-performance Rb atomic clock

The atomic clock setup is shown schematically in figure 2(a). The main clock hardware components are: 1) the Laser source, 2) the Physics Package (PP) and 3) the local oscillator (LO). In the following, we briefly describe the laser source and the PP. The LO was described in details in [17].

3.1. The Laser source

A distributed-feedback (DFB) laser diode emitting at 780 nm (Rb D2 transition) is placed in a compact laser head (LH) used as the laser source. Moreover, an evacuated ^{87}Rb reference cell which is thermally controlled and an AOM are integrated in the LH assembly [18]. Figure 3 shows the schematics of the optical setup of the LH. The optical components in the LH occupy 1 dm^3 only while its overall volume is 2.4 dm^3, including space for control electronics. The total power consumption for the laser supply and heaters (laser, reference cell and baseplate) is less than 1.5 W at ambient temperature. The frequency of the laser is stabilized to the sub-Doppler absorption lines using an evacuated ^{87}Rb reference cell. The AOM that is used for fine tuning of the laser frequency and also as a fast switch, with rise and fall times < 5 μs to control the durations and output powers of optical pumping and optical detection pulses. The laser has an emission linewidth of < 2MHz and shows a fractional frequency stability below 1×10^{-11}. All these characteristics make the LH appropriate for use in the DR atomic clock setup.

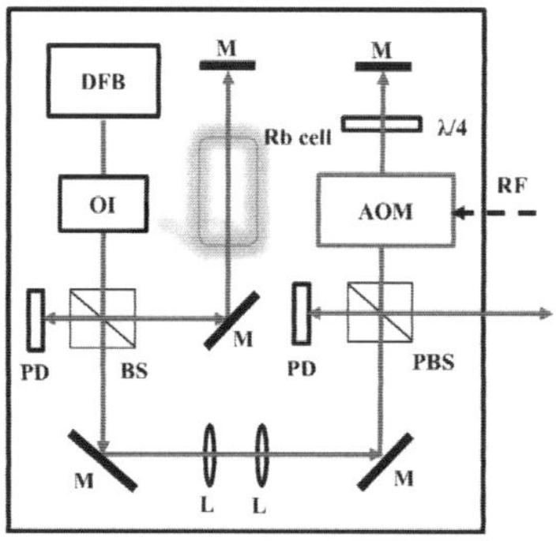

Figure 3. Schematics of the optical setup of the LH.

3.2. The Physics Package

The clock Physics Package (PP), with overall volume of 0.8 dm^3, contains a vapour cell, the magnetron type cavity, a C-field coil, a magnetic shield and an optical detector. The vapour cell is made of glass and is developed in-house. It is filled with ^{87}Rb atomic vapour and a mixture of Argon and Nitrogen buffer-gases of 26 mbar total pressure with the ratio of $P_{Ar}/P_{N2}=1.6$. The vapour cell has a cylindrical shape with a diameter of 25 mm and a height of 25 mm. The cell has a stem acting as a reservoir for the Rb atoms. It is placed in a compact so-called magnetron-type cavity based on a loop-gap resonator approach. The cavity has a total external volume of 45 cm^3 which resonates at the ^{87}Rb clock's transition frequency of ≈6.835 GHz, in a TE_{011}-like mode with a relatively low loaded quality factor Q≈200 [19]. The cavity is mounted inside a C-field coil that creates a static magnetic field parallel to the laser propagation vector (z direction) to split the respective Zeeman levels of ^{87}Rb hyperfine ground states. The field orientation factor (FOF), ξ, –which describes the fraction of useful magnetic field driving the clock transition with respect to the total field energy over the entire cell volume– can characterize the uniformity of the microwave magnetic field distribution inside the cavity [19]. In our magnetron-type cavity ξ was measured up to 0.95% that guarantees a strong clock signal. Moreover, the achieved high-contrast Ramsey signal and Rabi oscillations prove the homogeneity of the microwave magnetic field distribution [14]. A two-layer magnetic shield surrounds the whole PP to shield against external magnetic field fluctuations. A photodetector collects the transmitted light behind the cell.

3.3. Clock signals and short-term stability

The recorded clock signals for both CW-DR and Ramsey-DR schemes applied to the same Rb atomic clock setup are shown in figure 4. For CW-DR mode (the solid red line in figure 4), the signal has a contrast of C=26% and its linewidth of $\Delta\nu_{1/2}$=334 Hz is limited by optical and microwave power broadening here [11]. In the case of Ramsey-DR scheme (the dashed blue line in figure 4) the central fringe has a contrast of C=40% and a linewidth of $\Delta\nu_{1/2}$=160 Hz, limited by the 3 ms Ramsey time, T_{Ramsey} (duration between two $\pi/2$ microwave pulses, see figure 2(c)).

The measured short-term stability of the clock is limited by the shot-noise [11], the signal to noise ratio [20] and the microwave noise from the LO through the Dick effect [21]. This short-term stability is measured on the levels of $1.4\times10^{-13}\ \tau^{-1/2}$ and $2.4\times10^{-13}\ \tau^{-1/2}$ for CW-DR and Ramsey-DR schemes, respectively [22]. In the case of CW-DR scheme, the shot-noise limit contribution is at the level of $4.9\times10^{-14}\ \tau^{-1/2}$. It is about three times higher than the one from Ramsey-DR ($1.7\times10^{-14}\ \tau^{-1/2}$ [22]) which is explained by the lower light intensity during the optical detection in the Ramsey scheme. In both CW-DR and Ramsey schemes, the dominant instability contribiution in the short-term stability is due to technical laser noise and FM-to-AM noise conversion in the vapour cell [14, 19]. In the case of Ramsey-DR, additional noise from the AOM [14] results in slight degradation of the short-term stability of the clock compared to CW-DR case.

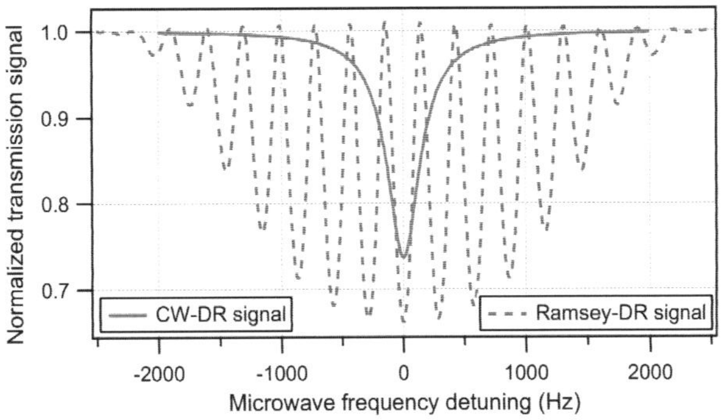

Figure 4. Double resonance signal in the case of the CW-DR scheme (solid red line) compared to Ramsey fringes in the case of Ramsey-DR (dashed blue line) [11, 14].

3.4. Limitations to long-term stability

Generally, in DR vapour-cell atomic clocks the long-term stability at $\tau = 10^4$ s to one day is limited by systematic variations of various parameters like the cell temperature and the input powers and frequencies of the light and microwave that can perturb the clock frequency. On the long-term clock stability of our Rb atomic clock, the contributions from the temperature shift and the light shifts (LS) are the dominating contributors, for both CW-DR and Ramsey-DR schemes [23]. The geometry of the vapour cell affects the temperature shift [24]: for example, the temperature coefficient (TC) improved by an order of magnitude for a newly-designed vapour cell with a shorter stem to 5.5×10^{-15}. The LS intensity and frequency coefficients α and β, respectively, were measured in the same Rb clock setup where the laser frequency was stabilized to the sub-Doppler cross-over transition CO11-12 for both CW-DR and Ramsey-DR schemes. The results are shown in table 1. It can be seen that α and β are suppressed about 5 and 125 times, respectively, from CW-DR to Ramsey-DR scheme as a consequence of the separation of the light and microwave interaction in the case of Ramsey-DR. The LS can be completely suppressed in Ramsey-DR, potentially, if no light passes through the cell unless during the optical pumping and the optical detection pulses. In our setup, the fraction of the residual light intensity to the optical pumping intensity was measured about 0.0005 which thus gives rise to the residual LS measured.

Table 1. LS intensity and frequency coefficients for both CW-DR and Ramsey-DR Rb clocks.

	CW-DR	Ramsey-DR
LS intensity coeff. α (/%)	3.5×10^{-13}	-6.6×10^{-14}
LS frequency coeff. β (/MHz)	-1.0×10^{-11}	-8.0×10^{-14}

4. Miniature clock

In parallel to the studies on high-performance atomic clocks described in section 3, we have conducted several studies towards further miniaturization of atomic clocks, notably in view of mobile applications and low power-consumption devices [25]. One focus of these studies concerns microwave-optical double resonance in micro-fabricated vapour cells [26], which constitutes a very promising alternative to the Coherent Population Trapping (CPT) approach widely pursued for chip-scale atomic clocks [27]. In fact, for an atomic vapour cell of same size the DR scheme allows reaching higher clock signal amplitude and thus better clock stability than the (CPT) scheme. With our collaboration partners we have developed several different designs and geometries of micro-fabricated alkali vapour cells, including e.g. a 4mm-size Rb vapour cell based on multi-stack anodic bonding of silicon and glass wafers [28] (Figure 5). Combined with a miniaturized microwave resonator cavity, the micro-loop-gap resonator [29], clock stabilities on the level of $7\times10^{-12}\ \tau^{-1/2}$ were obtained, which are of high interest for future miniature atomic clocks, for example in view of smart-grid applications. We also demonstrated double-resonance spectroscopy in a miniature vapour cell using a micro-fabricated discharge lamp as light source [30], as an alternative to laser-based chip-scale atomic clocks.

A potential alternative to existing approaches consists in covering the inner walls of the alkali vapour cell with an "anti-relaxation" coating [31]. This approach is particularly interesting in micro-fabricated vapour cells, where the high buffer-gas pressures required generally result in a loss of signal amplitude. However, the use of anti-relaxation wall coatings in micro-fabricated cells has been long time hindered by the incompatibility of temperature ranges for operational wall coatings and established cell-sealing techniques. After an initial study on the metrological properties of an atomic clock based on a cm-scale glass-blown wall-coated cell [32], we could demonstrate a first micro-fabricated vapour cell with an operational anti-relaxation wall-coating, based on a dedicated low-temperature sealing technique [33].

The pulsed Ramsey-DR clock scheme with its advantage of reduced light-shift demonstrated for the much bigger high-performance atomic clocks could also be applied to the realm of miniature atomic clocks. Our numerical simulations show that this approach should allow observation of high-quality Ramsey fringes from a micro-fabricated vapour cell, compatible with a clock stability on the level of $\leq 10^{-11}\ \tau^{-1/2}$ [22]. Combined with recently demonstrated pulsed laser interrogation for miniature atomic clocks using laser current modulation only [34], this approach bears high promise for the realization of strongly miniaturized atomic clocks with reduced light-shift effect and thus improved long-term stability.

Figure 5. Micro-fabricated Rb vapour cell with 4mm-thick core and excellent control of the cell geometry, manufactured by multi-stack anodic bonding [28]. The photo shows the cell on top of a 1 euro-cent coin.

5. Conclusion

We have studied high-performance Rb vapour cell atomic clocks based on microwave-optical DR using a very small magnetron-type microwave cavity in view of highly compact clock device. The clock was operational in both CW-DR and Ramsey-DR schemes. An integrated AOM in the frequency-stabilized laser system allowed us to separate optical and microwave pulses Ramsey-DR scheme. In both schemes, the optical detection noise was the main limitation in short-term clock stability while on the long-term scales ($\tau \approx 10^4$ s) the temperature sensitivity from the vapour cell and light-shift effects dominated. The vapour cell TC can be controlled by the geometry and design of the cell and the LS can be suppressed by switching from CW-DR to Ramsey-DR scheme. Based on previous developments, application of the Ramsey interrogation scheme also to strongly miniaturized Rb DR clocks appears possible.

Acknowledgments

This work was supported by the Swiss National Science Foundation (SNSF grant no. 140712) and the European Metrology Research Programme (EMRP project IND55-Mclocks). The EMRP is jointly funded by the EMRP participating countries within EURAMET and the European Union. We also acknowledge previous support from the European Space Agency (ESA) and the Swiss Space Office (Swiss Confederation). We thank A. K. Skrivervik, C. Stefanucci, and A. E. Ivanov (all EPFL-LEMA) for their support on the microwave cavity, and C. Calosso (INRIM, Italy) for the microwave LO. We thank T. Bandi and M. Pellaton for their contributions to the early phases of the work, and our co-authors of our cited references for fruitful collaborations.

References

[1] Camparo J 2007 *Phys. Today*, **60** 33
[2] Kastler A 1950 J. Phys. Radium, **11** 255
[3] Cohen-Tannoudji C and Barrat J 1961 *J. Phys. Radium*, **22** 329
[4] Mathur B S, Tang H and Happer H 1968 *Phys. Rev.*, **171** 11
[5] Franzen W 1959 *Phys. Rev.*, **115** 850
[6] Dupuis R T, Lynch T J and Vaccaro J R 2008 *Proc. IEEE Int. Frequency Control Symposium* edited by Jadusliwer B. *(Honolulu, Hawaii, USA 19-21 May 2008)* pp. 655-660
[7] Waller P, Gonzalez S, Binda S, Sesia I, Hidalgo I, Tobias G and Tavella P 2010 *IEEE Trans. Ultrason. Ferroelct. Freq. Cont.*, **57** 738
[8] Chunhao H, Zhiwu C, Yuting L, Li L, Shenghong X, Lingfeng Z and Xianglei W 2013, *Int. J. Navig. Observ.* **2013** 371450
[9] Rochat P, *et al.* 2005 *Proc. IEEE Frequency Control Symp., Vancouver, BC, USA*, pp. 26–32
[10] Lecomte S, *et al.* 2007 *Proc. IEEE Frequency Control Symp. and 21st European Frequency and Time Forum, Geneva, Switzerland*, pp. 1127 – 1131
[11] Bandi T, Affolderbach C, Stefanucci C, Merli F, Skrivervik A K and Mileti G 2014 *IEEE Trans. Ultrason. Ferroelct. Freq. Cont.*, **61** 1769-1778
[12] Micalizio S, Godone A, Levi F and Calosso C 2009 *Phys. Rev. A.*, **79** 013403
[13] Kang S, Affolderbach C, Gruet F, Gharavipour M, Calosso C E and Mileti G 2014 *Proc. 26th European Frequency and Time Forum (EFTF) (Neuchatel, Switzerland, 23-26 June 2014)* pp. 544-547
[14] Kang S, Gharavipour M, Affolderbach C, Gruet F and Mileti G 2015 *J. Appl. Phys.*, **117** 104510
[15] Esnault F X, Rossetto N, Holleville D, Delporte J and Dimarcq N 2011 *J. Adv. Space Res.*, **47** (5) pp. 854–858
[16] Guéna J, Abgrall M, Clairon A and Bize S 2014, *Metrologia*, **51** 108–120
[17] Calosso C E, Micalizio S, Godone A, Bertacco E K and Levi F 2007 *IEEE Trans. Ultrason., Ferroelect. Freq. Control.* **54** 1731
[18] Gruet F, Pellaton M, Affolderbach C, Bandi T, Matthey R and Mileti G 2012 *Proc. Int. Conf. on Space Optics (ICSO) (Ajaccio, Corsica, 9-12 October 2012)* no. 48

[19] Stefanucci C, Bandi T, Merli F, Pellaton M, Affolderbach C, Mileti G and Skrivervik A K 2012 *Rev. Sci. Instrum.*, **83** 104706

[20] Micalizio S, Calosso C E, Godone A and Levi F 2012 *Metrologia*, **49** 425-436

[21] Deng J Q, Mileti G, Drullinger R E, Jennings D A, and Walls F L 1999 *Phys. Rev. A*, **59** 773

[22] Gharavipour M, Affolderbach C, Kang S, Bandi T, Gruet F, Pellaton M and Mileti G 2016 *Jour. of Phys.: Conference Series* **723** 012006

[23] Affolderbach C, Gharavipour M, Kang S, Gruet F and Mileti G 2015 *Proc. 5th International Colloquium on Scientific and Fundamental Aspects of the Galileo Programme, Braunschweig, Germany, October 27-29, 2015*

[24] Calosso C E, Godone A, Levi F and Micalizio S 2012, *IEEE Trans. Ultrason., Ferroelect. Freq. Control.* **59** No. 12

[25] Knappe S 2007 *Comprehensive Microsystems* vol 3, ed. Gianchandani Y, Tabata O and Zappe H (Amsterdam: Elsevier) p 571

[26] Pellaton M, Affolderbach C, Petremand Y, de Rooij N F and Mileti G 2012 *Physica Scripta* **T149** 014013

[27] Lutwak R, Emmons D, Riley W and Garvey R M 2002 *Proc. 34th Annual Precise Time and Time Interval (PTTI) Meeting* (Reston, VA, 3–5 Dec. 2002) pp 539–550

[28] Pétremand Y, Affolderbach C, Straessle R, Pellaton M, Briand D, Mileti G and de Rooij N F 2012 *Journal of Micromechanics and Microengineering* **22** 025013

[29] Violetti M, Pellaton M, Merli F, Zürcher J–F, Affolderbach C, Mileti G and Skrivervik A K *IEEE Journal of Sensors* **14** 9

[30] Venkatraman V, Kang S, Affolderbach C, Shea H and Mileti G 2014 *Appl. Phys. Lett.*, **104** 054104

[31] Robinson H G and Johnson C E 1982 *Appl. Phys. Lett.*, **40** 771

[32] Bandi T, Affolderbach C and Mileti G 2012 *J. Appl. Phys.*, **111** 124906

[33] Straessle R, Pellaton M, Affolderbach C, Pétremand Y, Briand D, Mileti G and de Rooij N F 2014 *Appl. Phys. Lett.* **105** 043502

[34] Ide T, Goka S and Yano Y 2015 *Proc. Joint Conf. IEEE Int. Frequency Control Symposium & European Frequency and Time Forum (EFTF) (Denver, Colorado, USA 12-16 April 2015)*, pp. p 167

An optical frequency standard based on ultracold magnesium atoms

A N Goncharov[1,2,3], **A E Bonert**[1], **D V Brazhnikov**[1,2], **O N Prudnikov**[1,2], **M A Tropnikov**[1], **S A Kuznetsov**[1], **A V Taichenachev**[1,2], **S N Bagayev**[1,2]

[1] Institute of Laser Physics SB RAS, 13/3 Acad. Lavrentyev ave., Novosibirsk 630090, Russia
[2] Novosibirsk State University, 2 Pirogova str., Novosibirsk 630090, Russia
[3] Novosibirsk State Technical University, 20 K. Marx ave., Novosibirsk 630073, Russia

E-mail: gonchar@laser.nsc.ru

Abstract. This paper presents the recent experimental results on development of an optical frequency standard based on ultra cold magnesium atoms with relative frequency uncertainty and long term stability at the level of $\Delta\nu/\nu < 10^{-16}$. We stabilized the frequency of our clock laser system at 655 THz to narrow Ramsey fringes in a time separated laser fields interacting with cooled Mg atoms localized in a magneto-optical trap (MOT). The intercombination line $^1S_0 \rightarrow {}^3P_1$ was used as the reference for frequency stabilization. The results of stabilization were studied with femtosecond comb based on Ti:Sa laser.

Optical frequency standards based on cooled and localized atoms play an important role in both fundamental research and various metrological and navigational applications. Alkaline-earth and alkaline-earth-like atoms such as Yb [1], Ca [2], Sr [3–5], Hg [6] and Mg [7, 8] are the main candidates to create frequency standards of a new generation. Today, Yb and Sr optical-lattices frequency standards demonstrate a relative instability and uncertainty of extremely low levels $10^{-17} \div 10^{-18}$ [1, 4]. Mg frequency standard at the moment much less developed compare to Sr and Yb optical standards. At the same time, magnesium atoms have some advantages compared to the other candidates in terms of the frequency standard. Thus, the blackbody radiation (BBR) shift of the clock transition for magnesium is much smaller than for Yb, Ca, Sr. The simplified level scheme of Mg atoms is shown at figure 1.

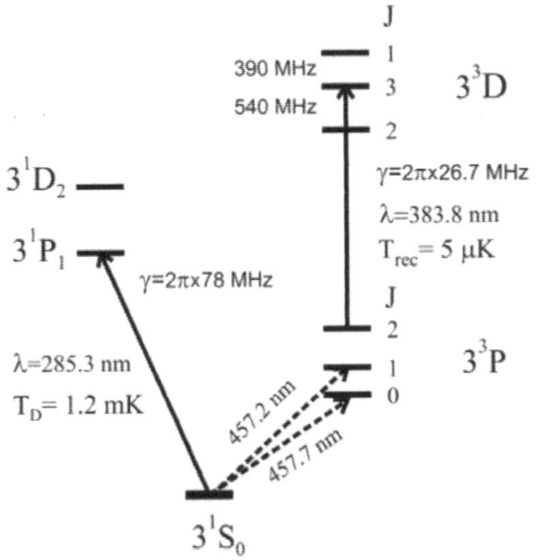

Figure 1. Simplified term diagram of ^{24}Mg atoms. Solid lines denote the cooling transitions with corresponding temperature limits, while dashed lines denote possible "clock" transitions, which can be used for laser frequency stabilization.

The resonant strong transition $^1S_0 \rightarrow {}^1P_1$ is a very suitable for an effective laser cooling and trapping of Mg atoms in a magneto-optical trap (MOT) directly from a thermal atomic beam. The ^{24}Mg $^1S_0 \rightarrow {}^3P_1$ transition natural width is only 31 Hz and it could be used to develop an optical frequency standard with high long term frequency stability. Strongly forbidden $^1S_0 \rightarrow {}^3P_0$ transition is of particular interest to develop an "optical lattice" standard with relative frequency uncertainty of $10^{-17} \div 10^{-18}$. To observe $^1S_0 \rightarrow {}^3P_0$ transition the magnetic-field-induced spectroscopy method could be used [9].

Despite a number of difficulties for deep cooling of Mg atoms down to a temperature less than 100 μK due to a large value of recoil energy, studies aimed at creating an optical frequency standard based on Mg atoms are carried out at the Institut für Quantenoptik, Hannover, Germany [7] and the Institute of Laser Physics, Novosibirsk, Russia [8]

In our previous studies we localized a cloud of cooled Mg atoms in MOT and have observed narrow Ramsey resonances in time separated laser fields at the intercombination transition $^1S_0 \rightarrow {}^3P_1$ [8]. This paper presents preliminary results on a clock laser frequency stabilization using Ramsey resonances and a frequency stability measurements with a frequency comb generator based on a femtosecond Ti:Sa laser. A schematic diagram of Mg optical frequency standard is shown at figure 2.

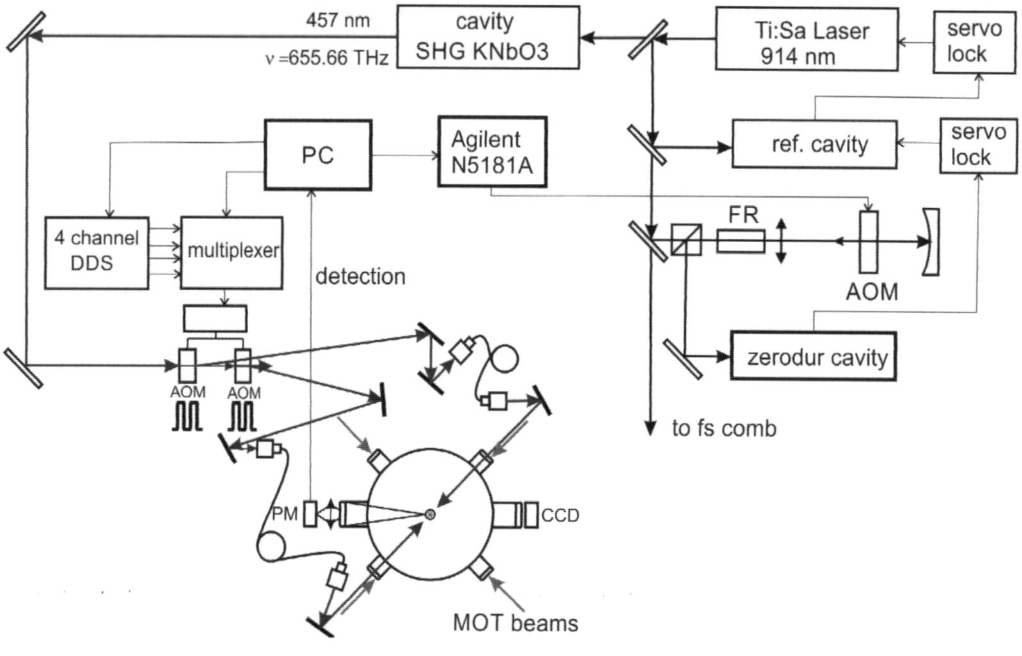

Figure 2. Simplified schematic of the experimental setup, showing the 914 nm Ti:Sa laser, the cavity with $KNbO_3$ crystal for frequency doubling to the "clock" wavelength of 457 nm, the reference cavity for Ti:Sa laser linewidth preliminary narrowing, the zerodur cavity with a two pass acousto-optical modulator (AOM) for frequency stabilization and fine tuning of the clock laser frequency, the MOT with a detection system.

The frequency of clock laser system was stabilized using a highly stable zerodur cavity and was tuned with the generator Agilent N5181A controlled AOM. A digital version of the third harmonic detection method is used to generate an error signal for frequency stabilization of the clock laser system to a central Ramsey fringe [10]. The four radio frequencies were generated with four channel DDS generator. These frequency were $\Delta_1 = f_0+\Delta/4$, $\Delta_{-1} = f_0-\Delta/4$, $\Delta_3 = f_0+3\Delta/4$, $\Delta_{-3} = f_0-3\Delta/4$, were Δ is a value of the Ramsey fringes period, f_0 = 80 MHz is a central frequency of AOM, that formed time separated pulses for cooled Mg atoms excitation. A calculated error signal $dS=(S_{\Delta-3}-3S_{\Delta-1}+3S_{\Delta 1}-S_{\Delta 3})$

was used for frequency stabilization. Figure 3 shows detected signal dS with a time separation of excitation pulses equal to 58 μs and with the pulses duration of 4 μs.

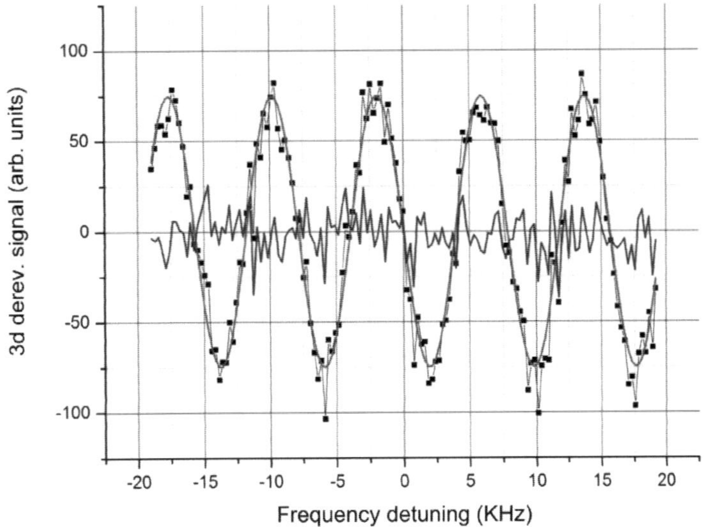

Figure 3. The error signal (a third harmonic of Ramsey fringes) that used for frequency stabilization (black line and symbols), red line is a sinusoidal fit of the error signal, blue line is a difference between the error signal and the fit. Blue line characterizes a signal/noise ratio S/N= 6, an averaging time is τ = 0.2 s for a point (total curve comprises 144 points).

Femtosecond frequency comb based on Ti:Sa laser [11] and referenced to Yb:YAG/I_2 optical frequency standard [12] was used to measure frequency stability of clock laser system stabilized to central Ramsey fringe. The scheme of the setup is shown at the figure 4.

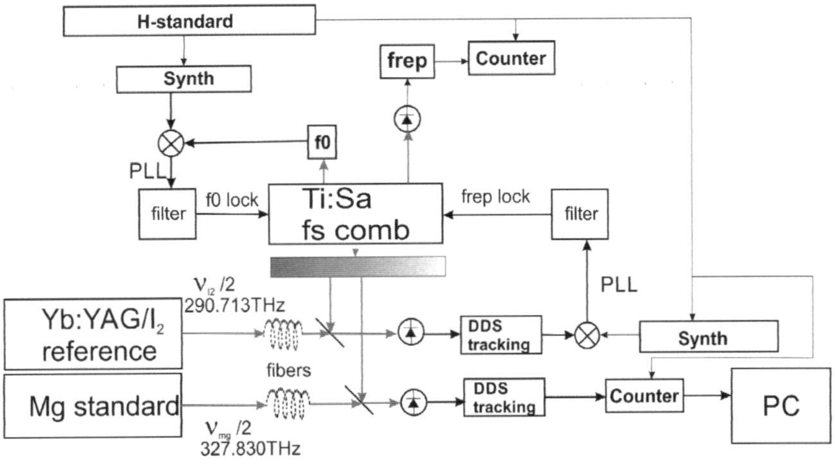

Figure 4. Scheme of the experimental setup for a frequency stability measurement of the clock laser system.

The precision of a frequency stability measurements was limited by the stability of the reference Yb:YAG\I_2 frequency standard. Figure 5 shows Allan deviation of our Mg optical frequency standard measured with femtosecond comb generator.

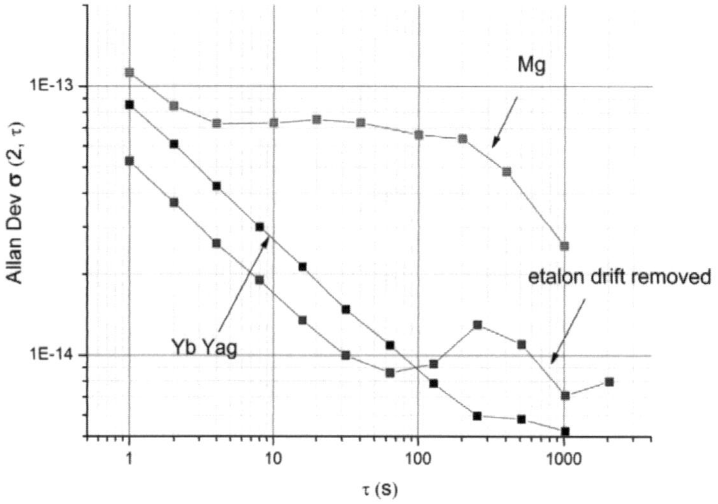

Figure 5. Allan deviation for the Mg frequency standard: locked to Ramsey fringes – red square, blue square –laser frequency lock to reference cavity only with a frequency drift removed from the measured data. Black square – Allan deviation for the Yb:YAG/I_2 reference laser measured with a beat signal of two Yb:YAG/I_2 lasers. The pick at the curve with blue square at 200 ÷ 400 s is due to frequency perturbations of the reference cavity that differ from a polynomial low and can't be removed from the data.

Preliminary experiments on frequency stability measurement demonstrate that at the moment the main limitation factors are large frequency drift of our zerodur reference cavity due to it temperature variation and relatively small value of S/N. The frequency stability of Mg standard will be improved substantially in future with an improvement of reference cavity thermal stabilization system and an optimization of Ramsey fringes detection.

In a future we plan to realize sub-Doppler laser cooling of magnesium atoms using $^3P_2 \rightarrow ^3D_3$ transition [13, 14] and localize atoms in an optical lattice for Mg optical frequency standard with relative frequency uncertainty of $10^{-17} \div 10^{-18}$.

Acknowledgments
The work was partially supported by the Russian Science Foundation (project no. 16-12-00054). A.N. Goncharov and D.V. Brazhnikov thank the Russian Foundation for Basic Research for the support (grant nos. 15-02-06087, 15-32-20330-"mol_a_ved"). D.V.B. was also supported by the Russian Presidential Grant no. NSh-6689.2016.2.

References

[1] Hinkley N *et al.* 2013 *Science* **341** 1215
[2] Sterr U *et al.* 2004 *C. R. Physique* **5** 845–55
[3] Falke S *et al.* 2014 *New J. Phys.* **16** 073023
[4] Bloom B J *et al.* 2014 *Nature* **506** 71

[5] Akatsuka T *et al.* 2010 *Phys. Rev.* A **81** 023402
[6] McFerran J J *et al.* 2012 *Phys. Rev. Lett.* **108** 183004
[7] Kulosa A P *et al.* 2015 *Phys. Rev. Lett.* **115** 240801
[8] Goncharov A N *et al.* 2014 *Quantum Electron.* **44** 521
[9] Taichenachev A V *et al.* 2006 *Phys. Rev. Lett.* **96** 083001
[10] Degenhardt C *et al.* 2005 *Phys. Rev.* A **72** 062111
[11] Matyugin Yu A *et al.* 2012 *Quantum Electron.* **42** 250–7
[12] Ignatovich S M *et al.* 2017 *J. Phys.: Conf. Ser.* **this volume**
[13] Prudnikov O N *et al.* 2015 *Phys. Rev.* A **92** 063413
[14] Prudnikov O N *et al.* 2016 *Quantum Electron.* **46** 661–7

IOP Conf. Series: Journal of Physics: Conf. Series **793** (2017) 012009 doi:10.1088/1742-6596/793/1/012009

Ultracold atoms for simulation of many body quantum systems

David A. W. Hutchinson

Dodd-Walls Centre, Department of Physics, University of Otago, Dunedin, New Zealand

E-mail: David.Hutchinson@otago.ac.nz

Abstract. Feynman famously proposed simulating quantum physics using other, better controlled, quantum systems. This vision is now a reality within the realm of ultracold atomic physics. We discuss how these systems can be used to simulate many body physics, concentrating the Berezinskii-Kosterlitz-Thouless transition in 2D physics and the role of disorder.

1. Introduction

It took 70 years from the prediction of Bose-Einstein condensation (BEC) in 1925 to the experimental realization [1] in 1995 in ultracold atomic gases. Whilst phenomena such as superfluidity in liquid helium, superconductivity of Cooper Pairs and even the macroscopic occupation of an individual cavity mode in the laser are all related to BEC, the experiments in 1995 where the first to directly show this condensation directly. In time of flight images – where the trapped cloud of atoms are allowed to expand and are then detected via an absorption image – the momentum distribution of the atoms is directly inferred showing the macroscopic occupation of the lowest momentum state.

The first experiments were then an exercise in characterization and developing an understanding of wave interference in these new systems. We were involved in a large part in the characterization of collective excitations at finite temperatures in these early systems, with the key characteristic being the demonstration of precision control both over internal and external degrees of freedom. For example, experiments by Debbie Jin [2] explored the temperature dependence of the collective excitation spectrum of early condensates revealing anomalous high temperature (here "high" means a hundred nano-Kelvin or so) shifts in the monopole and quadrupole modes. The theory to explain this, developed by us[3], involved a detailed understanding of the many body scattering properties of dilute gases in the presence of a condensate as well as the interplay between the dynamics of the condensate and that of the thermal atoms. The latter was characterized in a phenomenological manner by Zaremba [4] and co-workers and later by us in a microscopic treatment [5] based upon a Beliaev [6] type approach.

These early examples formed the foundation for studies of coherence including interference phenomena [7], over which it is now hard to believe there was such controversy and early demonstrations of matter wave beams, or atom lasers [8] as was the buzzword. Applications of matter wave interference and the atom laser have begun to find practical applications in the realms of inertial guidance [9–11] and even gravity sensing [12, 13] where the massive nature of the atoms yields an advantage over optical equivalents, but we believe it is in the field of quantum simulation that cold atoms will play their most important role.

2. Cold gases for quantum simulation

The ultracold gases were saved from being a purely demonstration experiment by the advent of further control over the inter-particle interactions through the use of so-called Feshbach [14] resonances. These resonances vary the relative energy between open and closed channels in two-particle scattering by the imposition of a uniform magnetic field. The differing Zeeman shift for the different levels then leads to a change in the position of the last bound state, varying the effective *s*-wave scattering length. Interactions could therefore be tuned from attractive through zero (non-interacting) to repulsive, including also the divergent scattering length in the unitary regime.

Optical lattices [15] were then introduced in to the experimentalist's arsenal of techniques. This uses two counter-propagating laser beams to create a standing wave. The ac Stark shift then leads to an effective harmonic potential for the atoms. One, two and three dimensional lattices can be formed in this manner together with more exotic potentials through the use of spatial light modulation techniques. By making one or more of these lattices very strong, with tight confinement in that direction, even the dimension of the system can be manipulated to create two-dimensional sheets, one dimensional wires, or dot-like structures.

3. Simulation of two-dimensional systems

Beautiful experiments by the group of Jean Dalibard in Paris then used optical lattices to produce two-dimensional sheets [16] of degenerate Bose gases. Theory of the non-interacting Bose gas shows that there is no BEC transition in the uniform two-dimensional Bose gas, but predicts a finite degeneracy temperature for a harmonically trapped gas – like those in the Dalibard experiments. The interacting two-dimensional gas can however undergo a superfluid phase transition to the so-called Berezinskii-Kosterlitz-Thouless (BKT) Phase. In this state fluctuations in the form of thermally activated vortices are bound in to vortx-antivortex pairs. Whilst global phase coherence – the characteristic of BEC – is destroyed, the system retains a local order that leads to a superfluid phase. The superfluid phase is then destroyed at higher temperatures by the vortex-antivortex pairs unbinding. The critical temperature is therefore related to the interparticle interactions. Cross over from the BEC to BKT phase and through to the thermal phase was investigated and characterized in these experiments and supporting theoretical work from ourselves[17–20] and others at Otago [21–23] as well as elsewhere [24].

Whilst this area is especially timely due to the recent award of the 2016 Nobel Prize in Physics to Kosterlitz and Thouless, as well as Duncan Haldane (also two-dimensional physics), much has been added to our understanding of two-dimensional physics in finite systems by studies in the cold gases. This continues to be the case, not least through the new experiments by Perrin reported elsewhere in this volume.

4. Disordered systems

The next piece in the quantum simulation jigsaw was the introduction of disorder. Disorder can be introduced in to a cold atom system via either a new, weak optical lattice incommensurate [25] with the primary optical lattice or through laser speckle [26]. The former, while not strictly disorder, provides an effectively random shift to the bottoms of the primary lattice potential. Laser speckle consists of light shone through a diffusive plate to provide a random optical potential shift. Both mechanisms have been used to demonstrate localization and effective superfluid to insulating phase transitions in ultracold gas systems. Work is now ongoing investigating the effects of disorder in ultracold gases with the hope of shedding light on the role of disorder in more complicated solid state systems.

One of the beauties of the cold atom systems, as stated above, is the ability to tune interactions. The interplay between localizing effects of disorder and the role of inter-particle interactions is a fascinating one and many body localization [27] is now one of the fashionable areas of cold atom research. Our main area of focus however has been on the role of magnetic fields in the presence of disorder. Simple arguments can be used to highlight how time reversal symmetry leads to enhanced localization in the presence of disorder. The imposition of a magnetic field breaks time reversal symmetry and hence suppresses localization leading to *negative magnetoresistance*. We have demonstrated theoretically [28] how this may be achieved in ultracold atomic gases, but how does one simulate a charged gas in a magnetic field using neutral atoms?

5. Synthetic gauge fields

The final element to be introduced to our world of ultracold atomic emulation of solid state systems are gauge fields. These ultracold atomic systems are by definition uncharged. They therefore do not couple to electric or magnetic fields in the sense of, say, electrons in solids. We need to introduce effective potentials in to the system Hamiltonian via other means. For example rotating the system introduces a term in the kinetic energy which mirrors the effects of a magnetic field. Quantised vortices and vortex lattices have been observed using rotation. Analogues of the integer and fractional quantum Hall effect based upon rotation have been proposed but are limited. Rotation also introduces an effective centripetal force which, at a rotation frequency equal to that of the harmonic trap confinement leads to instability of the gas through failure of the trapping.

A scheme first implemented by Ian Spielman's group at NIST used a Raman scheme to couple hyperfine states [29] in a spatially dependent manner. This lead to a minimum in the kinetic energy term centred at a momentum $\boldsymbol{p}\neq 0$. Effectively this introduces a spatially dependent vector potential $\boldsymbol{A}$ and hence a synthetic magnetic field. Spielman was able to demonstrate that implementing this coupling led to vortices in his condensate analogous to rotation. This scheme has since been extended to other gauge fields [30, 31] including non-abelian fields and also spin-orbit coupling.

6. Conclusions

In this very brief review we have concentrated on our own contributions to the field as this is what we talked about in Novosibirsk. I would however like to point the interested reader to the much better review article in *Nature Physics* [32].

What we would like to stress though is that ultracold atoms really offer a platform for studying many phenomena from solid state physics. They provide excellent control, through precision laser and atomic physics, overall system parameters. Model Hamiltonians can really be engineered in the spirit of Richard Feynman's vision [33] of quantum simulation from the 1980s.

References

[1] Cornell E A and Wieman C E 2002 *Rev. Mod. Phys.* **74** 875
[2] Jin D S, Matthews M R, Ensher J R, Wieman C E and Cornell E A 1997 *Phys. Rev. Lett.* **78** 764
[3] Hutchinson D A W *et al.* 2000 *J. Phys.* B **33** 3825
[4] Jackson B and Zaremba E 2001 *Phys. Rev. Lett.* **87** 100404
[5] Morgan S A, Rusch M, Hutchinson D A W and Burnett K 2003 *Phys. Rev. Lett.* **91** 250403
[6] Beliaev S T 1958 *Sov. Phys. JETP* **34** 323
[7] Andrews M R *et al.* 1997 *Science* **275** 637–41
[8] Helmerson K *et al.* 1999 *Physics World* **12** 31
[9] Shin Y *et al.* 2004 *Phys. Rev. Lett.* **92** 050405
[10] Scott R G, Judd T E and Fromhold T M 2008 *Phys. Rev. Lett.* **100** 100402
[11] Scott R G *et al.* 2009 *Phys. Rev.* A **79** 063624
[12] Dickerson S M *et al.* 2013 *Phys. Rev. Lett.* **111** 083001
[13] Biedermann G W *et al.* 2015 *Phys. Rev.* A **91** 033629
[14] Vogels J M *et al.* 1997 *Phys. Rev.* A **56** R1067
[15] Morsch O and Oberthaler M 2006 *Rev. Mod. Phys.* **78** 179–215
[16] Hadzibabic Z *et al.* 2006 *Nature* **441** 1118
[17] Gies C, van Zyl B P, Morgan S A and Hutchinson D A W 2004 *Phys. Rev.* A **69** 023616
[18] Gies C and Hutchinson D A W 2004 *Phys. Rev.* A **70** 043606
[19] Gies C, Lee M D and Hutchinson D A W 2005 *J. Phys.* B **38** 1797–809

[20] Simula T P, Lee M D and Hutchinson D A W 2005 *Phil. Mag. Lett.* **85** 395–404
[21] Simula T P and Blakie P B 2006 *Phys. Rev. Lett.* **96** 020404
[22] Bisset R N *et al.* 2009 *Phys. Rev.* A **79** 033626
[23] Foster C J, Blakie P B and Davis M J 2010 *Phys. Rev.* A **81** 023623
[24] Hadzibabic Z and Dalibard J 2011 *Rivista del Nuovo Cimento* **34** 389
[25] Roati G *et al.* 2008 *Nature* **453** 895–8
[26] Billy J *et al.* 2008 *Nature* **453** 891–4
[27] Choi J-Y *et al.* 2016 *Science* **352** 1547
[28] Towers J, Cormack S C and Hutchinson D A W 2013 *Phys. Rev.* A **88** 043625
[29] Lin Y-J *et al.* 2009 *Nature* **462** 628
[30] Dalibard J, Gerbier F, Juzeliunas G and Ohberg P 2011 *Rev. Mod. Phys.* **83** 1523–43
[31] Huo M-X *et al.* 2014 *Sci. Rep.* **4** 5992
[32] Bloch I, Dalibard J and Nascimbene S 2012 *Nature Physics* **8** 267–76
[33] Feynman R P 1982 *Int. J. Theo. Phys.* **21** 467

Yb:YAG/I_2 optical frequency standard at 515 nm with instability at the level 10^{-15}

S M Ignatovich[1], M N Skvortsov[1], V I Vishnyakov[1], D V Brazhnikov[1,2] and N L Kvashnin[1]

[1] Institute of Laser Physics SB RAS, 13/3 Acad. Lavrentyev ave., Novosibirsk 630090, Russia
[2] Novosibirsk State University, 2 Pirogova str., Novosibirsk 630090, Russia

E-mail: sign@laser.nsc.ru

Abstract. We present the results of development of an optical frequency standard based on Yb:YAG laser with second harmonic at 515 nm. The laser frequency is locked to the saturated-absorption resonance in a gas cell filled with molecular iodine. Final instability of the standard is 1.3×10^{-15} at 10^4 s.

1. Introduction

Optical frequency standards, involved saturated-absorption resonance in molecular iodine as a reference, have been developing many years [1-3]. The best value of instability, which was reached previously, is equaled to ~ 6×10^{-15} at 10^4 s. We developed optical standard of frequency based on Yb:YAG laser with intracavity second harmonic at 515 nm. As reference for stabilization resonance of iodine R(66)44-0 was chosen. It resonance has biggest ratio amplitude to width inside frequency tuning range of laser. Width of resonance was Γ = 310 kHz. For achieving of maximum stability optical scheme with detection of resonance in luminescent light was developed (see figure 1). Such scheme produce better signal/noise ratio relatively schemes with detection in transmitting light due to reducing short noise of photo detector from background light. Also luminescent detection make possible use for receiving of signal bigger photosensitive area, what allowing to use massive of photodiodes with next adding all signals together. So as every photodiode produce shot noise signal/noise ratio increased in $\sqrt{N}$ times, where N is a quantity of photodiodes in the massive. We are using 24 FDS10X10 Thorlabs photodetectors in every standard, it produce of signal/noise in 5 times higher in comparison with one photodiode and allow to observe resonance with signal/noise $>10^4$ in bandwidth 10 Hz.

Conventional locking scheme based on the third harmonic detection of a signal [4] has been used in our case. Functional diagram of the optical frequency standard is presented at figure 1.

2. Setup

We have quantitatively estimated the influence of various destabilizing factors on performance of optical frequency standard and, as the result, made several key modifications:

1. Electronic digital systems for precise laser beam modulation and for detecting of reference resonance have been developed and implemented. Modulation was produced by acousto-optic modulator (AOM) with signal from direct digital synthesizer (DDS). For achieving high stability we apply OCXO quartz with instability 10^{-10} in 1 s as clock signal for DDS. It allows tuning frequency of laser with accuracy 4×10^{-3} Hz. Also this clock signal was used in digital lock-in amplifier (DLock) for

detection of resonance. Digital lock-in was built at FPGA chip. Main advantage of digital lock-in is absence of output drifts of zero level, but for realization of this property we needed to program PID regulator inside of FPGA.

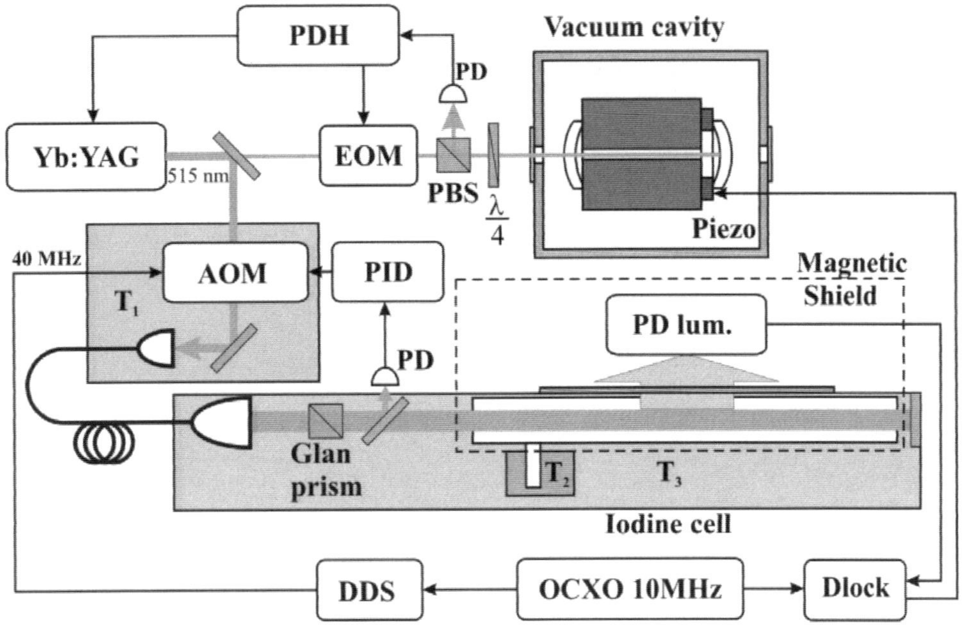

Figure 1. Functional diagram of optical frequency standard.

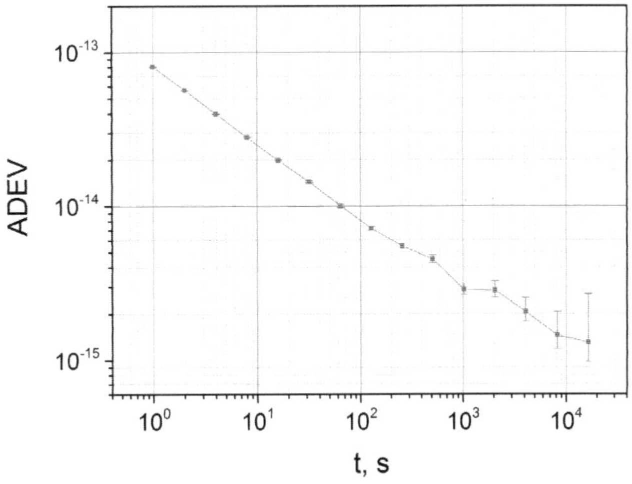

Figure 2. Allan deviation.

2. Stabilization of residual amplitude noises in modulated laser beam has been ensured up to 10^{-8} from total power. It was done due to mixing irregularity distribution of noises over the beam diameter in the polarization maintaining optical fiber and suppression of noises by PID which controlled power of signal at AOM [5].

3. Two-stage system of iodine cell thermostabilization has been designed for accurate stabilization of iodine vapors pressure. So as vapors pressure shift is ≈ –800 Hz/°C, instability in temperature should be better than 1 mK. It was achieved with two stages. External stage was built around all cell and has accuracy 20 mK. Internal stabilize temperature only finger of cell with instability <1 mK.

4. High-quality beam profile has been provided, allowing us to avoid relevant frequency shift due to wave-front curvature. It was done due to polarization maintaining optical fiber with high quality collimator.

Stability of the constructed optical frequency standard has been retrieved from stability of the beat signal of two identical setups. So, instability equals to 1.3×10^{-15} at 10^4 s (see figure 2).

Acknowledgments
This work was partially supported by the Russian Foundation for Basic Research (grant nos. 15-32-20330, 16-32-00159) and Russian Presidential Grant no. NSh-6689.2016.2.

References

[1] Skvortsov M N, Okhapkin M V, Nevsky A Yu and Bagaev S N 2004 *Quantum Electronics* **34** 1101
[2] Hall J L, Ma L-S, Taubman M, Tiemann B, Hong F L, Pfister O and Ye J 1999 *IEEE Trans. Instrum. Meas.* **48** 583
[3] Korel I I, Nyushkov B N, Denisov V I, Pivtsov V S, Koliada N A, Sysolyatin A A, Ignatovich S M, Kvashnin N L, Skvortsov M N and Bagaev S N 2014 *Laser Physics* **24** 074012
[4] Nyholm K, Merimaa M, Ahola T and Lassila A 2003 *IEEE Trans. Instrum. Meas.* **52** 284
[5] Denisov V I, Ignatovich S M, Kvashnin N L, Skvortsov M N and Farnosov S A 2016 *Quantum Electronics* **46** 464

Deep laser cooling of strontium atoms on $^1S_0 \rightarrow {}^3P_0$ transition

R Ya Ilenkov[1,2], A V Taichenachev[1,2], V I Yudin[1,2,3] and O N Prudnikov[1,2]

[1] Institute of Laser Physics SB RAS, 13/3 Acad. Lavrentyev ave., Novosibirsk 630090, Russia
[2] Physics Department, Novosibirsk State University, 2 Pirogova str., Novosibirsk 630090, Russia
[3] Physical Engineering Department, Novosibirsk State Technical University, 20 K. Marx ave., Novosibirsk 630073, Russia

E-mails: ilenkov.roman@gmail.com

Abstract. Momentum distributions of strontium atoms are characterized by means of bimodal functional dependences, which properties depend on the laser field parameters. It is shown that ultracold fraction of atoms ($T \leq \hbar\gamma/k_B$) can reach 60 %. The calculations are based on the quantum treatment of the problem with taking into account the recoil effect in full. The results can be used for ultradeep laser cooling of strontium atoms.

1. Introduction

Invention such precise and powerful tool as a laser has opened before scientists a wide range of capabilities for atom manipulation: acceleration, deceleration, localization, deflection, and focusing. So laser cooling has become an integral part of both fundamental science and many practical applications (high-precision frequency and time standards, nanolithography, quantum information etc.)

The theoretical description of the kinetics of neutral atoms in the polarized light fields with all the atomic levels, the coherence, the recoil effect is both important and challenging problem. The first step toward understanding mechanisms of interaction between atoms and light was called quasi-classical approach [1, 2]. It lies in the fact that the equations for the density matrix can be reduced to the Fokker-Planck equation for the Wigner function in the phase space. Simplicity of this approach has allowed to understand many of cooling mechanisms in the usual and ordinary terms of force and diffusion. However, this approach can only be applied in certain cases. First, the small recoil frequency parameter compared to the rate of spontaneous decay, and secondly, the momentum of a light field photon should be much smaller than the width of the momentum distribution of the atoms. Later quantum methods were developed [3, 4], for example, the secular approach which describes cooling and localization of atoms in the optical potential. In this approximation distance between the energy bands in the optical potential is greater than their broadening caused by optical pumping. At a fixed depth of the optical potential this approximation is valid in the limit of large detuning, and thus, for a given configuration is disrupted in a deep optical potential. Moreover, even when this condition, the secular approximation is valid only for the lower vibrational levels, and fails for the higher, where the distance between the levels becomes smaller due to the effects of anharmonicity. The more secular approximation is not applicable to atoms undergo above-barrier motion.

Therefore, to describe the laser cooling on forbidden transitions, such as intercombination transitions of Yb, Mg, Sr, used in experiments for ultradeep laser cooling [5–8] it is necessary to use methods of quantum effects into account the effect of the impact and spatial localization of atoms.

2. Laser cooling of strontium atoms

The system of equations on density matrix elements in two-point representation describing the interaction of two-level atom with a field of a standing light wave with taking into account of the effect of recoil and spatial localization:

$$\begin{cases} \left(-\frac{i\hbar}{m}+\frac{\gamma}{2}-i\delta\right)\rho_{21}(1,2)=-\frac{i}{\hbar}\left(V(1)\rho_{11}(1,2)-\rho_{22}(1,2)V(2)\right) \\ \left(-\frac{i\hbar}{m}+\frac{\gamma}{2}+i\delta\right)\rho_{12}(1,2)=-\frac{i}{\hbar}\left(V^{*}(1)\rho_{22}(1,2)-\rho_{11}(1,2)V^{*}(2)\right) \\ \left(-\frac{i\hbar}{m}+\frac{\gamma}{2}-i\delta\right)\rho_{22}(1,2)=-\frac{i}{\hbar}\left(V(1)\rho_{12}(1,2)-\rho_{21}(1,2)V^{*}(2)\right) \\ \left(-\frac{i\hbar}{m}+\frac{\gamma}{2}-i\delta\right)\rho_{11}(1,2)-\gamma f(q)\rho_{22}(1,2)= \\ =-\frac{i}{\hbar}\left(V^{*}(1)\rho_{11}(1,2)-\rho_{22}(1,2)V(2)\right), \end{cases} \quad (1)$$

there m – atomic mass, $\delta=\omega-\omega_0$ – detuning of light field of atomic transition frequency, γ – frequency of spontaneous radiative decay, $x=(x_1+x_2)/2$, $q=x_1-x_2$ – spatial coordinates, ρ_{11},ρ_{22} – the population of the excited and ground states, ρ_{12},ρ_{12} – optical coherence between levels, $f(q)$ - function which characterizes the recoil effect in the spontaneous emission:

$$f(q)=\frac{3}{2}\left(\frac{cos\,(q)}{q^2}-\frac{sin\,(q)}{q^3}+\frac{sin\,(q)}{q}\right), \quad (2)$$

and $\hat{V}$ – interaction operator with field of a standing light wave

$$\hat{V}=-2dE_0 cos\,(kx)\begin{pmatrix}0 & 1\\ 1 & 0\end{pmatrix}, \quad (3)$$

There d is the matrix element of atomic dipole moment, E_0 is the electric field amplitude.

We have developed an own quantum method [9] to obtaining the stationary distribution of two-level atoms in a standing wave of arbitrary intensity, allowing full account the recoil effect. The method used is to decompose the density matrix elements in the Fourier series for the spatial harmonics, which is possible due to periodicity of the light wave. Thus we obtain a system recursively coupled equations, where each harmonic is expressed through the previous one, and starting from free selected one (in our calculations, usually twenty or more) all the harmonics are equal to zero. Using this method kinetics of atoms in light fields of varying intensity was investigated. The new and most important result was mode which we called the anomalous localization. In strong standing wave (Rabi frequency greater than the constant spontaneous relaxation) was detected a anomalous behavior of atoms, namely, the concentration at the peaks of the optical potential [10].

The next important step was to study the quantum modes for different parameters of the problem. It is known that the quasi-classical approach gives Gaussian shape momentum distributions of and is completely inapplicable to the quantum regimes ($w_r \approx \gamma$, with $w_r = \hbar k^2/2m$ the recoil frequency and k the wave vector). Laser cooling of strontium atoms at the transition $^1S_0 \rightarrow {}^3P_0$ is a vivid example of such a regime. It is studied in this paper. The results of quantum calculation are shown in figure 1. There is a bimodal distribution: a narrow central peak and a broad substrate (background). Such a clearly expressed bimodal structure with high accuracy can be approximated by two Gaussian functions corresponding to the two-speed groups of atoms. Atoms are essentially non-equilibrium momentum distribution, which is a good approximation can be described by two Gaussian functions. For these parameters "cold" fraction of atoms has temperature $T_1=0.67\hbar\gamma$, and "hot" fraction of atoms $T_2=26.2\hbar\gamma$.

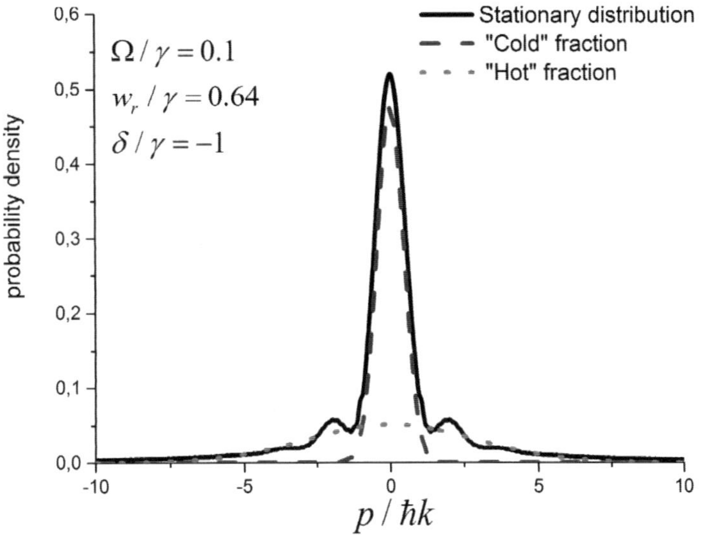

Figure 1. Bimodal momentum distribution of cold strontium atoms. The distribution has narrow structure of cold atoms and wide wings. Parameters of the calculation are written on the figure.

The proportions of "hot" and "cold" fractions of atoms figure 2b, as well as their temperatures figure 2a depend on the parameters of light fields (the Rabi frequency $\Omega = dE_0/\hbar$). Temperature of "cold" fraction is significantly lower than the average temperature defined as $<p^2>/m$ (with $<p^2>$ the average square of atomic momentum) and reaches the limit values $< \hbar\gamma$. The main trend is the general heating. It increases as the temperature of cold fraction and fractions hot temperature. The temperature of the hot fraction has two behaviors violate the monotony of growth. They both can be attributed to distortions of the form of pulse distributions of atoms. The number of atoms in the "cold" fraction maximizing in a weak field, and reaches 60 %. In strong fields the contribution of both factions aligned and it is about 50 %.

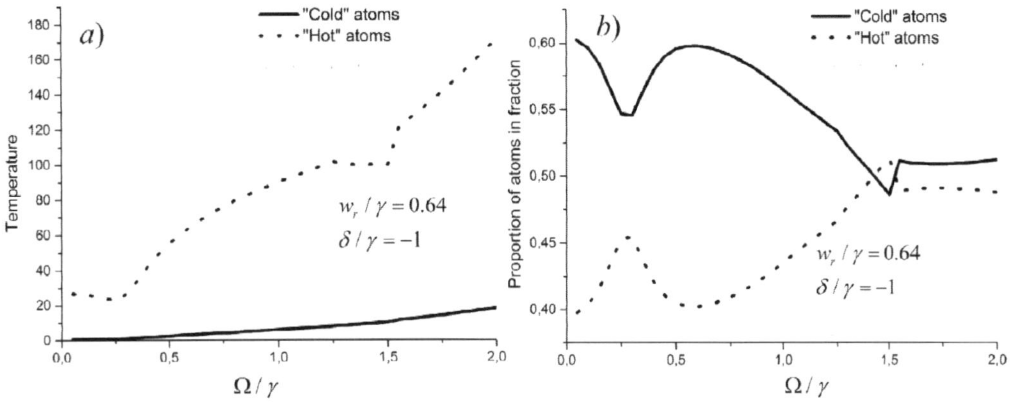

Figure 2. Dependences of temperature (a) and proportion (b) of atoms for "cold" and "hot" fractions of atoms of Rabi frequency. Parameters of the calculation are shown on the plots.

3. Conclusion

Investigation of the possibility of cooling of strontium atoms on the weak transition has been carried out. It has been shown that the atoms can exhibit the essentially non-equilibrium momentum distribution, which can be well approximated by two Gaussian functions. The possibility of obtaining the temperature of cold fraction below $\hbar\gamma$ has been demonstrated. The proportion of cold atoms fraction can be sufficiently high and reach 60 %. If we put atoms with bimodal distribution into the shallow dipole trap, it will lead to cutting off the hot atoms and we get the final momentum distribution of deeply cooled atoms, maintaining a significant number of cold atoms.

Acknowledgments

The studies presented in this paper were supported by the Russian Scientific Foundation (grant no. 16-12-00054).

References

[1] Minogin V G, Letokhov V S 1987 *Laser Light Pressure on Atoms* (Gordon and Breach, New York) p 248
[2] Kazantsev A P, Surdutovich G I and Yakovlev V P 1990 *Mechanical Action of Light on Atoms* (World Scientific, Singapore) p 392
[3] Aspect A, Arimondo E, Kaiser R *et al.* 1988 *Phys. Rev. Lett.* **61** 826
[4] Yoo S M and Javanainen J 1991 *J. Opt. Soc. Amer.* B **8** 1341
[5] Barber Z W, Stalnaker J E, Lemke N D *et al.* 2008 *Phys. Rev. Lett.* **100** 103002
[6] Poli N, Barber Z W, Lemke N D *et al.* 2008 *Phys. Rev.* A **77** 050501
[7] Poli N, Drullinger R E, Ferrari G *et al.* 2005 *Phys. Rev.* A **71** 061403
[8] Brazhnikov D V, Bonert A E, Goncharov A N *et al.* 2014 *Laser Phys.* **24** 074011
[9] Prudnikov O N, Il'enkov R Ya, Taichenachev A V *et al.* 2011 *JETP* **112** 939
[10] Brazhnikov D V, Il'enkov R Ya *et al.* 2012 *JETP Lett.* **95** 399

Mid-infrared tunable, narrow-linewidth difference-frequency laser based on orientation-patterned gallium phosphide

G Insero[1,2], C Clivati[3], D D'Ambrosio[3], P De Natale[1,2], G Santambrogio[1,2,3], P G Schunemann[4], S Borri[1,2], and J J Zondy[5]

[1]Istituto Nazionale di Ottica, INO-CNR, and European Laboratory for Nonlinear Spectroscopy, LENS, via N. Carrara 1, 50019 Sesto Fiorentino, Italy
[2]Istituto Nazionale di Fisica Nucleare, INFN, Sezione di Firenze, via G. Sansone 1, 50019 Sesto Fiorentino, FI, Italy
[3]Istituto Nazionale di Ricerca Metrologica, INRIM, Strada delle Cacce 91, 10135 Torino, Italy
[4]BAE Systems, Inc., MER15-1813, P.O. Box 868, Nashua, NH, USA 03061-0868
[5]Nazarbaev University, School of Science and Technology, Physics Department, Kabanbay Batyr 53, 010000 Astana, Kazakhstan

E-mail: jeanjacques.zondy@nu.edu.kz

Abstract. We report on the first characterization of orientation-patterned gallium phosphide (OP-GaP) crystals used to generate narrow-linewidth, coherent mid-infrared (MIR) radiation at 5.85μm by difference frequency generation (DFG) of continuous-wave (cw) Nd:YAG laser at 1064nm and diode-laser at 1301nm. By comparison of the experimental absolute MIR efficiency versus focusing to Gaussian beam DFG theory, we derive an effective nonlinear coefficient $d = (1/2)\chi^{(2)}_{eff}$ =17(3) pm/V for first-order quasi-phase-matched OP-GaP at the generated DFG wavelength. Using $d = (2/\pi)d_{14}$ and taking into account Miller's delta rule, we retrieve an absolute value of the d_{14} quadratic nonlinear susceptibility coefficient of GaP of $d_{14} = 27.2(3)$ pm/V at 5.85 μm, in good agreement with the latest absolute measurement of this nonlinear coefficient from non-phase-matched second-harmonic generation (1.32 μm ⟶ 0.66 μm) taking into account multiple reflection effects [Shoji *et al* 1997 *J. Opt. Soc. Am.* B **14** 2268]. The temperature and signal-wave tuning curves are also in qualitative agreement with a recently proposed temperature-dependent Sellmeier equation for OP-GaP when focusing effects are taken into account.

1. Introduction

Narrow-linewidth (<100 kHz), single-frequency and powerful laser sources are needed for precision molecular spectroscopy in the mid-IR range above 5 μm. The 6-μm range is particularly interesting for high-precision spectroscopy of cold CO molecules [1], and the only available sources, with linewidths in the few MHz to tens of MHz range and ~ 0.1W maximum power, are quantum cascade lasers (QCL's) [2]. For precision spectroscopy, μW-level of narrow-linewidth coherent radiation must be produced to either phase-lock or injection-lock such QCL lasers [3, 4].

Continuous-wave (cw) OPOs would be interesting alternative coherent sources combining single-frequency and powerful emission, however except with silver gallium sulfide ($AgGaS_2$) at 2.5 μm [5, 6], genuinely cw mid-IR OPO's have not yet been demonstrated with the existing mid-IR birefringent nonlinear materials (among which $AgGaS_2$, $LiInSe_2$, $CdSiP_2$ [7]) due to their larger residual absorption loss compared to oxide materials ($LiNbO_3$, $KTiOPO_4$ ...) whose mid-IR transparency range cut-off however lies at ~ 4.5 μm. Deeper (> 5 μm) mid-infrared alternatives to oxide quasi-phase-matched (QPM) nonlinear crystals (PPLN, PPLT,...) have appeared since a couple of decade [8], in the form of orientation-patterned III-V semiconductors such as gallium arsenide (OP-GaAs) for which periodical domain inversion is realized during the growth process using molecular beam epitaxy and hydride vapor phase epitaxy (MBE-HVPE) for the re-growth of thicker (~ mm) samples [9, 10]). OP-GaAs cannot be pumped at the popular laser wavelength of 1064nm due to its strong absorption and low energy bandgap. Despite a lower nonlinearity, OP-GaP exhibits a larger bandgap [11] and thus should tolerate 1-μm cw laser pumping, making it a potential candidate for cw OPO emitting above 5 mm. To date however none of them have yet demonstrated *genuinely* cw parametric oscillation. Only *quasi*-cw parametric oscillation at 4.7 μm has been reported for OP-GaAs by chopping the cw pump laser so has to avoid deleterious thermal effects, despite a pump wavelength of 2.1 μm (Ho:YAG laser) where residual absorption is much lower than at 1064 nm [12].

The remaining alternative to cover the 5 – 6 μm range with μW to mW power range from convenient ~1 μm lasers is then to use difference frequency generation (DFG) with a ~ 1.3 μm diode laser in a high-nonlinearity mid-IR crystal. For cw DFG quasi-phase-matched OP-GaAs or OP-GaP are preferable to birefringence phased-matched chalcogenides because the first ones are free from spatial walkoff which dramatically reduces the nonlinear interaction length. While down-conversion schemes using OP-GaAs has been extensively investigated, mostly in the pulsed regime [13,14], to date still few work has been devoted to the newly developed orientation-patterned gallium phosphide (OP-GaP) [15]. A recent cw DFG experiment making use of OP-GaAs pumped by a 40 W Tm: fiber laser at 2.01 μm and a tunable (2.4 – 2.9 μm) cw OPO idler-wave laser with up to 3 W power has yielded multi-tens of mW of 6.4 – 7.6 μm radiation [16]. In the present case where the target MIR range is 5.8 – 6 μm, and given the much lower of the signal-wave diode laser at ~ 1.3 μm, the choice of OP-GaP rather than OP-GaAs is, in addition to its larger thermal conductivity, dictated by its larger bandgap energy which is better suited for a pumping at 1064nm [7].

In this Letter, we provide the first characterization of the linear, thermo-optic and nonlinear properties of OP-GaP via cw DFG of a single-frequency Nd:YAG laser (λ_p = 1064nm, *Mephisto* MOPA, Coherent Inc, linewidth Δν ~ 10kHz, maximum power 50W) and an extended-cavity tunable diode laser, called here after the signal laser (λ_s = 1301.1nm, DL 100, Toptica Photonics AG, Δν ≈100kHz, maximum power 50 mW). Up to 65 μW of single-frequency idler at λ_i = 5.85 μm have been generated from ~10W of Nd:YAG laser and ~ 45mW of diode laser in a 24.6mm long QPM structure, limited by thermal dephasing effects arising from the non-negligible absorption at the pump and signal lasers' wavelengths, despite the larger bandgap of GaP. From the absolute measurement of DFG conversion efficiency versus focusing and comparison with cw Gaussian beam DFG theory [17] an effective first-order QPM nonlinear coefficient $d = (2/\pi)d_{14}$ =17pm/V (±15%) has been evaluated. From the experimental spectral and temperature tuning curve bandwidths, we could also validate a recently proposed temperature-dependent Sellmeier equation for GaP [15]. Additionally, we have highlighted an OP-GaP/GaAs QPM layer thickness limitation in tightly focused down-conversion processes yielding long-wavelength idler: a minimum QPM layer thickness of ≧ 1 mm is required so as to prevent severe loss of the highly diffracting long-wave idler as soon as it expands outside of the periodically-poled layer, due partly to the higher absorption in the (unpoled) bulk semiconductor substrate.

2. DFG experimental setup

In this section we describe the experimental laser set up, the samples of OP-GaAs used and experimental methods aimed at retrieving the absolute value of the effective nonlinear coefficient of OP-GaP.

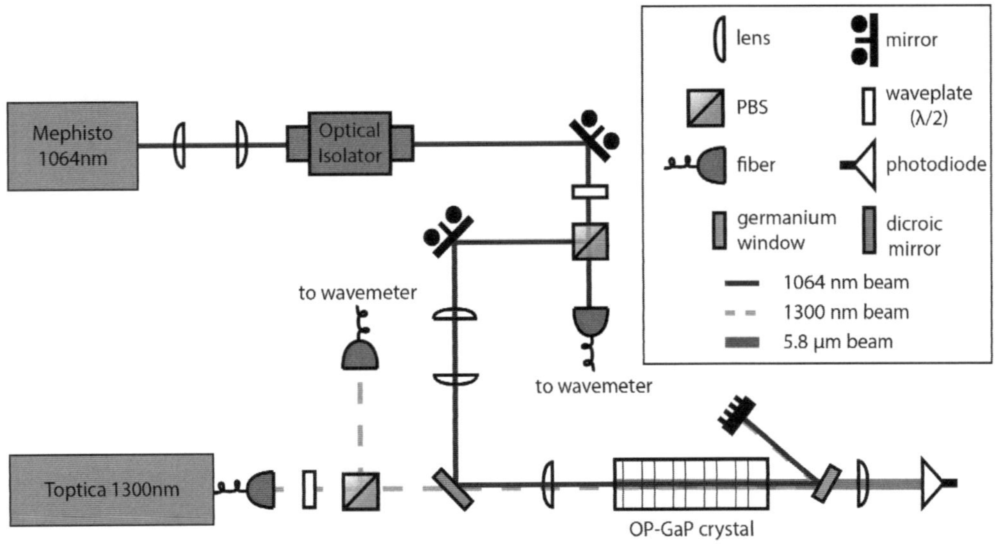

Figure 1. Experimental DFG laser setup. The external-cavity diode laser is fiber-outcoupled. The pump and signal waves are polarized vertically (along the [001] crystallographic axis) and horizontally (along [110]), which yielded horizontally polarized idler radiation. The CaF_2 collecting lens after the Ge filter is actually made of one f = 5 cm lens collimating the diverging idler at the output of the crystal and a second one focusing it on the thermo-electrically cooled 200 μm diameter active area MCT photovoltaic detector. The OP-GaP crystal is hosted in a temperature- regulated oven.

2.1. The laser setup

The layout of the single-pass difference-frequency laser setup is shown in figure 1. The pump laser (*Mephisto* MOPA) is a high-power master-oscillator power-amplifier Nd:YAG laser. The master oscillator is made of a non-planar ring oscillator (NPRO) monolithic cavity, conferring to the emission a narrow short-term linewidth (Δν = 10 kHz). It injection-locks a subsequent Nd:YAG solid-state power amplifier, capable of delivering up to 50 W of laser power. The *Mephisto* MOPA pump laser output power was however limited to P_p < 10 W during the experiment, due to strong absorption of the samples at 1064 nm and thermal saturation of the DFG efficiency. The signal-wave laser (DL 100 Toptica) is a grating-tuned external-cavity diode laser emitting at ~1301 nm. Its output power is limited to 40 mW, so that the pump drive exceeds by far the signal drive. Both near-IR laser beam transverse mode profiles are nearly TEM_{00}.

A variable telescope allows to match the pump beam diameter to the signal beam diameter (D ~ 2mm), and to longitudinally superimpose the two waits inside the crystal during the idler power optimization. The pump and signal beams are superimposed by a dichroic beam-splitter and then focused inside the crystal by several glass (BK7) lenses with various focal lengths (f=150, 100, 75, 50 mm) in order to investigate the dependence of the DFG efficiency on focusing. For tightly focused cw DFG, it is essential to insure a perfect spatial overlap of the two input waists. The common location of the two waists inside the crystal is achieved by fine longitudinal translation of the first telescope lens, while lateral overlap is achieved with slight angular tilts of the beamsplitter (for the pump) and of another (not shown in figure 1) signal laser steering mirror. These fine adjustments are facilitated once enough idler beam signal is detected by the thermo-electrically-cooled mercury cadmium telluride (MCT) photo-voltaic detector equipped with a trans-impedance pre-amplifier (VIGO System). To facilitate idler wave voltage optimization on the oscilloscope monitor, the signal laser is mechanically chopped during the alignments. The generated idler beam at λ_i ~ 5.85 μm is filtered from the near-IR pump and signal beams using a thick Ge window. The photodetector absolute power responsivity in V/W was calibrated with tens of microwatt radiation form a 6-μm attenuated QCL light that was simultaneously measured by a calibrated powermeter and an absolute radiometer (both giving a consistent power reading to ±5%). All the MIR power reported further has been corrected for all the propagation loss from the OP-Gas output to the MCT detector.

2.2. The OP-GaP samples

The OP-GaP QPM structure was grown by polar-on-nonpolar molecular beam epitaxy (MBE), lithographically patterned, reactive ion-etched and regrown by MBE to yield tens of nanometer-thick templates for subsequent re-growth by low-pressure hydride vapor phase epitaxy (HVPE) [11,15]. The crystal samples carried a 400μm-thick bulk GaP substrate, over which the patterned QPM layer extends for another 400μm thickness. An unpatterned GaP layer extends further above the QPM layer, as a result of the termination of the HVPE regrowth process after the domain walls have coalesced. The duty-cycle quality of the domain period is quite good, showing excellent domain propagation as shown in ref.[15]. The overall thickness of the chips is ~1.5mm, with a ~ 6mm wide QPM layer carrying a 50%-duty cycle periodic domain reversal (grating period $\Lambda = 24\mu m$).

In this experiment, we have used two OP-GaP samples ($l_c = 15.3$ and 24.6 mm) cut from the $\Lambda = 24$ μm channel. Only the base of the crystals (corresponding to the original wafer base) is optically polished, their upper surface being left rough. Their input/output facets are anti-reflection-coated at the 3 wavelengths ($R_{p,s,i} < 1\%$). The overall thickness of the chips is ~1.5mm, with a ~ 6mm wide QPM layer carrying a 50%-duty cycle periodic domain reversal (grating period $\Lambda = 24\mu m$).

The expected quasi-phase-matching temperature for the downconversion process 1064.3 nm – 1301 nm ⟶ 5.85 μm, calculated from the OP-Gas temperature Sellmeier equation [15] is $T \sim 43^{\circ}$ C. Therefore, the samples were housed in a massive copper-block oven whose temperature was regulated by a P-I servo-controller, and the oven was mounted on a XYZ-θφ positioner for DFG efficiency optimization. No crystal clamping to the polished copper base plate was used, relying on the sole flatness of the polished crystal base to insure thermal conduction (the top groovy facet of the samples did not touch the oven cap, preventing thus heat removal by conduction from this facet). The experimental phase-matching temperature was found to be around $T = 43^{\circ}$C as expected. One disadvantage of oven-heating as compared to the use of a thermos-electrically-cooled Peltier element is that while heating process is fast, cooling down the crystal in order to compensate for the thermally-induced radial phase mismatch $\Delta k(T,r)$ arising from the strong pump absorption. Typically we estimate the on-axis ($r = 0$) temperature gradient due to absorption heating to be as large as $\Delta T = 4 - 5^{\circ}$C at a pump power of ~ 6–7 W. Each time the pump is ramped, one has to lower gradually the oven set-point temperature in order to optimize the DFG efficiency by partly compensating for the thermally-induced phase-mismatch. For the largest pump power used (P_p = 8–10 W) the oven temperature could no more be decreased below $T \sim$ 35-36°C due to the thermal inertia of the massive copper block. In future experiments, a thermo-electric cooling method in better suited for the dynamic phase-matching control of the thermally-loaded sample.

Prior to the DFG experiment, the absorption of the OP-GaP chips at both the pump and signal laser wavelengths were cautiously characterized, as the retrieval of the absolute nonlinear coefficient from Gaussian beam DFG theory requires the knowledge of the absorption loss at the near-IR wavelengths (the loss at 6 μm measured from additional FTIR absorption spectrum data is $\alpha_i \sim 0.01$ cm^{-1}). The near-IR loss within the QPM layer was found somewhat less than the absorption loss in the bulk GaP substrate (the HVPE growth technique is almost free from the point defects found in bulk Czochralski-technique growth process). Because the absorption coefficients in the near-IR range as measured by FTIR spectroscopy are less accurate, the absorption coefficient of the samples was measured using the laser sources, by measuring the power transmittance $P_{out}/P_{in} = \exp(-\alpha l_c)$ of the focused laser beams. While inside the 400 μm-thick QPM layer, average measured absorption coefficients amounted to $\alpha_p \approx 0.17$ cm^{-1} and $\alpha_s \approx 0.12$ cm^{-1}, much larger coefficients were found within the bulk substrate (α_p

$\approx 0.58 cm^{-1}$ and $\alpha_s \approx 0.53 cm^{-1}$). Such amount of absorption at 1.3 μm and 1.064 μm within the QPM layer does actually not reflect the intrinsic expected absorption coefficient, which should be much lower according to the larger UV bandgap of GaP compared with GaAs. Further investigation to understand the origin of these absorption bands are being carried out.

2.3. *Experimental methods: Waist measurements*

In order to analyze the DFG conversion efficiency using Gaussian beam theory [17], it is important to measure as accurately as possible the pump and signal laser waists. The waists $w_{p,s}$ were accurately measured using either the knife-edge technique or the calibrated pinhole transmission technique [19], and were found, as expected for Gaussian beams with equal diameter focused by the same lens, to satisfy the diffraction-limited formula $w = (4/\pi)\lambda f/D$ where f is focal length of the common focusing lens located in front of the sample. As a consequence, all waist ratios were further found to satisfy $w_s/w_p \approx \lambda_s/\lambda_p$. Table 1 summarizes the result of the waist measurements corresponding to the different focusing lenses used. The focused beam DFG theory states that optimal conversion efficiency $\Gamma = P_i/P_pP_s$ (in unit of W/W^2) is achieved when the equal pump/signal confocal parameter condition (ECP) is achieved [17], that is when the two *internal* Rayleigh range $z_{p,s} = \frac{1}{2}k_{p,s}w_{p,s}^2$ are equal ($k = 2\pi n/\lambda$ is the internal wavevector accounting for the index of refraction $n = n(\lambda)$). Although our waist ratio $w_s/w_p \approx \lambda_s/\lambda_p$ does not correspond to the ECP condition ($z_p = z_s$), the experimental ratio is sufficiently close to it. In calculating the theoretical efficiency $\Gamma(w_p, w_s)$ versus the focusing parameter $l = l_c/z_d$, the experimental waist ratio condition must be imposed to confront theory and experimental result. In the definition of the focusing parameter $l = l_c/z_d$, $z_d = \frac{1}{2}k_i w_d^2$ and $w_d = w_p w_s/[w_p^2 + w_s^2]^{1/2}$ are respectively the idler Rayleigh range and idler effective waist (which is smaller than either the pump or signal waist, meaning that the generated idler is highly diffracting all the more than the generated MIR wavelength is long).

Table 1. Measured experimental waists w_p and w_s (±10% uncertainty) and calculated idler waist $w_d = w_p w_s/[w_p^2 + w_s^2]^{1/2}$ and idler Rayleigh range $z_d = (1/2)k_i w_d^2$

f(mm)	w_p (μm)	w_s (μm)	w_d (μm)	z_d (mm)
150	67.5	82.5	52.24	4.67
100	45	55	34.83	2.07
75	27	33	20.90	0.75
50	19	23	14.65	0.37

Among the input beams, the signal wave diffracts more than the pump due to its longer wavelength. However, with the waist values given in table 1, all near-IR input beam modes remains confined with the QPM layer thickness t = 400 μm. For instance, for the strongest focusing conditions (with f = 50 mm) corresponding to a pump and signal confocal parameters $b_p = 2z_p$ = 6.4 mm and $b_s = 2z_s$ = 7.7 mm the pump/signal mode diameters at the exit crystal facet (assuming a waist located at the center of the chip) are not clipped by the QPM layer aperture. However, for the two strongest focusing cases (f = 50 and 75 mm) the output diameter of the generated idler mode that strongly diffracts as it propagates to the output chip facet can be as large as twice the thickness of the QPM layer, meaning that part of the idler mode will experience some additional severe absorption loss from the bulk substrate. Furthermore, for the idler mode volume located outside the QPM layer the QPM condition $\Delta k = 2\pi[n_p - n_s - n_i - 1/\Lambda] = 0$ [8] is no longer satisfied. This propagation of part of the idler energy outside the QPM layer volume may also alter the phase coherence of the DFG process, leading to imperfect constructive interference of each idler wavelet generated in consecutive slabs of the nonlinear material (for downconversion $\varphi_p - \varphi_s - \varphi_i$ depends on the residual phase-mismatch

$\Delta kl_c/2$ which is ~ 0 only within the QPM layer). Indeed we shall observe a drastic reduction of the DFG efficiency for these tightly focused conditions in section 3.

3. Experimental results for DFG conversion efficiencies

Figure 2 displays the absolute MIR power generated by the longest sample (l_c = 24.6 mm) as function of the pump power measured in front of the OP-GaP sample, for a fixed signal wave power P_s ~ 40mW. A similar plot was obtained from the shorter sample. The displayed power has been corrected for all spurious loss in the collecting optics (Ge filter and CaF_2 collecting lenses). Each curve corresponds to the focusing conditions summarized in Table 1. For each data point, the temperature of the oven has been slightly optimized (decreased) as the pump power is ramped so as to partly cancel the thermally-induced phase mismatch. The maximum idler power P_i ~ 65 μW is generated at P_p ~ 10 W at the strongest focusing (f = 50 mm) although the curve for f =75 mm almost overlaps the f = 50 mm curve, meaning that these focal lengths must be around the optimum focusing condition to saturate the DFG efficiency.

The limited transparency of GaP at 1064 nm results in heating of the crystal; the thermo-optic effect consequently changes the refractive index, affecting the QPM conditions. The latter is then responsible for two observed effects. First, at the largest powers, the pump beam heats the crystal by several degrees above the oven copper base temperature, shifting the crystal temperature out of the QPM condition. In our measurements we took into account this shift by decreasing the oven temperature, so that each data point in Figure 2 corresponds to the nearly optimal QPM condition, once thermal equilibrium is reached. Second, the finite thermal conductivity of GaP yields a transverse temperature gradient. This affects the QPM condition in a radially-dependent fashion. While the QPM condition is well satisfied for weak pump powers, for substantial thermal load it cannot be satisfied throughout the whole transverse beams volume. This accounts for the observed idler power saturation at larger pump powers, whereas a linear behaviour would be expected. For the largest pump power (P_p > 6–7 W) it was no longer possible to decrease the oven temperature below T ~ 32°C because of its thermal inertia and the absence of an active cooling mechanism.

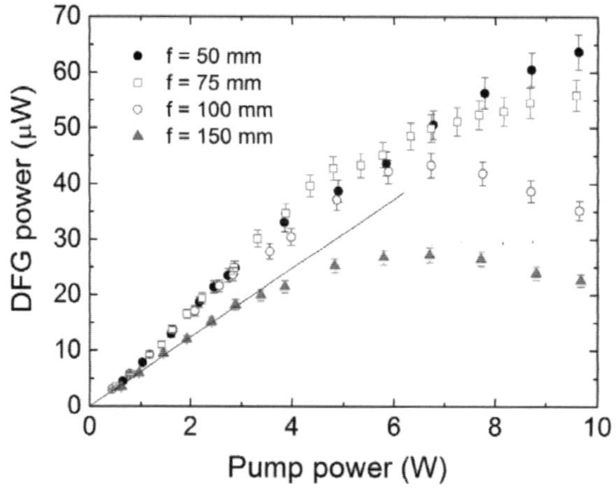

Figure 2. Idler power at λ_i = 5.85 μm versus 1064 nm pump power at maximum ~ 40 mW signal power at 1301 nm.

A striking feature in figure 2 also regards the earlier onset of saturation for loose rather than for strong focusing, while intuitively one would have expected the opposite. The explanation has to do with the poor thermal management (cooling or heating flux occurs through the crystal base only, the upper groovy chip facet not being in contact with the copper cover) and to the relatively large full transverse diameter ~ 4 w_p of the pump beam respective to the thin QPM layer thickness. For the f = 150 mm data in figure 2 for instance corresponding to cylindrical plane-wave focusing (pump confocal parameter $b_p = 2z_p$ = 85.6 mm >> l_c), the on-axis part of the cylindrical Gaussian beam along which the power density is largest cannot be cooled efficiently (the radial extent of the temperature gradient $\Delta T(r)$ is more pronounced for larger transverse beam diameter). By moving vertically the sample position so as to bring the pumped volume toward the colder oven base, beam aperture effect by the thin QPM layer occurs. For stronger focusing instead (e.g. f = 50 and 75mm) the pumped volume in the vicinity of the smaller wait can be moved closer to the oven base which is colder, allowing for a more efficient cooling process by conduction. The same striking feature was also observed for the shorter (l_c=15.3 mm) crystal.

Finally the straight line overriding the loosest focusing data in figure 2 is a fit of the initial linear trend of the conversion efficiency, before the occurrence of thermal load effects. Such line slopes provide the thermal-effect-free conversion efficiency $\Gamma = P_i / P_p P_s$ (plotted as symbols in figure 3) that will be used to retrieve the absolute value of the effective nonlinear coefficient of OP-GaP.

4. Gaussian beam DFG conversion efficiency analysis

In order to evaluate the effective nonlinear coefficient d only the linear data slope of each $P_i = \Gamma\, P_s\, P_p$ curve in figure 3 was considered to discard any thermal effects from the analysis. The conversion efficiency Γ (in W^{-1}), obtained by dividing the slope with the constant input power of the signal laser, is plotted against the focusing parameter. The experimental data are then compared with results from the Gaussian beam DFG analysis [17] taking into account the effect of absorption and which further goes beyond the ECP assumption (i.e. valid for any values of $w_{p,s}$). The conversion efficiency Γ from focused Gaussian beam DFG theory taking into account the effect of absorption is given by [17]

$$\Gamma = K\exp(-\alpha_i l_c) l_c \left(k_s^{-1} - k_p^{-1}\right)^{-1} h(a, l, f, \sigma) \tag{1}$$

$$h(a,l,f,\sigma) = \frac{\kappa_e}{2l} \iint_{-f}^{l-f} d\tau d\tau' \frac{\exp[-a(\tau+\tau'+2f) + i\sigma(\tau-\tau')]}{(1+i\kappa_e\tau')Q(\tau) + (1-i\kappa_e\tau)Q^*(\tau')} F_\beta(\tau,\tau').$$

In Equation (1), $K = 8\omega_i^2 d^2 / \pi\epsilon_0 c^3 n_p n_s n_i$ is the scaling factor proportional to the square of the effective nonlinear coefficient d, and h is the DFG Gaussian beam focusing function (also called *aperture* function) which depends on the focusing parameter $l = l_c/z_d$, on the normalized phase-mismatch parameter $\sigma = \Delta k z_d$, on the normalized input beam foci location $f = \eta l$ measured from input facet (when absorption cannot be neglected, the relative location η moves ahead from mid-crystal position $\eta = 0.5$ to $\eta < 0.5$ at strong focusing) and on the absorption normalized parameter $a = (\alpha_p + \alpha_s - \alpha_i) z_d / 2$. The definition of all parameters in the integral definition of h can be found in [17]. Here we just note that the spatial walkoff function $F_\beta(\tau,\tau')$ in the kernel is set to unity in the case of a quasi-phase-matched down-conversion (scaled walkoff angle $\beta = 0$).

For a given value of the focusing parameter l, and given waist location η, the focusing function $h(a,l,f,\sigma)$ must be optimized (maximized) over the phase-mismatch term σ first. Once this optimum $< h(\eta) >_\sigma$ found, η is incremented (decremented) and a new value of $\sigma_{opt}(\eta)$ is sought until the optimal couple $(\eta_{opt}, \sigma_{opt})$ is found, corresponding to the optimal value of h, noted $< h >_l$ for the focusing parameter l. Then l is incremented and the same multi-variable optimization procedure is repeated so as to construct the theoretical curve $\Gamma(l)$ shown in the next section. Such an optimization of the focusing function, performed using an adaptive step-size FORTRAN algorithm code to repetitively compute the double-integral, actually mimics the experimental optimization of the idler power by optimizing experimentally these parameters by using the sample's XYZ-θφ positioner. The computational optimization procedure yields the expected results that for the rather strong pump and signal absorption loss, the optimal waist location shifts ahead of the sample input facet (η_{opt} ~0.4) instead of being located at mid-sample (η_{opt} = 0.5) for the two strongest focusing cases. Additionally, the retrieved values of σ_{opt} are not nil as it is for plane-wave DFG but have negative non zero values as a result of focusing [17]. In performing the computational optimization leading to the curves $\Gamma(l)$, the experimental condition $w_s/w_p \approx \lambda_s/\lambda_p$ has been imposed. Once $< h(l) >$ is evaluated, the scaling factor K in (1) is then adjusted via d to match the curves to the experimentally determined $\Gamma(l)$'s.

5. Experimental DFG efficiencies and retrieval of the effective nonlinear coefficient

The comparison between experimental Γ's (symbols) and DFG theory (solid lines) is shown in figure 3 for both sample lengths, where the nonlinear effective d was adjusted to the two loose focusing data rather than the strong focusing ones for the reasons stated at the end of subsection 2.3, related to the efficiency limitations arising from the limited QPM layer thickness under strong focusing conditions (the fact that the generated idler mode expands then outside the QPM layer, in the severely absorbing bulk substrate). For those reasons we have chosen to match the theoretical $\Gamma(l)$ curves to the two loose focusing data point only for which all pump, signal *and* idler modes remain confined within the QPM volume. It thus appears that the two strong focusing data tend to depart from the predicted theoretical optimum for both samples, and such a departure is not due to thermal effects since the experimental Γ's are derived from the slope of the low-power linear data in figure 2.

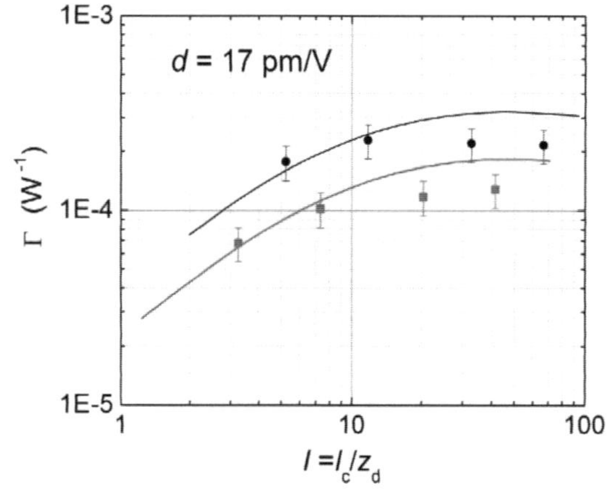

Figure 3. Plot of the absolute DFG conversion efficiency Γ as function of the focusing parameter l (crystal length expressed in units of the idler Rayleigh length). Experimental data are plotted as symbols (squares for the l_c = 15.3 mm OP-GaP sample and circles for l_c = 24.6 mm sample). The solid lines represent the optimized theoretical efficiency.

The best curve match to the loose focusing experimental data is found for d =17(3) pm/V. The ± 3 pm/V uncertainty originates from the waist measurement uncertainty and absolute idler power uncertainty. The fact the departure of the two strong focusing data from the predicted optimum focusing is more pronounced for the longer crystal corroborates our assumption that the QPM layer thickness must be increased to ~ 0.8 – 1 mm so as to confine the highly diverging ($\delta = \lambda_i/\pi n_i w_d$) idler wave within the QPM layer volume over the total sample length.

It is interesting to compare the measured absolute effective nonlinear coefficient $d = (2/\pi)d_{14}$ to values of the $d_{14}(\equiv d_{36})$ coefficient of GaP derived from non-phase-matched second-harmonic generation in bulk isotropic GaP. Levine reported $d_{14} = 37(2)$ pm/V from Maker-fringe technique at 10.6 μm (5.3 μm SH), $d_{14} = 49(9)$ and 47(10) pm/V from wedge technique at 1.32 μm (0.63 μm

SH) and 2.12 μm (1.06 μm SH), respectively [20]. More recently, Shoji *et al.* accurately measured $d_{14} = 36.8(4)$ pm/V from wedge technique at 1.32 μm taking for the first time into account multiple reflection effects that may play an important role in the previous measurements given the high index of refraction of GaP ($n \sim 3$ in the mid-IR) [21]. Note that the measurements by Levine have been rescaled to lower values following the standardization recommendation by Roberts [22]. Applying the Miller's rule factor $M_{14} = \prod_{i=1}^{3} (n^2(\lambda_i) - 1)/(n^2(\Lambda_i) - 1)$ to take account of $\chi^{(2)}$ dispersion, where the Λ_i are Shoji's reference second-harmonic generation wavelengths and using a recently determined temperature-dependent Sellmeier equation to account for dispersion [15], one finds $M_{14} = 0.82$ for our DFG wavelengths, and thus a nonlinear coefficient $d_{14} = 30.2$ pm/V for 5.85 μm generation using the present down-conversion pump and signal wavelength, from Shoji's most reliable measurement. For first-order QPM, this yields $d = 19(2)$ pm/V, close to our experimental value $d = 17(3)$ pm/V. Our d value translates into $d_{14} = 27.2(3)$ for GaP at ~ 6 μm.

6. Angular and spectral tuning curves

We have also investigated the temperature and signal wavelength DFG tuning curves, and compared the experimental acceptance curves with the theoretical curve given by $h(\sigma)$ where the wavevector-mismatch parameter $\Delta k(T, \lambda_i)$ in the reduced parameter $\sigma = \Delta k z_d$ has been evaluated with the temperature-dependent Sellmeier equation of Ref. [15] that was derived from the temperature-tuned wavelengths of a nanosecond Nd:YVO$_4$ pumped OP-GaP mid-IR OPO. Figure 4 compares the experimental and theoretical temperature tuning curve and figure 5 the spectral tuning curve when the Toptica diode laser was tuned around the phase-matched signal wavelength. Both the cylindrical focusing (f = 150mm, black circles) and strong focusing (f = 75mm, hollow circles) cases are plotted, and compared with simple plane-wave prediction ($\propto \text{sinc}^2(\Delta k l_c/2)$, dashed curve) and the Gaussian beam DFG theory prediction (solid curves).

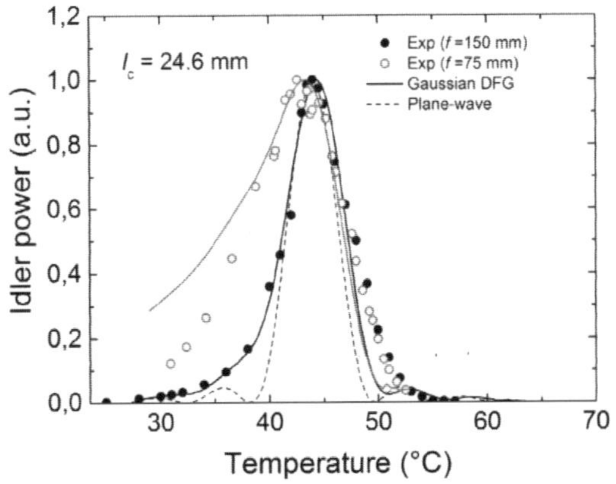

Figure 4. Temperature tuning curves for loose and strong focusing conditions (long sample). The dashed line is a plane-wave pattern.

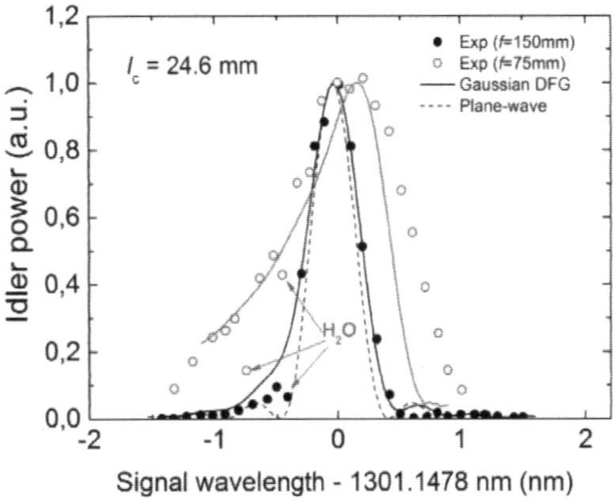

Figure 5. Signal-wave tuning curves for loose and strong focusing conditions. The arrows point to water line absorption as predicted by atmospheric databases near 5.85μm [23].

It can be seen that the broadening and asymmetry of the tuning curves at strong focusing are qualitatively accounted by the focused-beam DFG theory. It can also be noticed that beam aperture effects wash out the usual side lobes of the plane-wave tuning curve. At looser focusing there is a good match between both experiment and theory as to the temperature and spectral acceptance bandwidths, validating thus the thermo-optical dispersion relation given in Ref. [15].

7. Conclusions

In conclusion, we report on the first complete characterization of the cw difference-frequency generation process in an OP-GaP crystal, and demonstrate the first DFG using OP-GaP. The maximum generated idler power of ~ 65 μW is limited by thermal effects arising from the substantial absorption at 1064 nm by the OP-GaP samples. As the absorption coefficients of OP-GaP for wavelengths beyond 1.5 μm are about one order of magnitude lower than at 1064 nm, a pump laser in the telecom band around 1550 nm would allow to increase significantly the idler power (diode lasers for the signal around 2 μm are commercially available). The generated radiation can be referred to an absolute frequency standard by locking the pump and signal lasers to a near-IR comb, and the narrow-linewidth DFG radiation is sufficient to phase-lock a broader-linewidth QCL.

Funding. The ELI European project; Instituto Nazionale di Fisica Nucleare (INFN) (SUPREMO); Horizon 2020 (EMPIR-15SIB05-OFTEN); Participating Sates.

Acknowledgments
EMPIR-15SIB05-OFTEN has received funding from the EMPIR programme co-financed by the Participating States and from the European Union's Horizon 2020 research and innovation programme.

References

[1] Borri S and Santambrogio G *Adv. in Phys.*: X 2016 doi: 10.1080/23746149.2016.1203732
[2] Bartalini S, Borri S, Galli I, Giusfredi G, Mazzotti D, Edamura T, Akikusa N, Yamanishi M. and De Natale P 2011 *Opt. Expr.* **19** 17996
[3] Borri S, Galli I, Cappelli F, Bismuto A, Bartalini S, Cancio P, Giusfredi G, Mazzotti D, Faist J and De Natale P 2012 *Opt. Lett.* **37** 1011
[4] Galli I, Siciliani de Cumis M, Cappelli F, Bartalini S, Mazzotti D, Borri S, Montori A, Akikusa N, Yamanishi M, Giusfredi G, Cancio P and De Natale P 2013 *Appl. Phys. Lett.* **102** 121117
[5] Douillet A and Zondy J J 1998 *Opt. Lett.* **23** 1259
[6] Douillet A, Zondy J J, Yelisseyev A, Lobanov S and Isaenko L 1999 *J. Opt. Soc. Am.* B **16** 1481
[7] Petrov V 2015 *Prog. Quantum Electron.* **42** 1
[8] Hum D S and Fejer M M 2007 *C. R. Physique* **8** 180
[9] Koh S J, Kondo T, Ishiwada T, Iwamoto C, Ichinose H, Yaguchi H, Usami T, Shiraki Y and Ito R 1998 *Jpn. J. Appl. Phys.* **37** L1493
[10] Ebert C B, Eyres L A, Fejer M M and Harris J S 1999 *J. Crystal Growth* **201/202** 187
[11] Tassev V, Snurea M, Petersona R, Scheplera K L, Bedforda R, Manna M, Vangalab S, Goodhuec W, Lind A, Harris J, Fejer M and Schunemann P 2013 *Proc. SPIE* **8604** 86040V
[12] Pomeranz L, Schunemann P, Setzler S, Jones C and Budni P 2012 *CLEO-OSA Technical Digest*, paper JTh1I.4 doi:10.1364/CLEO_AT.2012.JTh1I.4
[13] Hildenbrand A, Kieleck C, Lallier E, Faye D, Grisard A, Gérard B and Eichhorn M 2011 *Proc. SPIE* **8187** 81870H
[14] Leindecker N, Marandi A, Byer R L, Vodopyanov K L, Jiang J, Hartl I, Fermann M and Schunemann P G 2012 *Opt. Express* **20** 7047
[15] Pomeranz L A, Schunemann P G, Magarrell D J, McCarthy J C, Zawilski K T and Zelmon D E 2015 *Proc. SPIE* **9347** 93470K
[16] Devi K, Schunemann P G and Ebrahim-Zadeh M 2014 *Opt. Lett.* **39** 6751
[17] Zondy J J 1998 *Opt. Commun.* **149** 181
[18] P. Kuo, PhD dissertation, http://nlo.stanford.edu/content/thick-film-orientation-patterned-gallium-arsenide-nonlinear-optical-frequency-conversion
[19] Zondy J J, Touahri D and Acef O 1997 *J. Opt. Soc. Am.* B **14** 2481
[20] Levine Z H 1994 *Phys. Rev.* B **49** 4532
[21] Shoji I, Kondo T, Kitamoto A., Shirane M and Ito R 1997 *J. Opt. Soc. Am.* B **14** 2268
[22] Roberts D A 1992 *IEEE J. Quantum Electron.* **28** 2057
[23] HITRAN, http://hitran.iao.ru/

Progress in optical frequency standards: ultracold Thulium, ions, and passive resonators

N Kolachevsky[1,2], K Khabarova[1,2], I Semerikov[1,2], I Zalivako[1,2,3] and A Borisenko[1,2,3]

[1]P N Lebedev Physics Institute, Russian Academy of Sciences, Moscow, Russian Federation
[2]Russian Quantum Center, Skolkovo, Moscow region
[3]Moscow Institute of Physics and Technology (State University), Dolgoprudnyi, Moscow Region

E-mail: kolachevsky@lebedev.ru

Abstract. We report on different types of optical clocks and passive frequency references which are under development in our laboratories: optical lattice clock based on the inner-shell transition in the Tm atom at $\lambda = 1.14\,\mu$m, optical ion clock on single $^{27}Al^+$ ion, and a family of lasers referenced to ultra-stable ULE and cryogenic silicon cavities.

1. Introduction

Frequency standards play an important role in a broad area of physics and many technical applications. For example, extremely stable and accurate frequency references are required for precision spectroscopy, tests of fundamental theories [1], measurements of fundamental constants [2] and provide a new approach to gravity sensing [3]. Progress in telecommunication [4], precision global positioning systems [5] and some other technologies are also strongly depend on performance of frequency standards. In the course of history fractional accuracy and stability of clocks were improving with increase in their operation frequency and the last revolution in the field happened with passage from microwave to the optical frequency domain in the beginning of the millenium. Active development of optical frequency standards began with advent of femtosecond frequency combs [6] which made it possible to transfer accuracy and stability from optical to radiofrequency domain. Also important role played advances in development of laser sources with sub-Hz spectral linewidths [7].

Narrowing of the "clock" laser spectral linewidth improves precision of the frequency standards and reduces integration time. For achieving sub-Hz performance, laser frequency is usually stabilized to an external high-finesse cavity. To minimize temperature influence, the body of the resonator is usually produced from materials with low temperature extension coefficients. In previous works we have reported on a family of laser systems based on diode lasers stabilized to ULE cavities [8] which show the fractional frequency instability of 1.5×10^{-15} in 1-100 s time interval approaching the thermal noise limit. Lasers are currently used as clock lasers for Sr, Tm and Al^+ optical clocks at P.N. Lebedev Institute and VNIIFTIRI. To reach better performance we are currently working at cryogenic Si cavities [9] with GaAs/InGaAs mirrors for $\lambda = 1.5\,\mu$m. The thermal noise will be lowered to 10^{-16} level which opens new perspectives both for applications in current optical frequency standards, and for relatively compact and simple ultrastable passive frequency references.

There are two main approaches to the development of an optical frequency standard: to use a narrow transition in a cloud of neutral atoms or in a single ion as a reference. Both ways have their own advantages and disadvantages.

In the first case a large number of interrogated atoms results in a high signal-to-noise ratio for the readout signal. Another advantage of these optical clocks is that the majority of atoms used for this purpose (Ca, Sr, Mg, H) [10] have "clock" and cooling transitions in visible region where they can be accessed with diode lasers. Drawbacks are relatively low lifetime of atoms in the cloud and strong trapping fields which cause shifts in energy structure of atoms and decreases performance of the clocks. The latter problem can be solved by trapping atoms in an optical lattice formed by laser fields at so-called "magic" wavelength [10]. Such lattice shifts energy levels of upper and lower clock levels equally leaving the transition frequency unperturbed. The best characteristics of accuracy and stability for neutral atoms clocks were achieved for Sr lattice frequency standards [11]. Their inaccuracy is on the level of 2.1×10^{-18} and instability reaches 2×10^{-18} at a few hours of averaging time.

The second possibility is to use a single ion in a radio-frequency trap. Single ions captured in Paul traps can be stored there for several hours or even months [12], subjected to weak trapping fields and do not interact with another particles. This enables one to build optical clocks with extremely high accuracy [13]. However signal-to-noise ratio for such clocks is inferior to lattice clocks which reduces stability and causes long integration times to reach the ultimate accuracy. Another disadvantage of single-ion optical clocks is that ions usually have cooling transitions in ultraviolet spectral domain and often can not be accessed by existing laser sources. In this case implementation of usual methods of direct laser cooling and electron shelving [14] for state readout becomes difficult or even impossible.

The method allowing to overcome these obstacles was demonstrated at NIST. Instead of direct laser cooling it was suggested to use sympathetic cooling [15] using a different axillary ion which has a convenient cooling transition. For state readout it was proposed to use quantum logic spectroscopy method [16] which enables one to map state of "clock" transition in the clock ion onto another metastable transition in auxiliary ion exploiting Coulomb interaction between particles. The state of auxiliary ion can be read with conventional electron-shelving method. Another advantage of quantum logic method is that after readout process reference ion remains in the initial state. This approach was demonstrated in an optical clock based on $^{27}Al^+$ with $^{9}Be^+$ [17] and $^{25}Mg^+$ [18] auxiliary ions. This clock with $^{25}Mg^+$ as an auxiliary ion showed fractional inaccuracy of 8.6×10^{-18} [18]. There are also proposals showing that performance of Al^+ clocks can be drastically improved [19, 20].

To implement a lattice optical clock we work on an optical frequency standard based on the inner-shell transition [Xe]$4f^{13}6s^2$ (J = 7/2, F = 4, m = 0) → [Xe]$4f^{13}6s^2$ (J = 5/2, F = 3, m = 0) in the Tm atom at $\lambda = 1.14\,\mu$m. It is considered as a candidate for an optical lattice clock. The transition wavelengths and probabilities for two clock levels J = 7/2 and J = 5/2 in the spectral range 250 – 1200 nm are calculated using the COWAN package which allows deducing of the differential dynamic polarizability and suggests that the magic wavelength is at around 807 nm with an attractive optical potential. The suggested clock transition demonstrates a low sensitivity to the BBR shift which provides a clock frequency instability at the low 10^{18} level competing with the best known optical clocks. We also evaluated other feasible contributions to clock performance (magnetic interactions, light shifts, van der Waals, and quadrupole shifts) which, after reasonable assumptions, can be lowered to the 10^{18} level. Together with the relative simplicity of laser cooling and trapping Tm atoms, our results demonstrate that Tm is a promising candidate for optical clock applications [21]. Experiments with direct excitation of the clock transition by spectrally narrow laser radiation at $\lambda = 1.14\,\mu$m set a lower limit for the upper clock level lifetime of 112 ms which corresponds to the natural linewidth of < 1.4 Hz. Modulating the trap depth and analyzing the corresponding parametric resonances frequencies, we deduced the scalar polarizability of the Tm ground state at 532 nm which shows reasonable agreement with our calculations.

To apply the second approach with a single trapped ion we are developing a single-ion-based optical frequency standard on base of the $^1S_0 \rightarrow {}^3P_0$ transition in Al^+ ion. This transition has a wavelength $\lambda = 267$ nm and the natural linewidth of 8 mHz [22]. The transition also has the lowest BBR shift [23] among all ions and atoms used in optical clocks so far and has low sensitivity to external fields which makes it a promising candidate for a frequency standard. Since the cooling transition at 167 nm cannot be accessed by existing laser sources we plan to implement methods of sympathetic cooling and quantum logic spectroscopy with $^{25}Mg^+$ ion.

In this paper we show our progress on trapping $^{24}Mg^+$ ions in a linear Paul trap and their Doppler laser cooling. This is a first step towards creation of an optical frequency standard based on $^{27}Al^+$ reference ion and $^{25}Mg^+$ as an auxiliary ion for sympathetic cooling and state readout.

2. Experimental setup

The main part of our setup is a linear quadrupole Paul trap [24] with four cylindrical tungsten electrodes. Diameter of those electrodes is $d_{el} = 2$ mm, the distance from the electrode surface to the trap centre of $r_0 = 1.475$ mm. Using cylindrical geometry instead of hyperbolically blade caused by better optical access. Radial confinement is realised by applying RF potential with frequency $\omega = 2\pi \times 18.25$ MHz and $V_{ac} = 600$ V peak to peak amplitude to the diagonal pair of cylindrical electrodes. Constant part of electrodes potential U_{dc} is set to zero. Axial confinement is provided by applying a DC voltage of 1000 V to four ring electrodes. Two inner rings have diameter of 12.5 mm two others are of 8 mm diameter. For $^{24}Mg^+$ axial frequency with described parameters is about $\omega_{ax} = 2\pi \times 80$ kHz, q Mathieu parameter is equal to 0.084, and radial secular frequency is $\omega_{rad} = 2\pi \times 1.54$ MHz.

Ions are loaded by electron impact ionisation of neutral Mg atoms directly in the trap, while trapping potentials are on. Mg atoms are produced by a resistive heating of a stainless-steel tube oven. The electron beam is provided by electron gun EGA-1012. Electron energy is about 500 eV, and current is 5 μA. We use such high electron energy rates due to electrons with lower energies are deflected by RF trap field and do not achieve its center. The number of stored ions can be measured with channeltron, situated on the trap axis.

All parts are mounted inside the stainless vacuum chamber with $P = 2\times10^{-10}$ mBar, pumped by ion-getter pump NexTorr D100-5. The scheme of the vacuum part of the setup is pictured in a figure 2

For Doppler cooling of $^{24}Mg^+$ we use cycling transition $^2S_{1/2} \rightarrow {}^2P_{3/2}$ with the natural linewidth of $\Gamma = 2\pi \times 41.4$ MHz and the saturation intensity of about $I_{sat} \approx 2.5\,\mathrm{mW/mm^2}$. This cycling transition requires single laser system at approximately 280 nm with stability and accuracy on a level of about 10 MHz. To meet this requirements we built a laser system which is shown on a figure 2. The ECDL laser with tapered amplifier Toptica TA pro provides 800 mW power at 1120 nm, which is converted to about 15 mW optical power at 280 nm with two stages of home-made bow-tie SHG cavities with LBO and BBO crystals respectively. Cooling laser frequency is locked to a wavelength meter WSU-10, which ensures long therm stability and accuracy on required level.

Ion luminescence is collected by high aperture lens assembly made of UV grade fused silica with anti-reflection coating. The lens covers about 3.6% of 4π solid angle. Photons are detected with Raptor Photonics Falcon Blue EMCCD camera with quantum efficiency of 28 % at 280 nm. Also fluorescence is measured by Hamamatsu H-8259 photomultiplier (PMT).

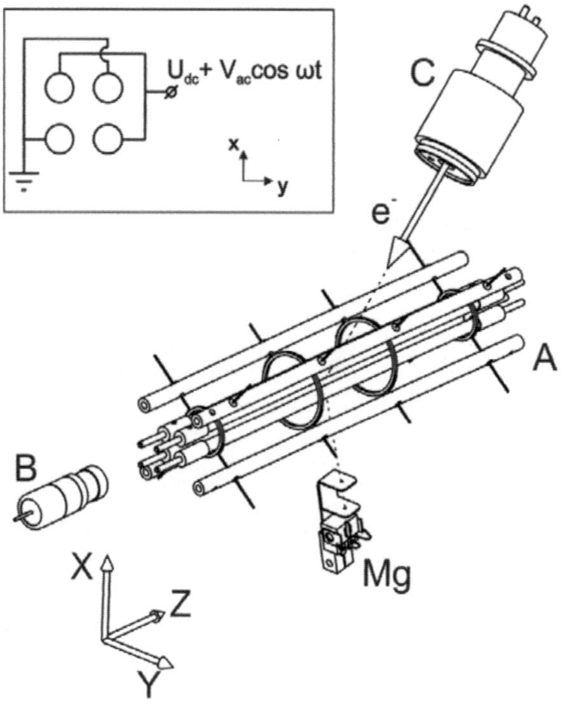

Figure 1. Sketch of the linear Paul ion trap, atomic source and detecting system. The inset shows the electric circuit of electrode connections. A – linear quadrupole Paul trap, B – channeltron, C – electron gun, Mg – magnesium atomic oven.

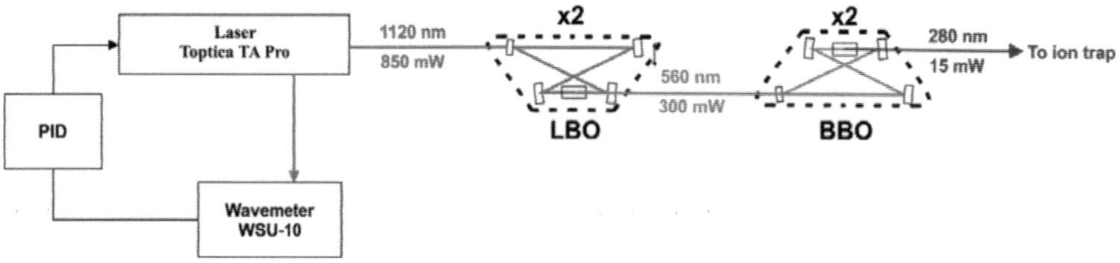

Figure 2. Laser system for Doppler cooling of $^{24}Mg^{+}$ ions.

3. $^{24}Mg^{+}$ ion cooling

To cool all degrees of freedom of ions we use two perpendicular laser beams. One of them propagates along the trap axis and another one is orthogonal to it. Due to very high initial ion temperatures beams are red detuned by $3\,\Gamma$ and and have an intensity $15\,I_{sat}$. Propagation directions of cooling beams and scheme of fluorescence detection system are shown in the figure 3.

In the beginning of an experimental cycle trapping fields, cooling beams, both atomic and electron guns are switched on and ions are being trapped and laser cooled. After approximately

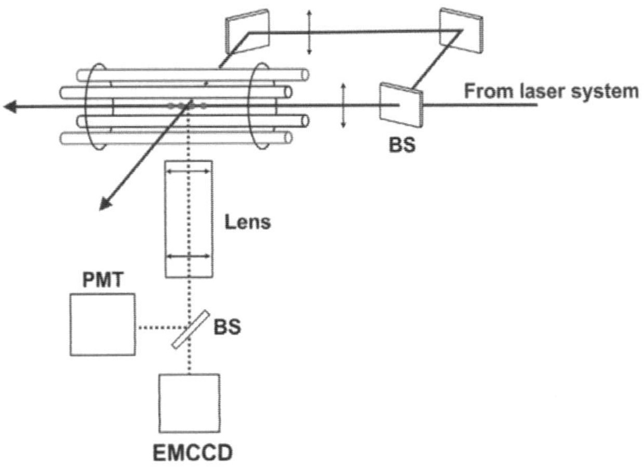

Figure 3. Orientation of cooling beams in the trap and fluorescence detection scheme. One of the laser beams propagates along trap axis, another one is orthogonal to it. Optical axis of the lens is perpendicular to both cooling beams. EMMCCD – Falcon Blue camera, PMT – Hamamatsu H-8259 photomultiplier, BS – beamsplitter.

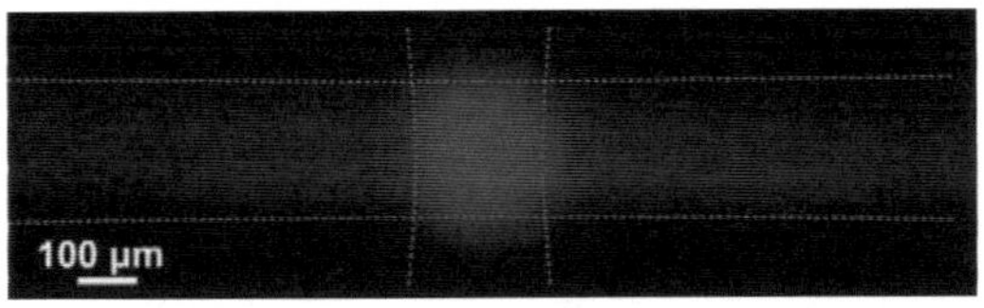

Figure 4. Picture of trapped cloud of $^{24}Mg^{+}$ ions. White lines show boundaries of cooling laser beams.

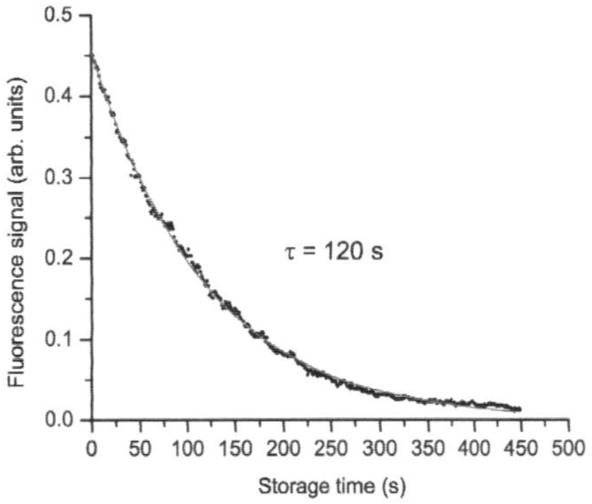

Figure 5. Dependence of ions cloud fluorescence signal on storage time. Lifetime of ions in the trap equals to $\tau = 120\,\mathrm{s}$.

10 seconds the trap is loaded and guns are turned off. Then we measure dependence of fluorescence signal on storage time using the camera reading. Image of a trapped cloud of ions is shown in figure 4 and dependence of fluorescence rate on storage time pictured in figure 5. Exponential approximation shows that lifetime of trapped ions is about 120 s, which is two order of magnitude higher than without cooling [25]. This time is limited by collisions with the rest gas.

Micromotion inducted by stray fields prevents ion crystallisation. To facilitate crystallisation we currently work on photo-ionisation system and compensation system for stray fields. It will allow to reduce influence of stray fields, caused by charging of electrodes with electron beam.

4. Outlook

Dopler cooling of $^{24}Mg^+$ ion is a significant step towards development of single Al^+ ion frequency standard. In this work we demonstrate detection of fluorescence signal from Mg^+ ions, and the lifetime of ions in the trap up to 120 s. Now we are working on reducing excess micromotion caused by stray fields from charging of trapping electrodes with electron beam. Also we are developing third laser system for Raman resolved sideband cooling of $^{25}Mg^+$ to achieve ground state cooling.

Acknowledgments

The work was supported by the Russian Science Foundation (Project No.16-12-00096).

References

[1] Rosenband T, Hume D B, Schmidt P O, Chou C W, Brusch A, Lorini L, Oskay W H, Drullinger R E, Fortier T M, Stalnaker J E, Diddams S A, Swann W C, Newbury N R, Itano W M, Wineland D J, Bergquist J C 2008 *Science* **319** 1808
[2] Bouchendira R, Cladé P, Guellati-Khélifa S, Nez F and Biraben F 2013 *Ann. Phys.* **525** 484
[3] Chou C W, Hume D B, Rosenband T, Wineland D J 2010 *Science* **329** 1630
[4] Gilbert S L, Swann W C and Dennis T 2001 *Proc. SPIE* **4269** 184
[5] Dow J M, Neilan R E and Rizos C 2009 *Journal of Geodesy* **83** 191
[6] Udem T, Reichert J, Holzwarth R and Hänsch T W 1999 *Opt. Lett.* **24** 881
[7] Alnis J, Matveev A, Kolachevsky N, Udem Th and Hänsch T W 2008 *Phys. Rev.* A **77** 053809
[8] Vishnyakova G A, Golovizin A A, Kalganova E S, Sorokin V N, Sukachev D D, Tregubov D O, Khabarova K Yu, Kolachevsky N N 2016 *Phys. Usp.* **59** 168173
[9] Matei D G, Legero T, Grebing Ch, Häfner S, Lisdat Ch, Weyrich R, Zhang W, Sonderhouse L, Robinson J M, Riehle F, Ye J and Sterr U 2016 *J. Phys.: Conf. Series* **723** 012031
[10] Poli N, Oates C W, Gill P and Tino G M 2014 *arXiv preprint* arXiv:1401.2378
[11] Nicholson T L, Campbell SL, Hutson R B, Marti G E, Bloom B J, McNally R L, Zhang W, Barrett M D, Safronova M S, Strouse G F, Tew W L and Ye J 2015 *Nature Commun.* **6** 6896
[12] Hua G, Yao H, Pei-Liang L, Wu B, Hu S and Ke-Lin G 2015 *Chin. Phys.* B **24** 054213
[13] Huntemann N, Sanner C, Lipphardt B, Tamm Chr and Peik E 2016 *Phys. Rev. Lett.* **116** 063001
[14] Nagourney W, Sandberg J and Dehmelt H 1986 *Phys. Rev. Lett.* **56** 2797
[15] Barrett M D, DeMarco B, Schaetz T, Leibfried D, Britton J, Chiaverini J, Itano W M, Jelenkovic B, Jost J D, Langer C, Rosenband T, Wineland D J 2003 *Phys. Rev.* A **68** 042302
[16] Schmidt P O, Rosenband T, Langer C, Itano W M, Bergquist J C, Wineland D J 2005 *Science* **309** 749
[17] Rosenband T, Hume D B, Schmidt P O, Chou C W, Brusch A, Lorini L, Oskay W H, Drullinger R E, Fortier T M, Stalnaker J E, Diddams S A, Swann W C, Newbury N R, Itano W M, Wineland D J, Bergquist J C 2008 *Science* **319** 1808
[18] Chou C W, Hume D, Koelemeij J C J, Wineland D J, Rosenband T 2010 *Phys. Rev. Lett.* **104** 070802
[19] Zhang J, Deng K, Luo J and Lu Z H 2016 *arXiv preprint* arXiv:1609.03667
[20] Leibrandt D, Brewer S, Chen J S, Hume D, Hankin A, Huang Y, Chou C W, Rosenband T, Wineland D 2016 *APS Division of Atomic, Molecular and Optical Physics Meeting Abstracts*
[21] Sukachev D, Fedorov S, Tolstikhina I, Tregubov D, Kalganova E, Vishnyakova G, Golovizin A, Kolachevsky N, Khabarova K and Sorokin V 2016 *Phys. Rev.* A **94** 022512
[22] Rosenband T, Schmidt P O, Hume D B, Itano W M, Fortier T M, Stalnaker J E, Kim K, Diddams S A, Koelemeij J C J, Bergquist J C, Wineland D J 2007 *Phys. Rev. Lett.* **98** 220801
[23] Rosenband T, Itano W M, Schmidt P O, Hume D B, Koelemeij J C J, Bergquist J C, Wineland D J 2006 *arXiv preprint* arXiv:physics/0611125
[24] Leibfried D, Blatt R, Monroe C, Wineland D 2003 *Rev. Mod. Phys.* **75** 281
[25] Semerikov I A, Zalivako I V, Shpakovskii T V, Borisenko A S, Khabarova K Yu, Sorokin V N, Kolachevsky N N 2016 *Quantum Electron.* **46** 935

IOP Conf. Series: Journal of Physics: Conf. Series **793** (2017) 012014 doi:10.1088/1742-6596/793/1/012014

Experimental investigations of 3 mm aperture PPLN structures

D Kolker[1,2,3], A Pronyushkina[2,3], A Boyko[2], N Kostyukova[2], S Trashkeev[1,2], B Nuyshkov[1] and V Shur[4]

[1] Research Laboratory of Quantum Optics Technology, Novosibirsk State University, 2 Pirogova str., Novosibirsk 630090, Russia
[2] Institute of Laser Physics SB RAS, 13/3 Acad. Lavrentyev ave., Novosibirsk 630090, Russia
[3] Novosibirsk State Technical University, 20 K. Marx ave., Novosibirsk 630073, Russia
[4] The Institute of Natural Sciences, Ural Federal University, 51 Lenin ave., Ekaterinburg 620000, Russia

E-mail: dkolker@mail.ru

Abstract. We are reporting about investigation of domestic 3 mm aperture periodically polled lithium niobate (PPLN) structures for cascaded mid-IR OPO. Wide aperture periodically poled MgO-doped lithium niobate (LiNbO3) structures at multigrating, fan-out and multi fan-out configuration were prepared at "Labfer LTD". Laser source based on such structures can be used for special applications. Four different PPLN structures were investigated and effective aperture for effective pumping was defined.

1. Introduction

Cascade optical parametric oscillator is a novel nonlinear device for consecutive transformation of pump laser energy at near IR to IR, MID IR and THz spectral region by utilizing of nonlinear nonoxide crystals or liquid crystals as a second step of primary pump energy conversion. The cascade OPO with intracavity pumping of initial signal wave possess flexibility for a choice of suitable second pump wavelength for conversion to MID IR and THz spectral region at second cascade. The theory of seldom OPO was presented at [1] by *F.Wong et al.* At [1] a parasitic OPO oscillation is addition component secondary signal and idler appeared at self-phase locked PPLN OPO when 3:2:1 regime was broken. Intracavity pumped optical parametric oscillators (OPOs) offer compact and robust design with higher pump power. However cascaded or tandem OPO schemes of this kind down conversion of laser radiation into the mid - IR spectral range have rarely been realized [2-3]. The second stage of this OPO based on nonoxide crystal $AgGaSe_2$ intracavity pumped by single grating PPKTP at first cascade at 1.85 μm. Extremely wide tuning range from 8 to 18 μm was demonstrated by using the angle tuning of type II $AgGaSe_2$ element into the OPO cavity. Lately at [4] the intracavity difference-frequency mixing (DFM) based on PPKTP-AGSe tandem was successfully demonstrated. The DFM tuning range 6.8-8.0 μm was driven by temperature tuning of PPKTP at the OPO cavity. At the present work, we are demonstrating a novel PPLN structures for effective intracavity pumping of cascade OPO.

2. Experimental setup

PPLN - OPO set-up is shown on figure 1. The pump source was a lamp-pumped and electro-optically Q-switched Nd:YAG laser LQ215 (Solar Laser Systems, Belarus) optimized for a repetition rate of 20

Hz. The linewidth of this laser is 1.5 cm^{-1}, $M^2<2$, and divergence <2.5 mrad. The laser generated pulse energy of 180 mJ and 5.6 ns (FWHM) pulses with an average power of 3.6W. The measured energy stability was ±2.5%. The Faraday isolator (Avesta 8AFI-1064C) installed to optical setup to avoid a feedback from OPO optical elements. The half-wave plate and polarizer used as an attenuator of the pump energy. Mirrors M1-M3 utilized for precision adjustment of the pump beam in to the OPO cavity. Fused silica lens f=300 mm with AR coating optimized for 1.064 μm was used for optimal focusing of pump radiation into the PPLN structure. M4 is dichroic mirror (Layertec model 103080). This mirror reflected for the pump radiation at 1.064 μm and transmitted for signal wave (spectral region 1.3-1.8 μm) and idler wave (spectra region 2.5 - 4.5 μm). The OPO cavity designed with two mirrors, Input coupler mirror M5 (Layertec 105804) was transmitting for the pump and idler wave, but R=98.5% for signal wave at the spectral region 1.350-1.70 μm. The back mirror M6 is Ag mirror (Thorlabs PF10-03-PO1) totally reflected pump, signal and idler waves. The beam splitter utilizing for separation idler and signal wave: idler wave reflected and energy measured by commercial power meter Thorlabs PM100D (sensor ES120C), signal wave transmitted and measured by wavelength meter Angstrom LSA L IR (High Finesse). Idler wavelength was recalculated by equation:

$$\lambda_{id}{}^{-1} = \lambda_p{}^{-1} - \lambda_s{}^{-1}.$$

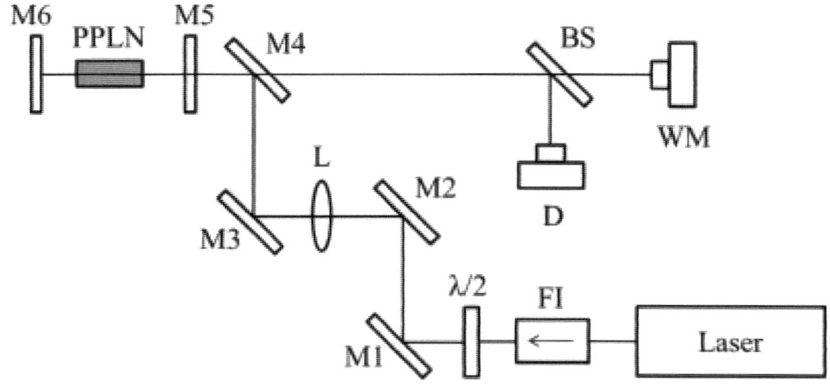

Figure 1. Experimental setup: Laser – LQ215 laser, λ = 1064 nm, τ = 6 ns; FI – Faraday isolator; λ/2 – half-wave plate; M1-M3 – flat mirrors; L – lens, f = 300 mm; M4 – dichroic mirror; M5 – input coupler; M6 – reflecting mirror; BS – beam splitter; D – pyroelectric sensor or power meter; WM – wavelength meter.

Four PPLN structures manufactured at LABER LTD for this experiment, and have 3 mm thickness. At figure 2 demonstrates the distributing of idler wave intensity for PPLN OPO. The PPLN structure was scanned by X-Y coordinate precision stage relatively of cavity axes of OPO cavity (signal wave in resonance, pump is double passed). Pump waist diameter was 200 μm. The pump energy limited by 70% of damage threshold energy of PPLN (200 MW/cm^2). Figure 2(a) is corresponding of fan-out structure (3×20×50 mm, Λ is varied 27.45 – 32.42 μm); 2(b) – multi fan-out structure (3×10×50 mm, four fan-out sections were placed at the same LN chip); 2(c) – multigrating structure with 3 symmetric sections: Λ_1 = 30,9 μm, Λ_2 = 31 μm, Λ_3 = 31,2 μm; 2(d) – multigrating structure with 3 symmetric sections Λ_1 = 31,3 μm Λ_2 = 31,4 μm, Λ_3 = 31,5 μm. The effective aperture (from full aperture) for fan-out structure is defined as 57 % (figure 2a), for multi fan-out is 37% and for multigrating 68% and 73% respectively (figures 2c,d).

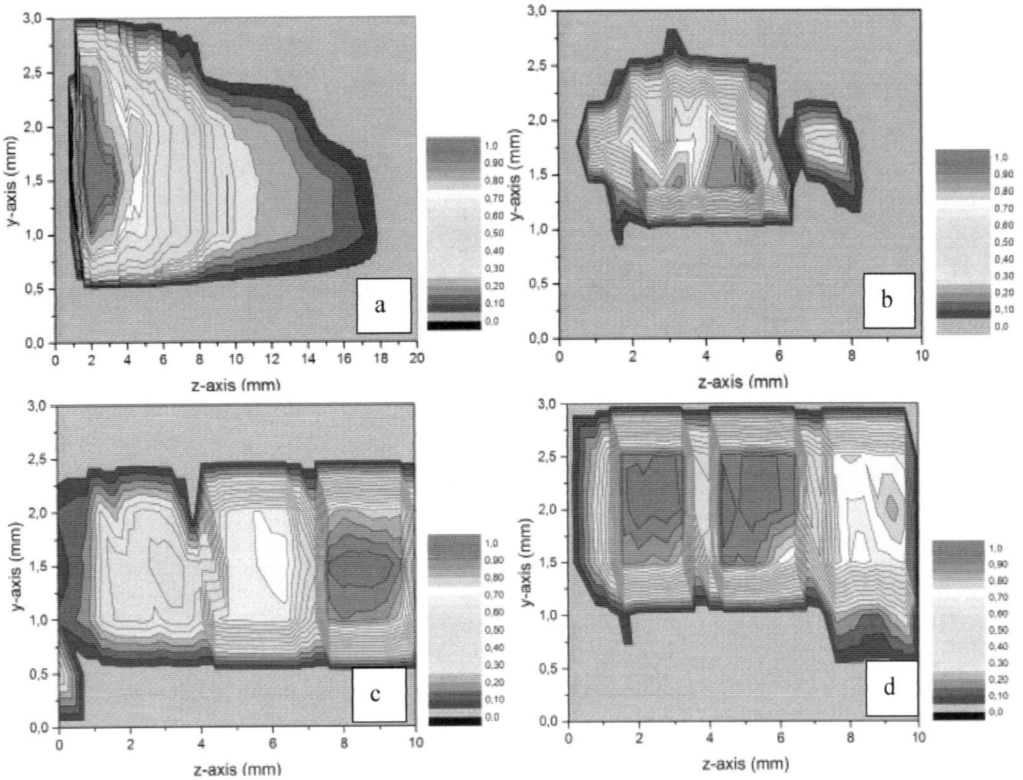

Figure 2. Idler energy distribution for four different PPLN structures.

3. PPLN OPO (3 mm aperture) experiment: performance and tuning curve

For the next experiment the lens L (figure 1) was changed by two lenses, L1 and L2, (figure 3) telescope (f1 = 200 mm f2 = -75mm). The lens L1 was mount at the precision X-coordinate table for the matching pump beam into the PPLN structure. The beam diameter was 2.5 mm horizontal and 2.7 mm at vertical direction.

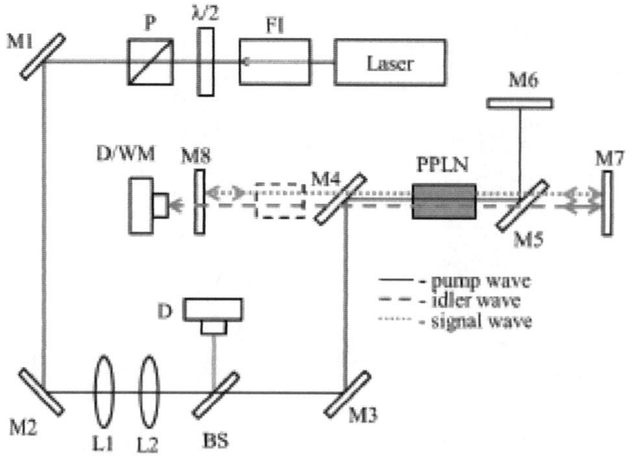

Figure 3. Experimental setup of 3 mm OPO. Laser Nd:YAG laser, l = 1064 nm, FI-Faraday isolator, λ/2 half wave plate, P-polarizer, L1, L2 –lenses, BS –quartz plate, D – power and energy meter, M4,M5 – bending mirrors, M6 – total reflecting mirror, M7.M8 – cavity mirrors for signal wave, WM-wavelength meter.

Figure 4 demonstrates tuning curve for idler wavelength versus pump energy for fan-out structure. Wide tuning range 2800-4500 nm for idler wave is demonstrated. Figures 5(a) and 5(b) demonstrate the performance of idler energy versus pump energy for λ_i=2.94 μm and λ_i=3.2 μm. The pump beam diameter was 2.7 mm for horizontal and 2.6 mm for vertical coordinate.

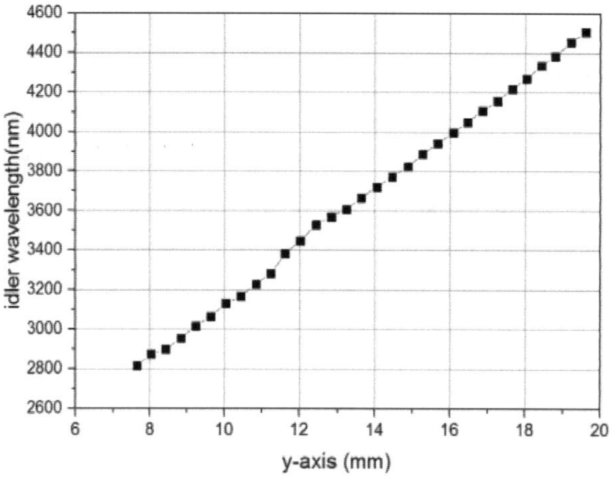

Figure 4. Tuning curve for 3 mm aperture fan-out PPLN structure.

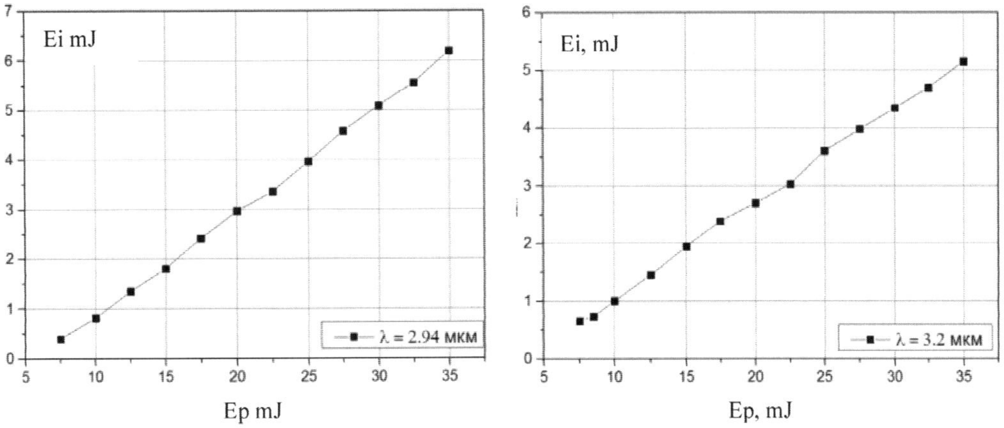

Fugure 5. (a) and (b) idler wave energy versus pump energy for two different grating.

For maximum, pump level, at figure 5(a): the corresponding pump-to-idler energy conversion efficiency is 17% and quantum efficiency is 48.9%, and the slope efficiency is 20.7%. At figure 5(b): pump to idler energy conversion efficiency is 14.8%, quantum efficiency is 41% and slope efficiency is 16.2%.

Conclusion

We are demonstrating efficiency of pump energy conversion (Ng:YAG laser at 1.064 μm) 17.7% for 2.94 μm and 14.8 % for 3.2 μm respectively. The estimated maximum signal intracavity energy for PPLN OPO (at 35 mJ primary pump by Nd:YAG laser) for 2.6 mm diameter of signal wave beam is promising to reach at least 5 times above secondary threshold by inserting second element II-type $AgGaSe_2$ into the OPO cavity for 8-18 μm oscillation.

Acknowledgment

This work was supported by the NSU program 5-100 of the Russian Ministry of Education and Science.

References

[1] Wong F, Gorelik P, Kolker D and Zondy J J 2006 *Optics Letters* **31** 2039–41
[2] Boyko A, Marchev G, Petrov V, Pasiskevicius V, Kolker D, Zukauskas A and Kostyukova N 2016 *Optics Express* **23** 33460–5
[3] Boyko A, Marchev G, Petrov V, Pasiskevicius V, Kolker D, Zukauskas A and Kostyukova N 2016 *High-Brightness Sources and Light-Driven Interactions* (OSA, EUV, HILAS, MICS)
[4] Boyko A A, Kostyukova N Y, Marchev G M, Pasiskevicius V, Kolker D B, Zukauskas A and Petrov V 2016 *Optic Letters* **41** 2791–4

Quantum-enhanced protocols with mixed states using cold atoms in dipole traps

K Krzyzanowska, M Copley-May, R Romain, C MacCormick and S Bergamini

Department of Physical Science, The Open University, Milton Keynes, United Kingdom

E-mail: Silvia.Bergamini@open.ac.uk

Abstract. We discuss the use of cold atoms in dipole traps to demonstrate experimentally a particular class of protocols for computation and metrology based on mixed states. Modelling of the system shows that, for a specific class of problems (tracing, phase estimation), a quantum advantage can be achieved over classical algorithms for very realistic conditions and strong decoherence. We discuss the results of the models and the experimental implementation.

1. Introduction

Entanglement is widely recognised as a key resource in quantum technology, however an advantage over classical computing can be achieved without it in the presence of non-classical correlations, quantifiable as *discord* [1]. Experiments using a few photonic qubits have shown that specific computational tasks can be efficiently solved even with no entanglement [2].

In the past years, there has been outstanding progress in the demonstration of quantum processing based on pure states with a limited number of qubits. However scalability remains an issue, mainly because of decoherence, which scales with the number of qubits. In pure-states quantum computation (QC) this problem can possibly be solved by error correction or using strategies as topologically protecting the qubits. Nevertheless, scaling up to a significant number of qubits and being able to perform a classically intractable calculation has been impossible so far.

As entanglement is very extremely vulnerable to decoherence, the investigation of protocols that are more robust against it is a promising route for progressing the field. One of such protocols is called Deterministic Quantum Computation with 1 clean qubit (or DQC1). This protocol relies on one pure-state control qubit together with a register of completely mixed state qubits, where non-classical correlations are created between the control and the register. DQC1 is a non-universal model of computation (essentially a phase estimation protocol) that can speed up some computational tasks for which no efficient classical algorithms are known. Whilst requiring only a single qubit with coherence, its power scales up with the number of mixed state register qubits. To date, experiments based on photonic implementation of DQC1 have evaluated the normalised trace of a two-by-two unitary matrix [2] and using an NMR implementation, performed the approximation to the Jones polynomial with a system of four qubits, thus experimentally demonstrating the principle of mixed state computation.

We present theoretical models of a cold-atoms based platform for the benchmarking of the protocol with a large number of qubits, thus extending the implementation to large Hilbert space. In comparison to the photonic or NMR experiments, a cold atoms approach offers scalability. Our scheme is based around a single, pure-state control qubit prepared in a tightly confining microscopic optical dipole trap and a register of N mixed state qubits prepared in a neighbouring trap with more relaxed confinement, as shown in figure 1. Conditional operation between the control and register qubits is implemented via the long range interaction between highly excited atomic Rydberg states, which are accessed by laser driven transitions. The Rydberg-Rydberg interactions between the control qubit and register qubits of cold atoms in mixed state can activate the implementation of DQC1. An appropriate sequence of laser pulses individually addressing the two traps will implement an algorithm to calculate the trace of a $2^N \times 2^N$ matrix.

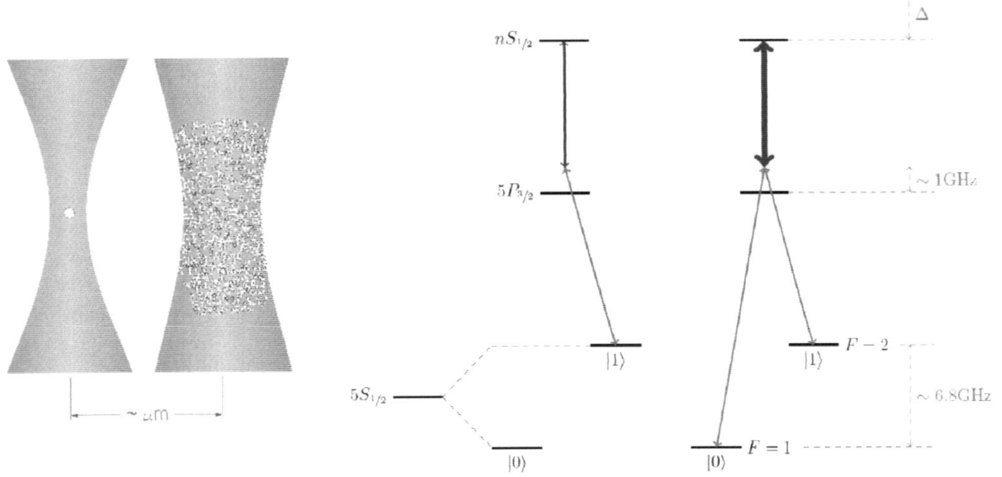

Figure 1. *Left panel:* Two dipole traps of different waist can be obtained and filled up with N atoms or a single atom respectively. If the two traps are positioned at a distance within the range of Rydberg-Rydberg interactions, conditional logic can be performed and a sensitive probe state can be created. *Right panel:* Laser scheme for Rydberg excitation/ Rydberg coupling and conditional logic.

We show that the protocol can be operated with many qubits using a cold atoms setting, and we explore the possibility of tackling non-trivial problems [3], such as many-body physics. The same scheme enables the preparation of quantum enhanced probes for phase estimation and promises high-precision measurement, without relying on quantum entanglement and using highly mixed states [4]. Modelling of this scheme, using cold atoms in dipole traps, demonstrates that the register of partly mixed qubits becomes a powerful resource for phase estimation when supplied with the coherence from the control qubit. A concrete mixed-state model for quantum sensing is also proposed. The scheme can achieve quantum-enhanced precision scaling with the size of the atomic register [5]. We will finally present the full design for this test and our progress in the experimental setup and implementation.

2. Computation and metrology with mixed states

DQC1 is a non-universal model of computation based on mixed states. For specific tasks the protocol can offer a quantum advantage that scales with the number of qubits in a maximally mixed state, and can exponentially speed up the processing even with a limited amount of entanglement. DQC1 and DQC1-derived protocols present a remarkable advantage with respect to standard QC protocols, in that they require only a single qubit with non-zero coherence, whilst their power scales up with the number of qubits in mixed state. In sensing applications, DQC1 derived protocols can surpass the standard quantum limit (SQL).

This model of computation was first proposed by Knill and Laflamme [1] and recent work has tried to classify and demonstrate what quantum feature gives rise to the quantum advantage. Although it has been shown that this scheme contains little to no entanglement [9], non-classical correlations are present in the output state of the DQC1 which can be quantified in terms of quantum discord [10].

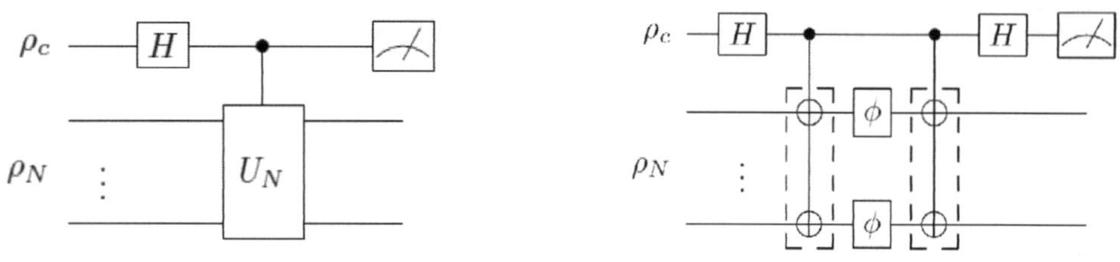

Figure 2. *Left*: Diagram of the DQC1 model of computation. An input single qubit (control qubit) is prepared in a coherent (or partly coherent) state ρ_C and a register of N qubits is prepared in the maximally mixed state $\rho_C = \mathbb{I}_N/2^N$. The protocol then performs a controlled unitary operation on the register qubits conditional on the state of the control. The normalized trace of the unitary is estimated by measurements on the control qubit only. *Right*: The metrology protocol. The CNOT gate between control and register qubits prepares a probe state that is then let to evolve under interaction with an unknown field. An interferometric scheme then allows to retrieve the phase accumulated during the interaction time. Measurement on the control qubit will retrieve information about the phase and hence recover the field with precision beyond the SQL even in the presence of strong decoherence.

The DQC1 equivalent circuit diagram is shown in figure 2 (left). An input single qubit (control qubit) is prepared in a coherent (or partly coherent) state ρ_C and a register of N qubits is prepared in the maximally mixed state $\rho_C = \mathbb{I}_N/2^N$. The protocol then performs a controlled unitary operation on the ensemble qubits conditional on the state of the control. The normalized trace of the unitary is estimated by measurements on the control qubit only.

It has been shown theoretically that this model of computation retrieves the trace of U_N with accuracy scaling with the number L of measurements on the control qubit, regardless of the number N of atoms in the mixed state register. In other words, this is a very efficient computational protocol, in that increasing the complexity of the computational task (size of U_N) does not increase the overhead in the calculation of its trace. It is remarkable that this protocol brings in a quantum advantage, even if no entanglement is present.

This protocol can be tested and the quantum resources tuned using cold atoms in dipole traps, enabling the conditional logic with Rydberg-Rydberg interactions. This platform has been discussed in [3] and we will review the experimental requirements in the next section.

Furthermore, it has been shown [4] that this protocol can be used effectively for phase estimation. We recently proposed a scheme, where cold atoms in dipole traps can be prepared in a probe state that can reach sensitivities beyond the SQL, even for moderate initial purity of the control and register states [4]. The circuit diagram representing this protocol is shown in figure 2 (right).

The concept that a significant quantum advantage can be achieved even with strongly mixed state qubits is potentially paradigm-shifting for quantum computation and for sensing. If we can understand thoroughly the role of quantum correlations (other than entanglement) in the powering up of quantum protocols, we will be able to bypass the stumbling block of decoherence by using appropriate algorithm (like DQC1 and DQC1-derived).

3. A cold atoms platform

We devised a cold-atoms scenario to test and benchmark these protocols. A rubidium magneto-optical trap (MOT) acts as a reservoir to load two optical dipole traps, one to serve as the (single) control qubit and one as a register of N qubit. In our scheme, the qubits are realised as the hyperfine doublet of the ground state of rubidium atoms. The population of the two traps will be tuned by controlling the trap loading rate (which depends on the number-density of the reservoir, common to both traps) and by individually controlling the trapping potentials. By these means, the traps will be adjusted so that a different average number of atoms can be achieved in the two traps [11].

The two dipole traps are to be individually laser-addressed so that high-fidelity stimulated Raman rotations [12] can be performed on the control atom and/or on the register atom to prepare a superposition of the ground hyperfine states (thus realising the Hadamard (H) gate in figure 2). It is to be noticed that the level of qubit purity can be tuned both on the control and the register atoms by activating spontaneous scattering on a closed transition (or transitions with different branching ratios), as discussed in [3]. This is a very important feature and allows us to tune the purity of the atomic qubits in each trap, which can range from fully coherent (within experimental limits) to maximally mixed. This tuning of resources is fundamental to understand the role of discord (as opposed to entanglement), the effects of decoherence and ultimately will allow us to explore the boundaries between purely classical to quantum enhanced behavior.

Quantum correlations are built between control and register to either implement a controlled transformation as in figure 2 (left) or a CNOT gate to prepare an N+1 qubit probe state for metrology, as in figure 2 (right). This is done exploiting Rydberg-Rydberg interactions and particularly the phenomenon of *Rydberg blockade* whereby if an atom is excited to Rydberg state, the simultaneous laser excitation of neighbouring atoms is blocked, provided the laser linewidth is small compared to the Rydberg-Rydberg interaction shift. In general for a given Rydberg-Rydberg interaction, a blockade radius is defined. Provided the displacement of all of the register atoms from the control atom fall well within this blockade radius, then a conditional logic scheme based on electromagnetically induced transparency can be used [13, 3].

In figure 1 (right) a level scheme showing the two-photon excitation of the Rydberg state and the EIT-based method of conditional rotation. The qubits are encoded in the hyperfine ground states F=1 ("0") and F=2 ("1") of ^{87}Rb. For the control qubit, state F=2 is coupled to a Rydberg state via 2-photon laser transition. The 0 and 1 states of the N register qubits are coupled by a 2-photon Raman transition (Λ configuration) scheme following [13], which drives rotations in the computational basis via the $5P_{3/2}$ intermediate state. Additionally, an intense laser beam coupling the $5P_{3/2}$ intermediate state to the Rydberg state is added so that the electromagnetically induced transparency condition is fulfilled and the interaction with the rotation beam is inhibited. However, the coupling of the control atom to Rydberg state leads to a strong Rydberg-Rydberg interaction and an additional shift of the Rydberg level that removes the condition for EIT, so that resonant Raman transfer between the 2 qubit states in the register atom is activated.

This scheme is particularly attractive as it allows for global addressing on the register to operate parallel conditional logic on the register qubits. It also maximizes the fidelity of the gates as it is not affected as much as other schemes by the spread in interaction strengths within the register (for a comprehensive discussion Ref. [13]).

We first modeled the performance of the protocol in figure 2 on a realistic cold atoms platform. We considered a register trap containing an average of 100 atoms, over which we perform a controlled Raman rotation. This is a trivial parallel operation on the register qubit, but one that would allow us to benchmark the protocol in high-dimensional Hilbert space. The Rydberg states are chosen so that the register can be fully blockaded by the control qubit. Imperfections in the controlled rotations are taken into account and are mainly due to spontaneous decay from the intermediate level (figure 1, right). For a separation between the traps of ~2 μm and interaction strength of 15 GHz (achievable at 2 μm with specific Rydberg states) the trace of U_N can be retrieved with accuracy of 10% with L=400 measurements on the control qubit. It is remarkable here that the principal source of uncertainty is due to the random loading of the register dipole trap [3], thus it becomes less significant for larger N.

Upon successful demonstration of the quantum advantage for trivial tasks, this platform offers the ability to implement a range of non-trivial unitaries [3].

Finally we modelled the metrology protocol (figure 2, right) for a particular scenario which will allow the probe state to detect gravity beyond the SQL. The C-NOT gate between control and register qubits prepares a probe state that is then let to evolve under interaction with the unknown field.
We consider N=25 atoms in the register (Poissonian loading), loaded in state-dependent trapping potentials, which allow for the atoms to be shifted vertically conditionally on their internal state. The potentials are initially overlapping to allow for initialization and probe state preparation. Subsequently the two state-dependent traps are adiabatically displaced along the z-axis so that the two states evolve a differential phase shift due to a variation in the gravitational potential experienced (state-dependent). In the case of Rubidium, with realistic trapping potentials (that we can achieve with the setup described in the next paragraph), and realistic purity levels (i.e. realistic decoherence effects) one can still gain a quantum advantage. With N = 25, and purity of control and register of 0.95 (well within what can be realistically achieved), the sensitivity of the probe exceeds the classical equivalent by a factor of 3.5. We also confirm that even in this scenario, the sensitivity of the protocol scales with the number of correlated atoms, thus demonstrating the quantum advantage even in the presence of strong decoherence.

4. The experiment: preparing the control and the register atoms

A magneto-optical trap (MOT) is formed in the area between two identical high N.A. lenses (see figure 3), which are ITO coated to ensure electrical conductivity. Electrodes are also placed around the body of the lens to control electric fields in the trapping region. Each lens has an effective focal distance of 7.03 mm so the distance between the two front faces of the lenses is 14.06 mm. Because of the restricted optical access, two of the MOT beams are crossed at a small angle of 40° (with the third one perpendicular to the plane of the first two) and have a diameter of 2.5 mm. The total MOT beam power is 8 mW and we choose a detuning of 1.1 $\Gamma = 2\pi\times6.7$ MHz. With these parameters, typical atomic densities are 10^8 atoms/cm^3 with a temperature of 300 μK (which can be reduced by increasing the laser detuning and simultaneously switching the quadrupole field off). A fluorescence image of the magneto-optical trap formed between the lenses is shown in figure 3 (right). In addition to the lenses are two sets of electric field plates, each set consisting of four electrodes. By applying voltages to these field plates, we are able to apply precise electric field gradients (including spatially constant fields) which allow fine tuning of the Rydberg-Rydberg interactions.

The MOT acts as a reservoir to load two dipole traps, one containing one atom and one containing a register of N atoms. As mentioned above, to control the populations of the traps we can adjust the reservoir number density to influence the trap loading rate, and individually, the trap depth and geometry [11]. The dipole traps are realised by imaging 852 nm laser beams at the focal plane of one of the high N.A. lenses. We have verified experimentally that the lenses allow us to obtain spot sizes of below 1 μm (measured across the $1/e^2$ points of the beam) over a large area of the focal plane. The same lens is part of an imaging system which observes the atoms which fluoresce at 780 nm. We separate the 780 nm fluorescence from stray 852 nm trap light using a dichroic beam splitter and a narrow band interference filter, and image the trapped atom on an intensified CCD camera.

4.1. Register

We prepare the register atoms by loading atoms from the background MOT into a tweezer obtained by focusing an 852 nm laser light down to a 2-4 μm waist. Two images of the dipole traps atoms are then taken, using a short 50 μs fluorescence pulse – one image with the dipole trapping beam on and one immediately after the dipole trapping beam has been switched off. The short pulse length ensures that the motion of the dense dipole trapped atoms is negligible during the imaging time, and that the expanding atomic density distribution matches the trapped atomic distribution. By comparing the atoms' fluorescence rate in the presence of resonant light it is possible to recover both the number of atoms loaded in the trap and the average light shift induced by the intense laser beam, which enables us to characterise the trap potential. An image of the dipole trap is shown in figure 4.

IOP Conf. Series: Journal of Physics: Conf. Series **793** (2017) 012015 doi:10.1088/1742-6596/793/1/012015

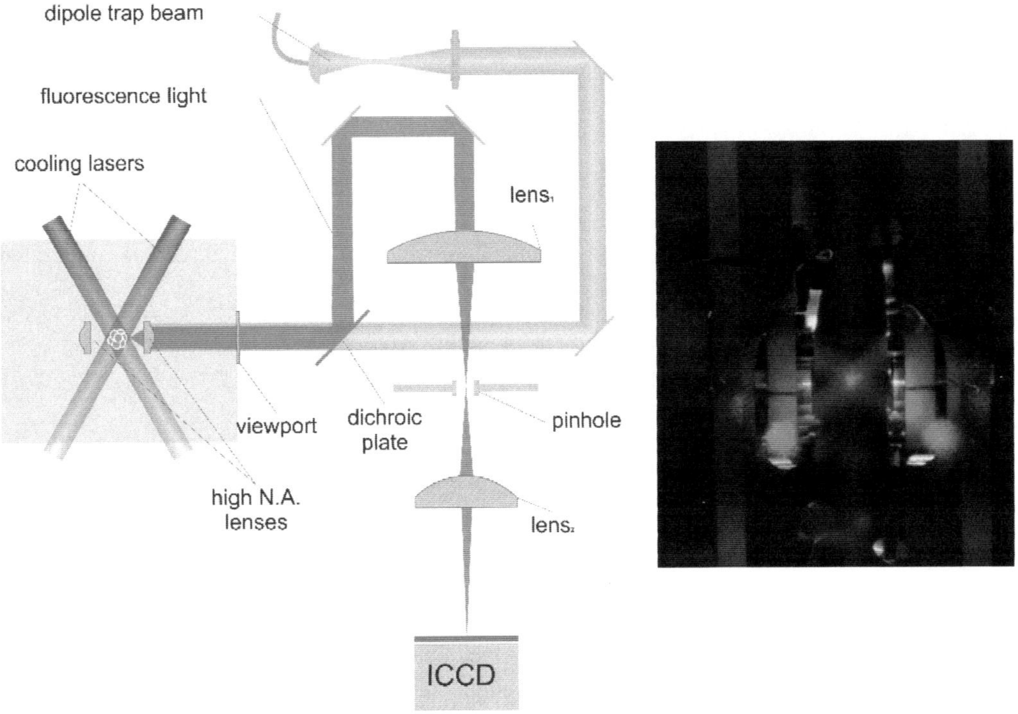

Figure 3. Experimental apparatus for tight dipole trapping of Rubidium atoms. *Left panel*: high numerical aperture lenses are used to produce a very tight (~μm waist) optical tweezer to load N atoms from a laser-cooled reservoir. Fluorescence light emitted from the atoms is also collected from the same lens and used to image the atoms. *Right panel*: image of the lenses inside the vacuum chamber.

With a trapping laser power of 110 mW we infer that the atoms in the dipole trap experience an average light shift of $2\pi\times80$ MHz, which is indeed compatible with a waist between 2 and 4 μm. We also determine the average number of atoms in the trap is N = 20.

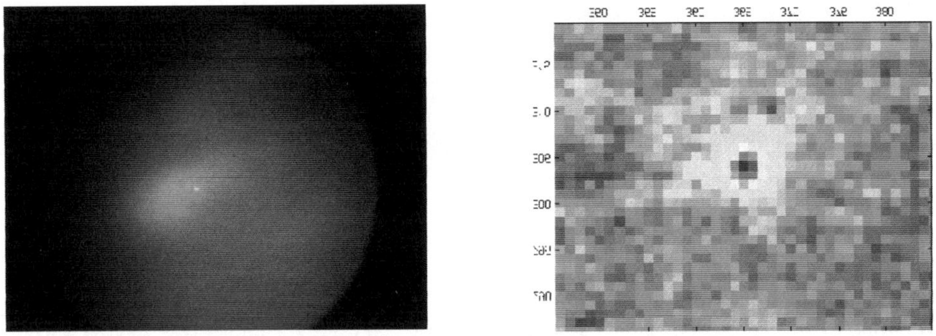

Figure 4. Dipole trap loading 20 atoms. *Left panel*: raw fluoresce image of the dipole trap (bright spot at the center) and the MOT reservoir *Right panel*: background subtracted image of a dipole trap containing 20 atoms. The magnification is such that 1 pixel on the ICCD corresponds to 1 μm at the lens focal plane.

4.2. Control

To isolate a single control atom, we need to create a much steeper potential nearby the register trap, so that inelastic laser-induced two body collisions ensure the filling of the trap with either 1 or 0 atoms with sub-poissonian statistics [14].

By changing the input diameter of the incoming beam on the high N.A. lens, we can reduce the transverse size of the beam (and by consequence the Rayleigh length that scales with the square of the waist). By changing the waist by a factor of ~4 we would so reduce the trapping volume, by a factor of $1/4^4$, thus allowing the trap to enter the regime of *collisional blockade* required for single atom trapping, as discussed in [14].

We have setup a second dipole trap potential and performed measurements of the irradiance spatial profile. In figure 5 we compare the measured profile to the theoretical models and we observe that the measured performance of the optical setup well in agreement with the expectations given by the models. We verify this way the much tighter potential for the trapped atoms that will allow us to achieve simultaneous loading of an ensemble cloud and single control qubit to perform mixed-states protocols.

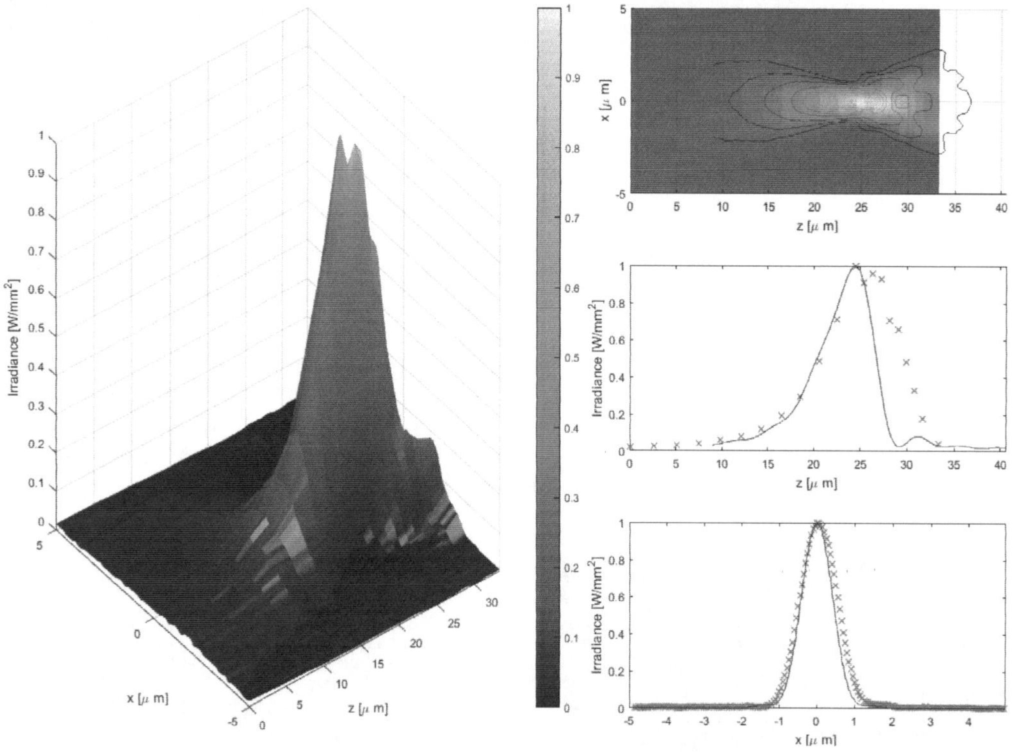

Figure 5. Measured irradiance profile for a ~1 μm waist trap and comparison with theoretical models

4.3. Excitation and conditional logic lasers

The excitation of Rydberg states and the EIT scheme for conditional rotation are achieved by the 2-photon laser schemes shown in figure 1. A commercial frequency doubled 960 nm laser (Toptica SHG) produces 270 mW of 480 nm light which we lock to a specific Rydberg state by extracting an error signal from an EIT based spectrometer [16]. The scheme has already been implemented and used to measure the transition strengths of the 480 nm 5P to nL Rydberg states via the Autler Townes effect [15]. We have excited n in the range 19 – 52 and the total orbital angular momentum quantum number

L can be either S or D. The 780 nm beam is locked by a modulation transfer scheme [17], and is used both for driving the 5S to 5P transition in the dipole trapped atoms and for locking the 480 nm laser.

5. Conclusions

We reported on progress on the design and implementation of a cold-atoms based experiment to test the quantum advantage to be gained in the presence of mixed states. At the present time we are able to study registers containing 20 ^{87}Rb atoms. In the future, we aim to increase the number of atoms in the register by tuning the MOT loading rate and the trap geometry. This method gives us the ability to tune the amount of quantum resources and decoherence present in the system and to study the boundaries between classical computing and quantum enhanced computing or sensing.

References

[1] Knill E and Laflamme R 1998 *Phys. Rev. Lett.* **81** 5672
[2] Lanyon B P, Barbieri M, Almeida M P and White A G 2008 *Phys. Rev. Lett.* **101** 200501
[3] C W Mansell, S Bergamini 2014 *New J. Phys.* **16** 053045
[4] Modi K, Cable H, Williamson M and Vedral V 2011 *Phys. Rev.* X **1** 021022
[5] MacCormick C, Bergamini S, Mansell C W, Cable H, Modi K 2016 *Phys. Rev.* A **93** 023805
[6] Datta A, Flammia S T and Caves C M 2005 *Phys. Rev.* A **72** 042316
[7] Datta A, Shaji A and Caves C M 2008 *Phys. Rev. Lett.* **100** 050502
[8] Schlosser N, Reymond G and Grangier P 2002 *Phys. Rev. Lett.* **89** 023005
[9] Saffman M, Zhang X L, Gill A T, Isenhower T and Walker T G 2011 *J. Phys. Conf. Ser.* **264** 012023
[10] Muller M, Lesanovsky I, Weimer H, Buchler H P and Zoller P 2009 *Phys. Rev. Lett.* **102** 170502
[11] Schlosser N, Reymond G, Protsenko I and Grangier P 2001 *Nature* **411** 1024–7
[12] Piotrowicz M J, MacCormick C, Kowalczyk A, Bergamini S, Beterov I I and Yakshina E A 2011 *New Jour. Phys.* **13** 093012
[13] Abel R P, Mohapatra A K, Bason M G, Pritchard J D, Weatherill K J, Raitzsch U and Adams C S 2009 *Appl. Phys. Lett.* **94** 071107
[14] McCarron D J, King S A and Cornish S L 2008 *Meas. Sci. and Technol.* **19** 105601

Highly efficient multimode diode-pumped Yb:KYW laser

S A Kuznetsov[1], V S Pivtsov[1, 2], A V Semenko[1, 2] and S N Bagayev [1, 3]

[1] Institute of Laser Physics SB RAS, 13/3 Acad. Lavrentyev ave., Novosibirsk 630090, Russia
[2] Novosibirsk State Technical University, 20 K. Marx ave., Novosibirsk 630073, Russia
[3] Novosibirsk State University, 2 Pirogova str., Novosibirsk 630090, Russian Federation

E-mail: clock@laser.nsc.ru

Abstract. Record high differential efficiency (53.2%) and full optical efficiency (48%) for a multimode diode-pumped Yb:KYW laser have been achieved. The characteristics of the laser and methods for improving its efficiency using a distributed Bragg reflector tapered diode laser (DBR TDL) are discussed.

1. Introduction

Optical frequency synthesizers based on femtosecond lasers are most important blocks of precision laser systems for measurement of absolute optical frequencies and their stability [1–3]. They are also used to solve the problem of transferring the frequency characteristics of optical frequency standards to the radio frequency range. As most of the master lasers are used femtosecond Ti:S laser or Er fiber lasers [4–6]. But in recent years, synthesizers on the basis of ytterbium-doped crystals are becoming more popular [7–10]. The absorption bands of ytterbium crystals coincide with the generation range of high-power semiconductor diode lasers (940 nm and 980 nm). Among ytterbium-doped media, the potassium tungstates Yb:KYW and Yb:KGW are most promising owing to large absorption cross-sections and induced radiation, wide luminescence band, and rather high heat conduction. In this paper, various schemes of small cavities of a Yb:KYW laser with high-power multimode diode pumping are investigated.

2. Experimental setup and results

As an active medium, we used Yb:KYW crystal with 5 at. % Yb3+ ions and a thickness of 1.5 mm. The crystals were cut in the direction of axis *c* (E || *a*, *b*). The temperature of the crystal was stabilized at 9°C using a Peltier element. The source of pumping was a LIMO25-F100-DL980 laser with a multimode fiber output. The central radiation wavelength was 981 nm, the maximum output power was 25 W, and the fiber core diameter was 105 μm. The pump laser radiation had circular polarization.

To eliminate the thermal lens astigmatism we have proposed a scheme of an ytterbium laser with symmetrical cooling of the gain medium (figure 1). The Yb:KYW crystal (tapering of 0.5°) with a highly reflective dielectric coating was glued to a copper heatsink. The other crystal surface had an antireflection coating. In this configuration, the thermal lens is symmetrical with respect to the axis of generation and may be compensated by changing the distance between the spherical mirror and the crystal. In fact, this is a modified disc laser, characterised in that the thickness of the crystal is much larger than the diameter of the pump and output beam waists, which were 110 and 40 mm, respectively. For this reason, their axes should coincide. Polarisation of the pump and output radiation was directed parallel to the crystal axis (p-polarisation). To compensate for the GVD we used a pair of GTI mirrors (Layertec) with dispersion D = – 900 ± 100 fs2.

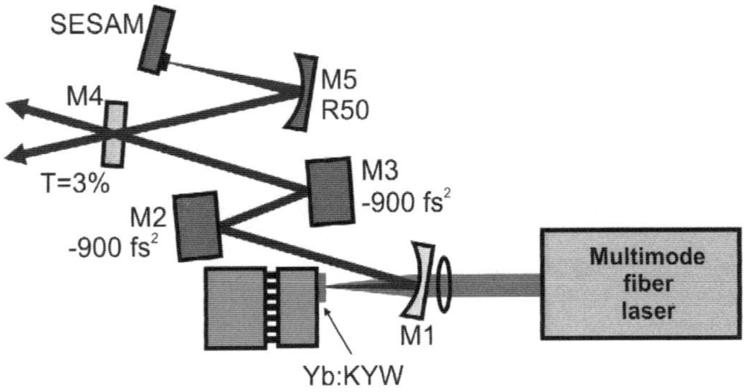

Figure 1. Scheme of a femtosecond ytterbium laser.

To start the mode locking we used a SESAM (BATOP GmbH). The intermode frequency was about 400 MHz. The output power as a function of the pump power for the free generation regime is shown in figure 2.The saturation of the dependence was not observed up to a pump power of at 14 W. We obtained a maximum output power of 3.0 W at 9.5 W of pump power. The lasing threshold was 0.8 W and η_{dif}= 35 %. We failed to obtain mode locking, apparently due to incomplete compensation for the GVD. The disadvantage of this scheme lies in the fact that the focusing of multimode pump radiation is not optimal.

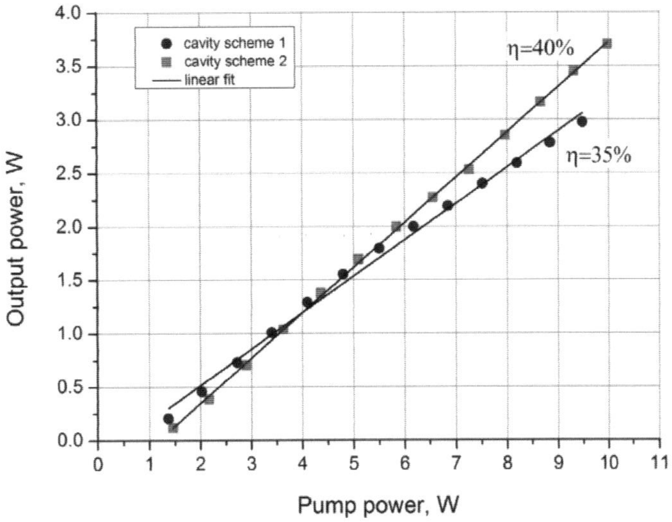

Figure 2. CW output power of the ytterbium laser with the resonator schemes given in figures 1 and 3 as a function of pump power.

In the scheme shown in figure 2, the above disadvantage is minimised. Here, the focal length of the lens can be reduced; therefore, the pump radiation will be focused to a spot smaller in size and the efficiency of the laser will be increased. We used a mirror M1 with a transmission of ~80 % and less than 0.05% for the pump radiation polarised along the *a* axis of the crystal (p-polarisation) and for the output radiation polarised along the *b* axis (s-polarisation), respectively.

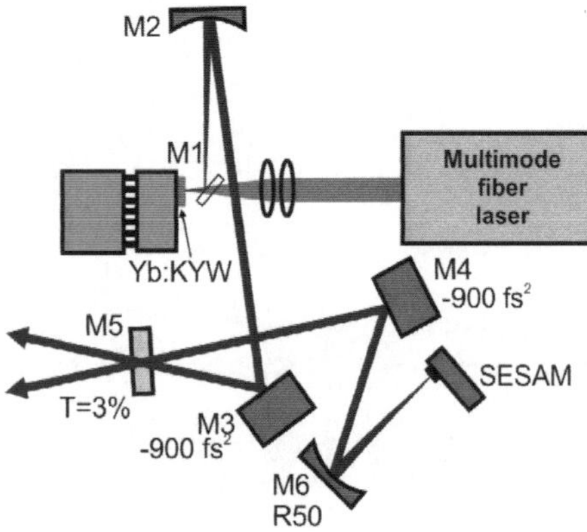

Figure 3. Scheme of a femtosecond ytterbium laser.

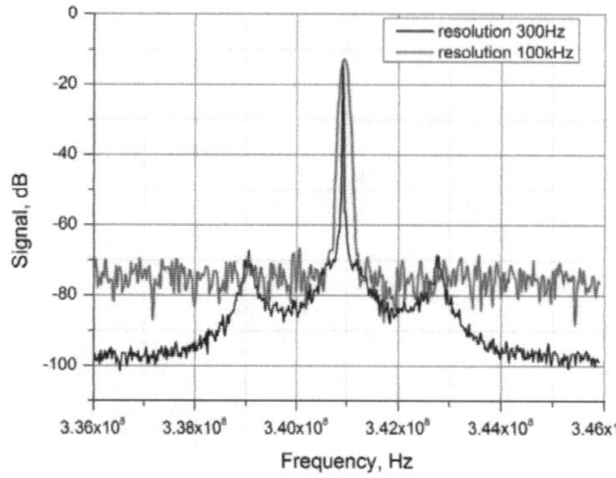

Figure 4. Emission spectra at the intermode frequency in the case of mode-locked operation, measured at a resolution of 0.3 and 100 kHz.

The lasing threshold was 1 W and η_{dif}= 40% (figure 2). At a pump power of 14 W the maximum output power reached 4.9 W, i.e., η_{dif}= 35 %, which is record high in the case of a multimode pump source. The pulse duration in the mode-locking regime was not measured; the intermode frequency was approximately 340 MHz, the output power in the mode-locking regime was 1.3 W at a pump power of 7 W, the radiation wavelength was 1041 nm and the spectral FWHM bandwidth was about 2 nm. In mode-locked operation there appeared a narrow intermode frequency comb with a high (greater than 60 dB) signal-to-noise ratio. The mode-locking spectrum at the intermode frequency is shown in figure 4.

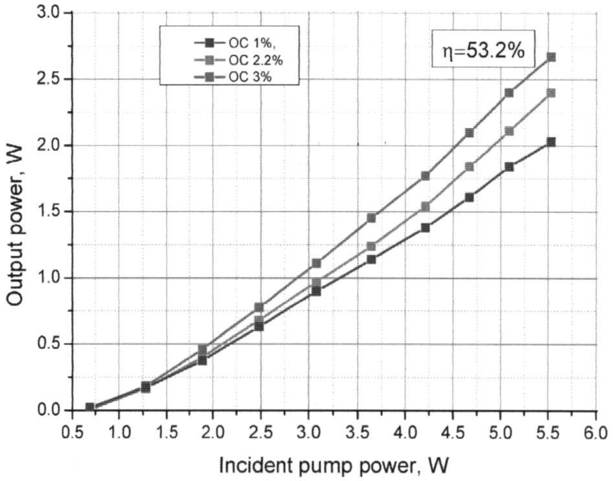

Figure 5. CW output power of the ytterbium laser for different output couplers.

Measurements for output mirrors with various transmission coefficients (1%, 2.2%, 3%) have been made (figure 5). The differential efficiency was 53.2%, the optical efficiency was 48%, and the maximal output power was 3.94 W at a pump power of 8.2 W. This is a record result when using a multimode pump source.

Maximum efficiency can be achieved by using single-mode diode pumping for this scheme when the pump and generation waists fully overlap. The pump source is a distributed Bragg reflector tapered diode laser [11].The pump diode consists of a 4-mm-long gain-guided tapered section and a 2-mm-long 4-μm-wide index-guided straight ridge-waveguide section containing a 1-mm-long surface Bragg grating. The DBR TDL emits at 981 nm with a spectral linewidth of less than 20 pm (FWHM).In addition, the device has a nearly diffraction limited output beam with a lateral beam propagation factor $M^2_{1/e2} = 1.3$ containing more than 72% of the power in the central lobe. The maximum output power is 7 W.

3. Conclusion

A promising small ytterbium laser with a diode pumping laser was designed. Record high efficiencies (a differential efficiency of 53.2% and a total optical efficiency of 48%) were obtained for a laser with an Yb:KYW crystal in the continuous-wave regime. Mode-locking laser operation with a pulse rate of ~340 MHz was demonstrated. This allows us to use the proposed design in the construction of small laser systems (e.g., mobile precision femtosecond synthesizers). The ultimate efficiency of the proposed laser could be achieved with single-mode diode pumping lasers and by optimizing the focusing system.

Acknowledgments

This work was partially supported by the Russian Foundation for Basic Research (project no. 16-02-00639-a) and Russian Presidential Grant (NSh-6689.2016.2).

References

[1] Bagayev S N, Chepurov S V, Klementyev V M *et al.* 2000 *Appl. Phys.* B **70** 375
[2] Udem Th, Holzwarth R and Hansch T W 2002 *Nature* **416** 233
[3] Bagayev S N, Denisov V I, Klementyev V M *et al.* 2004 *Laser Phys.* **14** 1
[4] Pivtsov V S, Nyushkov B N, Korel I I *et al.* 2014 *Quantum Electronics* **44** 507
[5] Zimmermann M, Gohle C, Holzwarth R *et al.* 2004 *Optics Lett.* **29** 310
[6] Korel I I, Nyushkov B N, Denisov V I *et al.* 2014 *Laser Physics* **24** 074012
[7] Pekarek S, Sudmeyer T, Lecomte S *et al.* 2011 *Opt. Express* **19** 16491
[8] Endo M, Ozawa A and Kobayashi Y 2012 *Opt. Express* **20** 12191
[9] Kuznetsov S A and Pivtsov V S 2014 *Quantum Electronics* **44** 444
[10] Kirpichnikova A A, Kuznetsov S A and Pivtsov V S 2015 *Bulletin of the Russian Academy of Sciences. Physics* **79** 169
[11] Fiebig C, Blume G, Uebernickel M *et al.* 2009 *IEEE J. of Selected Topics in Quantum Electronics* **15** 978

Quantum field theory and classical optics: determining the fine structure constant

Gerd Leuchs[1,2,3], **Margaret Hawton**[4] **and Luis L Sánchez-Soto**[2,5]

[1] Department Physik, Universität Erlangen-Nürnberg, Staudtstr. 7 B2, 91058 Erlangen, Germany
[2] Max Planck Institute for the Science of Light, Staudtstr. 2, 91058 Erlangen, Germany
[3] Department of Physics, University of Ottawa, 25 Tempelton Street, Ottawa, ON, K1N 6N5, Canada
[4] Department of Physics, Lakehead University, 955 Oliver Road, Thunder Bay, ON, P7B 5E1
[5] Departamento de Óptica, Facultad de Física, Universidad Complutense, 28040 Madrid, Spain

E-mail: gerd.leuchs@mpl.mpg.de

Abstract. The properties of the vacuum are described by quantum physics including the response to external fields such as electromagnetic radiation. Of the two parameters that govern the details of the electromagnetic field dynamics in vacuum, one is fixed by the requirement of Lorentz invariance $c = 1/\sqrt{\varepsilon_0\mu_0}$. The other one, $Z_0 = \sqrt{\mu_0/\varepsilon_0} = 1/(c\varepsilon_0)$ and its relation to the quantum vacuum, is discussed in this contribution. Deriving ε_0 from the properties of the quantum vacuum implies the derivation of the fine structure constant.

1. Introduction

At the starting point of the prototype quantum field theory, quantum electrodynamics (QED), are the equations of Maxwell and Dirac. Formulation of a consistent theory from scratch was delayed by the enormous difficulties encountered, until Feynman and Schwinger decided to postulate Maxwell's equations in vacuum and build the theory from there by adding the interaction between the electromagnetic field and the electron. When Maxwell formulated the equations carrying his name, the mathematical structure was motivated by the attempt to bring all laws of electromagnetism known hitherto into one consistent form. There are, however, parameters in Maxwell's equations, which had to be determined by experiment. These two parameters are, depending on your choice of the unit system: ε_0 and μ_0 or the speed of light in vacuum and the impedance of the vacuum, or whatever the choice of units is. The compatibility with special relativity requires Maxwell's equation to be Lorentz-invariant, which means that the speed of light in vacuum and the limiting speed in special relativity are identical. This related one parameter to the rest of physics, $c = 1/\sqrt{\varepsilon_0\mu_0}$, while the other parameter, the vacuum impedance $Z_0 = \sqrt{\mu_0/\varepsilon_0} = 1/(c\varepsilon_0)$ was still an experimental parameter unrelated to any other area in physics. This we challenge by postulating that this second parameter is intimately related to the property of the modern quantum vacuum. As will become clear, this also determines the fine structure constant.

2. A dielectric model for the quantum vacuum

In ultra high electric fields the vacuum is predicted to break down forming an electron–positron plasma. At lower field strength, virtual electron–positron pairs are polarized forming virtual ephemeral electric dipoles. This vacuum polarization provides the partial shielding of point charges. In analogy to the treatment of a dielectric medium, this vacuum polarization caused by a light field should appear in Maxwell's equations. The quantity to look for is the electric displacement field introduced by Maxwell. This is the first term on the right hand side of

$$\mathbf{D} = \varepsilon_0 \mathbf{E} + \mathbf{P} \,. \tag{1}$$

One might be tempted, as we are, to think of $\varepsilon_0 \mathbf{E}$ as of the polarization of the vacuum. The vacuum would then be a modern version of the aether, with all the quantum properties and with relativistic symmetry [1].

Along this line the response of the vacuum to an external electric field is the polarization of virtual particle–antiparticle pairs ubiquitous in the vacuum. And the precision with which Maxwell's equations describe low energy electro-magnetic experiments suggests that they already contain the linear response of the vacuum. This linear response can be calculated on the back of an envelope using a simple model [2, 3]. There it is argued that the dipole moment induced in a single virtual pair is inversely proportional to the third power of the rest mass of the particles. The polarization in equation (1) is a dipole moment density so we need to divide the induced dipole moment by the volume, a single particle pair occupies. A first guess for this volume is this particle's Compton wavelength cubed, which corresponds to a momentum cut-off of $p = mc$ when integrating over all momenta. In the resulting expression for the polarization of the vacuum due to virtual electron-positron pairs the mass of the electron drops out. The surprising consequence is that the electronic contribution depends only on the square of the electron charge and not on the mass. The same must then be true for all other types of elementary particles. They all contribute to the vacuum polarization in a similar way proportional to the square of their charge and one thus has to sum over all different types of elementary particles. Therefore, the parameters used in Maxwell's equations, determined experimentally so far, are related to the relativistic quantum vacuum and allow one to deduce information about all charged elementary particles known and unknown. According to this simple model the fine structure constant is given by the sum over all types of elementary particles. For ε_0 one obtains

$$\varepsilon_0 = \frac{1}{\hbar c} f \sum_j^{\text{e.p.}} q_j^2 \,. \tag{2}$$

The factor f in our model is of order unity, but depends on the details of renormalization. The index runs over all different types of charged elementary particle–antiparticle pairs (e. p.): $j = 1$ electron, $j = 2$ muon, $j = 3$ tauon, $j = 4$ to 21 the different versions of quarks, $j = 22$ the W^+ boson and what else is out there ($j > 22$?).

3. Implications of the result of model

The charge of the electron, which we see in low energy experiments, is to our current understanding the partially screened bare point charge. The screening is due to the polarization of the vacuum. The contribution to the point charge screening by a particular type of elementary particle happens at a distance from the point charge roughly equivalent to the Compton wavelength of the particular type of particles. The same holds for the other types of elementary particles [4]. As is well known from high-energy scattering experiments, for example between two electrons, the apparent charge of the electron increases as the scattering particles come closer. As a consequence, the screening by vacuum polarization is reduced when $(\omega_{\mathbf{k}}/c)^2$ of the photon exchanged by the scattering particles differs more and more from $\mathbf{k}^2$. This is quantified by the off-shellness $k^2 = \omega_{\mathbf{k}}^2/c^2 - \mathbf{k}^2$. Equivalently, the fine structure constant, which is proportional to e^2 likewise increases. In our model, however, it is the dielectric permittivity, which decreases as you start to leave out the low rest mass or high Compton wavelength particles contributing to the screening at larger distances. Inserting equation (2) into

$$\alpha = \frac{1}{4\pi\varepsilon_0}\frac{e^2}{\hbar c} \tag{3}$$

yields

$$\alpha^{-1} = 4\pi f \sum_j^{\text{e.p.}} \left(\frac{q_j}{e}\right)^2 . \tag{4}$$

This can be compared with the running fine structure constant routinely used in particle physics [5, 6]

$$\alpha^{-1}(k^2) \simeq \alpha^{-1}(0) - \frac{1}{3\pi}\sum_j^{\text{e.p.}} \left(\frac{q_j}{e}\right)^2 \ln\left(\frac{\hbar k^2}{m_j^2 c^2}\right) , \tag{5}$$

where $k^2 = \omega_{\mathbf{k}}^2/c^2 - \mathbf{k}^2$ and the limit $\hbar^2|k^2| \gg m_j^2 c^2$ is assumed. The value of k^2 for which the right-hand side of equation (5) vanishes, $k^2 = \Lambda_L^2$, is referred to as the Landau pole [7]. So, equation (5) gives

$$\alpha^{-1}(0) = \frac{1}{3\pi}\sum_j^{\text{e.p.}} \left(\frac{q_j}{e}\right)^2 \ln\left(\frac{\hbar \Lambda_L^2}{m_j^2 c^2}\right) . \tag{6}$$

It is obvious that all different types of virtual elementary particles contribute to α. This looks very much like equation (4) derived from the simple model, if one replaces the fudge factor f by

$$f = \frac{1}{12\pi^2}\left\langle \ln\left(\frac{\hbar \Lambda_L^2}{m_j^2 c^2}\right)\right\rangle_j . \tag{7}$$

Note that for the electron and the W^+ boson, the logarithmic term inside the angle brackets varies only by 23%: from 101 to 78, if one chooses $\hbar\Lambda_L/c$ be the Planck mass. The averaged result $f \simeq 0.7$ can be factored out of the sum. The relevance of the cut-off being determined by the Planck mass is that then $\alpha^{-1}(k^2 \to \infty) = \varepsilon(k^2 \to \infty) = 0$ in the bare vacuum. It is worth noting that in the beginning of quantum field theory several groups postulated a formula very similar to equations (4) and (5). However, this was before quarks were discovered at a time when one assumed that all charged elementary particles had the same electric charge e. In this case the sum simply turns into the number ν of different types of elementary particles [8, 9]

$$\alpha^{-1} = \frac{1}{3\pi}\nu \ln\left(\frac{\hbar c}{G m^2}\right) . \tag{8}$$

The main result here is the closed form in equation (6) for the fine structure constant, derived in a fully quantum field theoretical treatment and inspired by the simple dielectric model of the vacuum above.

[1] Laughlin R B 2005 *A Different Universe: Reinventing Physics from the Bottom Down* (New York: Basic Books)
[2] Leuchs G, Villar A S and Sánchez-Soto L L 2010 *Appl. Phys.* B **100** 9–13
[3] Leuchs G and Sánchez-Soto L L 2013 *Eur. Phys. J.* D **67** 57
[4] Gottfried K and Weisskopf V 1986 *Concepts of Particle Physics* vol II (New York: Oxford University Press)

[5] Peskin M and Schroeder D 1995 *An Introduction to Quantum Field Theory* (Redwood City, CA: Addison-Wesley)
[6] Hogan C J 2000 *Rev. Mod. Phys.* **72** 1149–61
[7] Landau L D, Abrikosov A A and Khalatnikov I M 1954 *Dokl. Akad. Nauk SSSR* **95** 497
Landau L D, Abrikosov A A and Khalatnikov I M 1954 *Dokl. Akad. Nauk SSSR* **95** 773
Landau L D, Abrikosov A A and Khalatnikov I M 1954 *Dokl. Akad. Nauk SSSR* **95** 1177
[8] Landau L D and Pomeranchuk I Y 1955 *Dokl. Akad. Nauk SSSR* **102** 489–93
[9] Zel'dovich Y B 1967 *JETP Lett.* **6** 345–7

IOP Conf. Series: Journal of Physics: Conf. Series **793** (2017) 012018 doi:10.1088/1742-6596/793/1/012018

Investigation of single defects created in crystals by laser emission and hard radiation

E F Martynovich[1,2], V P Dresvyanskiy[1], S V Boychenko[1], A L Rakevich[1], S A Zilov[1] and S N Bagayev[3]

[1] Irkutsk Branch of the Institute of Laser Physics SB RAS, 130A Lermontov str., Irkutsk 664033, Russia

[2] Irkutsk State University, 20 Gagarin blvd, Irkutsk 664033, Russia

[3] Institute of Laser Physics SB RAS, 13/3 Acad. Lavrentyev ave., Novosibirsk 630090, Russia

E-mail: filial@ilph.irk.ru

Abstract. The possibility of identifying radiation-created quantum systems via the characteristics of quantum trajectories of luminescence intensity measured on individual centers by confocal scanning fluorescence microscopy with the time-correlated single photon counting has been studied. Calculations of the quantum trajectories have been carried out by the density matrix method. Experimental studies have been carried out using a confocal microscope.

Exciton mechanism of radiation-induced defects formation is implemented to a large part of the known inorganic crystalline media. Exposure to particles and hard radiation quanta on the crystals leads to the formation of stable electron excitations: electrons, holes and excitons after a series of quick intermediate energy transfer processes. Then started the processes of interaction of electronic and nuclear subsystems of the crystals. Structural defects result from this interaction. The basic process of this kind is the decay of the excitons on the primary structural defects, i.e. Frenkel pairs with subsequent recharging, migration and aggregation.

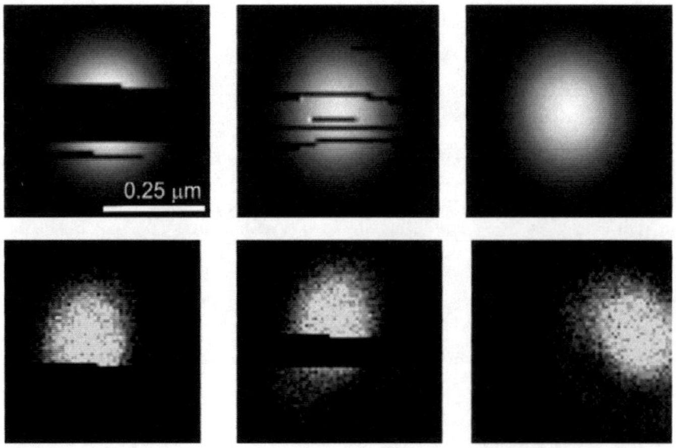

Figure 1. Confocal scanned fluorescent images of single color centers with different characteristics of quantum trajectories. Above – theoretically calculated, at the bottom – experimentally obtained.

In figure 1 the images of study single F_2 color centers in LiF crystal displays. Gaps in the images correspond to centers transitions in the non-emitting state during the scanning. One experimental trajectories presented in figure 2. In the process of recording this trajectory increased the intensity of the exciting radiation abruptly. Therefore, the trajectory consists of two parts: linear and non-linear.

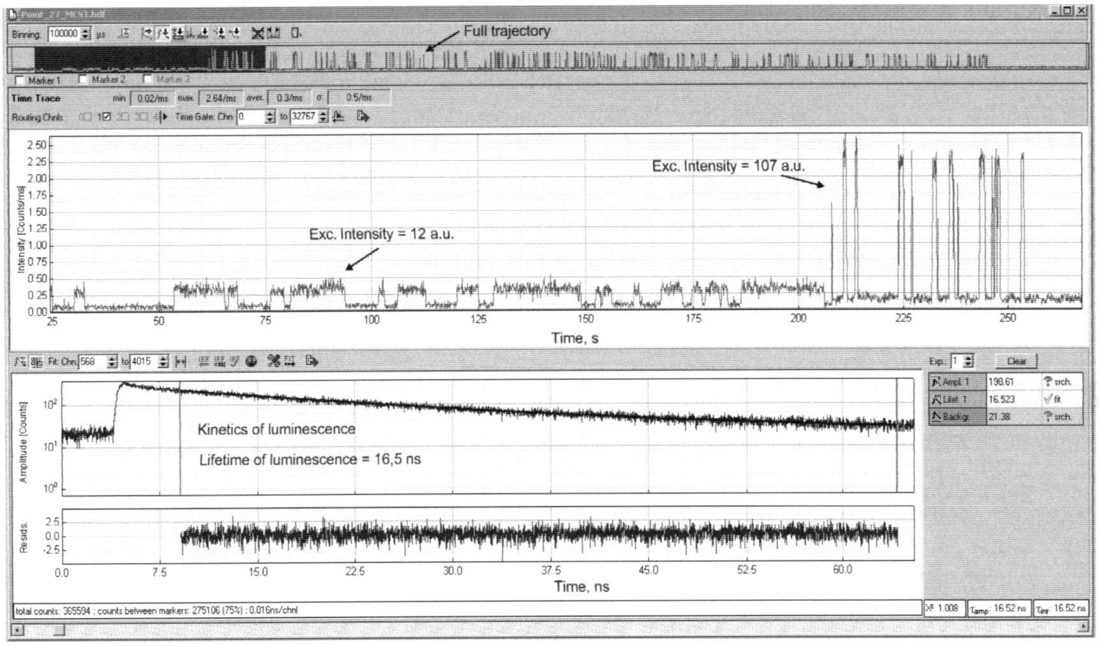

Figure 2. Linear and non-linear parts of the trajectory of the intensity of the same center when excited by continuous sequence of picosecond pulses. Switching the excitation intensity is on 208 seconds. Above – the full trajectory. In the mid – enlarged initial part. Below – the kinetics of the rise and decay of the luminescence intensity of the center.

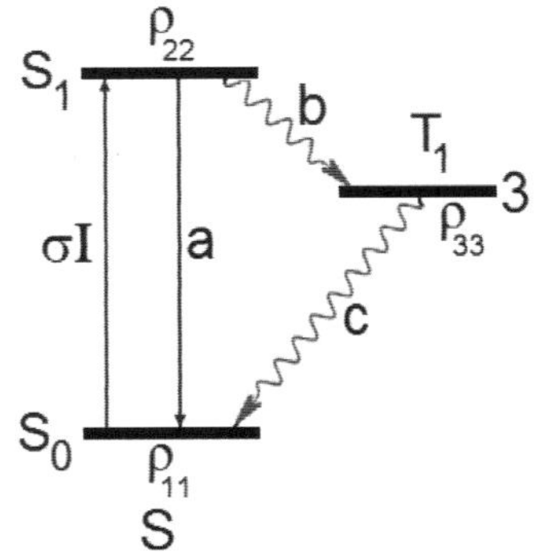

Figure 3. Scheme of transitions.

Trajectories can be called linear, if the processes occurring during excitation centers are limited to the scheme shown in figure 3. There are no induced transitions down under the influence of the exciting radiation. They should not be taken into account because the F_2 centers have a large Stokes shift and the cross section of stimulated emission in the spectral line of the pump is very little. In addition, in the diagram there are no stimulated transitions under the influence of the spontaneous emission, i.e. there is no superluminescence. In our experiments, it is confirmed by the fact that luminescence decay time does not depend on the intensity. Also on the scheme shown in figure 3, does not take into account the possible transitions on higher levels of singlet and triplet systems. For linear trajectories, the luminescence intensity increases with excitation intensity. Accordingly, the intervals of stay of the center in the radiating state S_0 decreases as increasing the probability σI.

With a substantial increase in the intensity of the exciting radiation observed decrease in the average time of stay in the center of a non-emitting state T_1. This is due to the transition to the nonlinear regime, due to the excitation overlying states and triplet-singlet conversion. The ratio of lifetimes in the emitting and non-emitting states at a fixed intensity of the exciting radiation can be used as a new individual "spectroscopic" characteristic of the centers.

Acknowledgments
This work was supported by the project 0307-2016-0004 of Fundamental Research Programs of State Academies of the Russian Federation, and the Russian Foundation for Basic Research (grant no. 16-52-44056).

References
[1] Martynovich E F, Dresvyansky V P, Zilov S A, Rakevich A L, Boychenko S V, and Bagayev S N 2015 *OSA Online Technical Digest of the Advanced Photonics 2015* (Optical Society of America, USA) paper SeW2B.5

IOP Conf. Series: Journal of Physics: Conf. Series **793** (2017) 012019 doi:10.1088/1742-6596/793/1/012019

Nonlinear optics in the mid-infrared: new morning

A V Mitrofanov[1,2,3,4,5], A A Voronin[1,2,3], D A Sidorov-Biryukov[1,2], A A Lanin[1,2], M V Rozhko[1,2], S Mitryukovsky[1], A A Ivanov[1], V Ya Panchenko[4,5], E A Stepanov[1,2], D V Meshchankin[1,2], E E Serebryannikov[1,2], A B Fedotov[1,2], V Shumakova[6], S Ališauskas[6], A Pugžlys[6], A Baltuška[6] and A M Zheltikov[1,2,3,4*]

[1] Russian Quantum Center, 100 Novaya str., Skolkovo, Moscow Region, 143025 Russia
[2] Physics Department, International Laser Center, M.V. Lomonosov Moscow State University, Moscow 119992, Russia
[3] Department of Physics and Astronomy, Texas A&M University, College Station TX 77843, USA
[4] Kurchatov Institute National Research Center, Moscow 123182, Russia
[5] Institute of Laser and Information Technologies, Russian Academy of Sciences, Shatura, Moscow Region, 140700 Russia
[6] Photonics Institute, Vienna University of Technology, 1040 Vienna, Austria

*E-mail: zheltikov@physics.msu.ru

Abstract. Recent breakthroughs in ultrafast photonics in the mid-IR help understand complex interactions of high-intensity mid-IR field waveforms with matter, offer new approaches for x-ray generation, enable mid-IR laser filamentation in the atmosphere, facilitate lasing in filaments, give rise to unique regimes of laser–matter interactions, and reveal unexpected properties of materials in the mid-IR range.

1. Introduction

Motivated and driven by numerous applications and long-standing challenges in strong-field physics [1, 2], molecular spectroscopy, semiconductor electronics, and standoff detection, ultrafast optical science is rapidly expanding toward longer wavelengths [3-16]. Experiments reveal unique properties of filaments induced by ultrashort laser pulses in the mid-infrared, where the generation of powerful supercontinuum radiation is accompanied by unusual scenarios of optical harmonic generation, giving rise to remarkably broad radiation spectra, stretching from the visible to the mid-infrared. Generation of few- and even single-cycle mid-infrared field waveforms has been demonstrated within a broad range of peak powers and central wavelengths. Below-the-bandgap high-order harmonics generated by ultrashort mid-infrared laser pulses are shown to be ideally suited to probe the nonlinearities of electron bands, enabling an all-optical mapping of the electron band structure in bulk solids.

2. Generation of ultrashort pulses in the mid-infrared

High-power ultrashort mid-IR driver pulses are delivered in our experiments by a laser system (figure 1) consisting of a solid-state ytterbium laser with an amplifier, a three-stage optical parametric amplifier (OPA), a grating–prism (grism) stretcher, a Nd:YAG pump laser, a three-stage OPCPA system, and a grating compressor for mid-IR pulses [6, 8]. The compressed-pulse idler-wave OPCPA output has a central wavelength of 3.9 μm, a pulse width of 80 fs, and an energy up to 30 mJ. Spectral measurements in the mid-IR range are performed with a homebuilt spectrometer consisting of a scanning monochromator and a thermoelectrically cooled HgCdTe detector. For the spectral measurements in the ultraviolet, visible, and near-IR ranges, OceanOptics HR4000 and NIRQuest spectrometers were employed. Temporal envelopes and phases of mid-IR pulses are characterized using frequency-resolved optical gating (FROG) based on second-harmonic generation (SHG) in a 0.5-mm-thick $AgGaS_2$ crystal.

Figure 1. (a) The laser facility with an OPCPA source of subterawatt ultrashort pulses in the mid-infrared at the Advanced Photonics Laboratory of the Russian Quantum Center, (b) ultrafast spectrochronography beamline, and (c) vacuum chamber for high-order harmonic generation.

3. Laser filaments in the mid-infrared

Laser filamentation in the mid-infrared involves a broad class of nonlinear-optical phenomena that do not show up in visible and near-infrared filaments. The most prominent examples include generation of multiple odd-order harmonics [8] and strong plasma refraction [6], which tends to give rise to ring-shaped beam patterns, especially well-resolved toward the back of the pulse, as well as a shift in the balance between the self-focusing due to the Kerr effect and plasma-induced defocusing. With all these effects manifested in the inherently off-axial beam dynamics [6, 12], angle-resolved studies on the supercontinuum filament output have to be included as an essential tool needed to understand laser filaments in the mid-infrared. Since such angle-resolved measurements have to be performed within a bandwidth of several octaves, stretching from the ultraviolet deep into the mid-infrared, multiple challenges need to be confronted in implementing experimental methods that would enable such studies.

Our experimental approach addresses these challenges through angle-resolved studies on a multioctavehigh-energy supercontinuum output of laser filaments induced in atmospheric air by 0.25-TW, 3.9-μm driver pulses. With optical harmonics up to the 15th order contributing to supercontinuum generation alongside Kerr-type and ionization-induced nonlinearities, the spectra of this supercontinuum radiation span over almost five octaves from the ultraviolet deep into the mid-infrared spectral range. As we demonstrate below, rewards for such angle-resolved measurements are numerous. Performed on an ultrabroadband beam, these studies reveal clearly resolved signatures of a complex off-axial beam dynamics, which is intrinsic to laser filamentation in the mid-infrared, as opposed to near-infrared filaments, thus offering important insights into an intricate and in many ways unusual spatiotemporal evolution of high-power mid-infrared pulses in the filamentation regime.

Our experiments demonstrate the generation of high-energy supercontinuum spanning 4.7 octaves, from 250 to 6500 nm, using a 0.25-TW, 3.9-μm output of a mid-infrared optical parametric chirped-pulse amplifier as a driver inducing a laser filament in the air. The high-frequency wing of the supercontinuum spectrum is enhanced by odd-order optical harmonics of the mid-infrared driver. Optical harmonics up to the 15th order are observed in supercontinuum spectra as overlapping, yet well-resolved peaks broadened, as verified by numerical modeling, due to spatially nonuniform ionization-induced blue shift.

Filamentation-assisted pulse compression in the gas phase is shown to enable the generation of subterawatt few-cycle pulses in the mid-infrared. With both spatial modulation instabilities and excessive plasma scattering of the mid-infrared beam prevented through a careful choice of the gas pressure and the input peak power, providing single-filament regime of pulse propagation, peak powers as high as 0.3 TW are achieved in a truly single-mode, almost diffraction-limited 35-fs output at a central wavelength of 4 μm.

4. Ultrashort pulses in the mid-infrared and new regimes of nonlinear optics

Because of the complexity of the spatiotemporal field waveform dynamics underlying filamentation, experiments designed to guide theoretical modeling toward a detailed physical understanding of filaments generate high-volume, high-variety data sets at high data-acquisition speeds. Ultrabroadband, supercontinuum radiation spectra have to be measured in a filamentation experiment simultaneously with extremely complex beam patterns, pulse shapes, and field phase profiles. For a meaningful interpretation of an experiment, all these measurements have to be done online in a multidimensional parameter space, which involves varying the input peak power, pulse width, pulse chirp, beam diameter, beam shape, focusing geometry, as well as changing both linear and nonlinear material response within a continuum of dispersion profiles, nonlinear susceptibilities, ionization potentials, band gaps, etc. Since the spectra of supercontinuum radiation generated in laser filaments often span over several octaves, spectral characterization of such waveforms cannot be performed using a single spectrometer, requiring a set of accurately cross-calibrated spectrometers enabling spectral measurements in partially overlapping frequency ranges. Pulse characterization of such supercontinua cannot be done by standard methods, calling for a combination of innovative approaches. Beam characterization has to combine measurements of transverse intensity profiles with radially resolved spectral measurements, often augmented with fast-camera recording. To make this already very intricate picture complete, optical nonlinearities in filaments tend to amplify noise in both space and time, giving rise to beam and pulse breakup due to spatial and temporal modulations instabilities. As a result, multiple uncontrolled factors, such as random field intensity fluctuations across the beam and along the pulse, are a part of the picture. Processing this diversified information with a need to include an enormous number of both controlled and uncontrolled physical factors inevitably requires big-data management solutions.

Due to the computation complexity of numerical analysis of filamentation, especially at the level of peak powers orders of magnitude higher than the self-focusing threshold, where the assumption of axial symmetry of a beam fails and the full (3 + 1)-dimensional problem has to be solved, some of the most interesting questions in the filamentation of high-intensity, high-peak-power ultrashort laser pulses open. Those include possible filamentation scenarios whereby high-intensity, high-peak-power ultrashort laser pulses available from the cutting-edge laser systems within a broad spectral range could be delivered over large distances, compressed in time, and coupled into regular or irregular arrays of quasistationary, quasisolitary spatiotemporal waveforms (Figure 2), including arrays of light bullets.

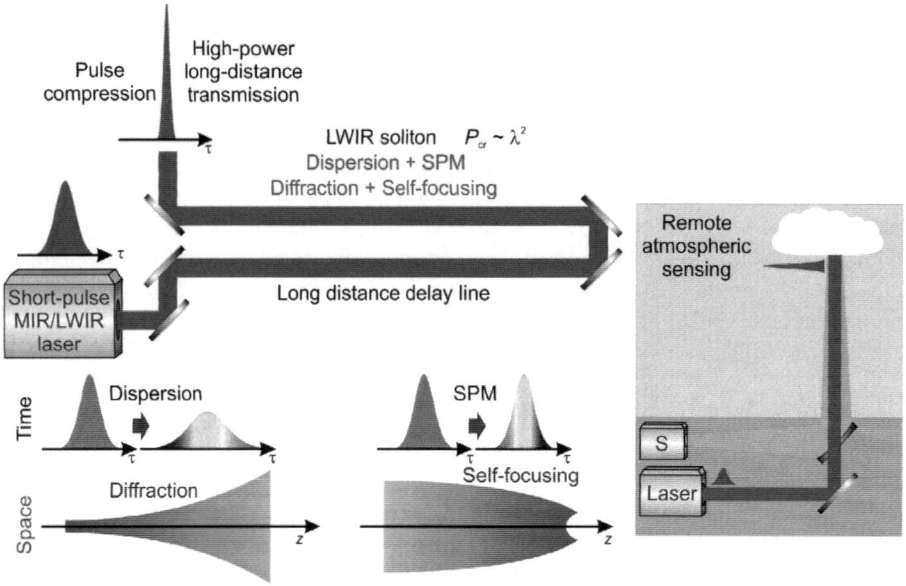

Figure 2. Envisaged remote sensing scheme using high-peak-power solitons in the mid- and long-wavelength infrared: SPM, self-phase modulation; S, spectrometer.

5. Conclusion

To summarize, our studies reveal unique properties of mid-infrared filaments, where the generation of powerful mid-infrared supercontinuum is accompanied by unusual scenarios of optical harmonic generation, giving rise to remarkably broad radiation spectra, stretching from the visible to the mid-infrared. Filamentation-assisted pulse compression in the gas phase is shown to enable the generation of subterawatt few-cycle pulses in the mid-infrared. Generation of few- and even single-cycle mid-infrared field waveforms with peak powers ranging from a few megawatts to hundreds of gigawatts has been demonstrated within a broad range of central wavelengths.

Acknowledgments

This research was supported in part by the Russian Foundation for Basic Research (14-29-07182, 16-02-00843, 16-32-60164, and 15-32-20713), Russian Science Foundation (project no. 14-12-00772), Welch Foundation (Grant No. A-1801), and ONR (Award No. 00014-16-1-2578).

References

[1] Colosimo P *et al.* 2008 *Nature Phys.* **4** 386–9
[2] Agostini P and Di Mauro L F 2008 *Contemp. Phys.* **49** 179–97
[3] Serebryannikov E E and Zheltikov A M 2014 *Phys. Rev. Lett.* **113** 043901
[4] Zhokhov P A and Zheltikov A M 2014 *Phys. Rev. Lett.* **113** 133903
[5] Serebryannikov E E and Zheltikov A M *Phys. Rev. Lett.* **116** 123901
[6] Mitrofanov A V *et al.* 2015 *Scientific Reports* **5** 8368

[7] Kartashov D, Ališauskas S, Pugžlys A, Voronin A A, Zheltikov A M and Baltuška A 2012 *Opt. Lett.* **37** 2268–70
[8] Mitrofanov A V *et al* 2015 *Opt. Lett.* **40** 2068–71
[9] Lanin A A, Voronin A A, Stepanov E A, Fedotov A B and Zheltikov A M 2015 *Opt. Lett.* **40** 974–7
[10] Stepanov E A, Lanin A A, Voronin A A, Fedotov A B and Zheltikov A M 2016 *Phys. Rev. Lett.* **117** 043901
[11] Balciunas T, Dutin C F, Fan G, Witting T, Voronin A A, Zheltikov A M, Frédéric G, Paulus G G, Baltuska A and Benabid F 2015 *Nature Communications* **6** 6117
[12] Shumakova V, Malevich P, Ališauskas S, Voronin A, Zheltikov A M, Faccio D, Kartashov D, Baltuška A and Pugžlys A 2016 *Nature Communications* **7** 12877
[13] Lanin A A, Voronin A A, Fedotov A B and Zheltikov A M 2014 *Scientific Reports* **4** 6670
[14] Serebryannikov E E and Zheltikov A M 2014 *Phys. Rev.* A **90** 043811
[15] Mitrofanov A V *et al.* 2016 *Optica* **3** 299–302
[16] Mitrofanov A V, Voronin A A, Sidorov-Biryukov D A, Mitryukovsky S I, Rozhko M V, Pugžlys A, Fedotov A B, Panchenko V Ya, Baltuška A, Zheltikov A M 2016 *Opt. Lett.* **41** 3479–82

Higher-order effects on uncertainties of clocks of Mg atoms in an optical lattice

V D Ovsiannikov[1], S I Marmo[1], S N Mokhnenko[1] and V G Palchikov[2,3]

[1] Voronezh State University, Voronezh 394006, Russia
[2] FGUP "VNIIFTRI", Mendeleevo 141570, Moscow Region,Russia
[3] National Research Nuclear University "MEPhI", Moscow 115409, Russia

E-mail: ovd@phys.vsu.ru

Abstract. Multipole, nonlinear and anharmonic effects on the optical-lattice-based clocks of Mg atoms are evaluated theoretically. Dipole polarizabilities, hyperpolarizabilities and multipolar polarizabilities for Mg atoms are calculated in the single-electron approximation with the use of analytical presentations for the wave and Green's functions in the modified model-potential approach. For comparison, the data are also given for atoms of the group IIb elements (Zn, Cd, Hg).

1. Introduction

The recent progress in detecting experimentally the $^3P_0 - {}^1S_0$ clock transition in neutral Mg atoms and determination of corresponding magic wavelength (MWL) λ_{mag} = 468.46 nm [1] paves the way to the development of a new time-frequency standard of Mg atoms in an optical lattice. To ensure the Mg clock fractional uncertainties below 10^{-17}, the role of nonlinear and non-dipole effects of interaction between trapped atoms and the magic-frequency lattice should be evaluated.

In this paper, we present theoretical considerations of the most appropriate operational conditions based on results of numerical calculations of electromagnetic susceptibilities of Mg atoms in their clock states. The single-electron Fues' model-potential (FMP) approach was used with modifications introduced for the most efficient account of contributions from the valence electrons of the $3s^2$ shell and from their interaction with the inner-electron shells [2, 3]. The modifications of the FMP parameters were performed, taking into account the contribution of the 3s electron to the polarizability of the upper clock state $3s3p(^3P_0)$. The modified approach improved significantly the agreement of the calculated MWL (471.6 nm) with the above-cited result of experimental measurements.

Together with electric-dipole (E1) frequency-dependent polarizabilities $\alpha^{E1}_{e(g)}(\omega)$ of the excited (e) and ground (g) clock states the magnetic dipole $\alpha^{M1}_{e(g)}(\omega)$ and electric quadrupole $\alpha^{E2}_{e(g)}(\omega)$ polarizabilities (M1 and E2) provide their specific contributions to the dynamic Stark shift. The E2-M1 contributions to the linear in the laser intensity I Stark shifts of the clock energy levels are added to the E1 shift in a traveling wave and subtracted in a standing wave, due to the on-phase and a quarter-period off-phase temporal and spatial distributions of corresponding interactions [4, 5].

The shift of the clock frequency in a traveling wave is determined by the difference of polarizabilities $\Delta\alpha^{\Sigma}(\omega) = \alpha^{\Sigma}_{e}(\omega) - \alpha^{\Sigma}_{g}(\omega)$:

$$\Delta\nu_{clock} = \Delta\alpha^{\Sigma}(\omega)\frac{I}{4} + \Delta\beta(\omega,\xi)\frac{I^2}{16}, \tag{1}$$

where $\alpha^{\Sigma}_{e(g)}(\omega) = \alpha^{E1}_{e(g)}(\omega) + \alpha^{qm}_{e(g)}(\omega)$, and $\alpha^{qm}_{e(g)}(\omega) = \alpha^{E2}_{e(g)}(\omega) + \alpha^{M1}_{e(g)}(\omega)$ is the sum of the E2-M1 polarizabilities, which in the far-off-resonance regions of frequencies is usually 6 to 7 orders smaller than the dipole polarizability $\alpha^{E1}_{e(g)}(\omega)$. The quadratic in I shift, determined by the difference of the clock-state hyperpolarizabilities $\Delta\beta(\omega,\xi) = \beta_e(\omega,\xi) - \beta_g(\omega,\xi)$, is also taken into account in (1). The second argument ξ indicates the dependence of hyperpolarizabilities on the circular polarization degree ξ [6], which for the clock states may be written as

$$\beta_{e(g)}(\omega,\xi) = \beta^{l}_{e(g)}(\omega) + \xi^2\left[\beta^{c}_{e(g)}(\omega) - \beta^{l}_{e(g)}(\omega)\right], \tag{2}$$

where two independent components of the hyperpolarizability tensor determine the quadratic shift of excited (ground-state) energy level in the field of linearly polarized ($\beta^{l}_{e(g)}(\omega)$) and circularly polarized ($\beta^{c}_{e(g)}(\omega)$) laser radiation.

The clock-frequency shift in a field of a lattice standing wave has quite different dependence on the lattice-laser intensity. In particular, this dependence includes in addition to linear and quadratic in I terms, those proportional to $\sqrt{I}$ and to $I^{3/2}$, which are related with the vibrational motion of an atom trapped in the lattice-induced dynamic Stark-effect energy well. Corresponding equations and their backgrounds are described in section 2 of this paper. In section 3 the numerical data of calculations on the basis of the model-potential approach in the single-electron approximation are presented for magnesium atoms in comparison with earlier calculated data [3] for some other atoms of the group II elements.

Atomic units $e = m = \hbar = 1$ are used throughout the paper, unless otherwise stated explicitly. The speed of light in atomic units $c = 137.036$ coincides numerically with the inverse fine-structure constant $\alpha = 1/137.036$.

2. Shifts of clock levels in the field of a lattice wave

The Stark effect on atomic energy levels enables trapping neutral atoms in the minima of thus created potential-energy wells in a lattice standing wave. But the Stark energies of the ground and excited clock states cause the clock-frequency shift which should be taken into account in evaluating the clock-frequency uncertainties.

The Stark energies are determined by the interaction of a trapped atom with a lattice wave of an electric-field

$$\mathbf{E}(X,t) = 2\mathbf{E}_0\cos(kX)\cos(\omega t), \tag{3}$$

and a magnetic field (a quarter-period off-phase, both in space and time)

$$\mathbf{B}(X,t) = 2\left[\mathbf{e}_x \times \mathbf{E}_0\right]\sin(kX)\sin(\omega t), \tag{4}$$

oscillating in time with frequency ω and in space along the incident laser beam of intensity $I = cE_0^2/8\pi$ and the wave vector $\mathbf{k} = k\mathbf{e}_x$, $k = \omega/c$; X is a displacement of an atom from the lattice standing-wave antinode in the lattice-laser-beam direction determined by the unit vector $\mathbf{e}_x$. The operator of atom-lattice interaction may be presented as $\hat{V}(X,t) = \mathrm{Re}\left\{\hat{V}(X)\exp(-i\omega t)\right\}$, where the spatial factor determines an X-distribution of electric dipole $\hat{V}_{E1} = (\mathbf{r}\cdot\mathbf{E}_0)$, electric quadrupole $\hat{V}_{E2} = \omega r^2\left(\{\mathbf{E}_0 \otimes \mathbf{n}\}_2 \cdot \mathbf{C}_2(\theta,\varphi)\right)/c\sqrt{6}$ and magnetic dipole $\hat{V}_{M1} = \left([\mathbf{n}\times\mathbf{E}_0]\cdot(\hat{\mathbf{J}}+\hat{\mathbf{S}})\right)/2c$ operators [3, 7]

$$\hat{V}(X) = \hat{V}_{E1}\cos(kX) + (\hat{V}_{E2} + \hat{V}_{M1})\sin(kX). \quad (5)$$

Note, that the operators are written with account of doubling the amplitude of an incident laser wave $\mathbf{E}_0$ in the amplitudes (3) and (4) of a standing wave.

With account of the second- and fourth-order in $\hat{V}(X)$ terms of perturbation theory, linear and quadratic in the lattice-laser intensity I correspondingly, the interactions (5) produce lattice potential wells for an atom in its ground (g) or excited (e) state, which may be presented, as follows [8]

$$U^{latt}_{g(e)}(X; I,\omega,\xi) \approx -D(I,\omega,\xi) + U^{harm}_{g(e)}(I,\omega,\xi)X^2 - U^{anh}_{g(e)}(I,\omega,\xi)X^4 \quad (6)$$

The separation of neighbour wells equals half wavelength λ of the lattice wave. To exclude tunneling atoms between potential wells, its location near the well bottom (at the equilibrium position $X=0$) should hold the inequality $|X| << \lambda/4$, where $\lambda/4$ is the distance between the top and bottom of the well. At this condition the ratio between absolute values of consecutive terms in the right-hand side of equation (6) is significantly smaller than the unit. Therefore, the motion of a trapped atom is described by the wavefunctions of an anharmonic oscillator. The depth of the potential well (6)

$$D_{g(e)}(I,\omega,\xi) = \alpha^{E1}_{g(e)}(\omega)I + \beta_{g(e)}(\omega,\xi)I^2, \quad (7)$$

is determined by the lattice-frequency-dependent electric dipole polarizability $\alpha^{E1}_{g(e)}(\omega)$ and hyperpolarizability $\beta_{g(e)}(\omega,\xi)$ of the ground (excited) state. At an operational intensity the quadratic term in the right-hand side of equation (7) is at least 6 orders of magnitude smaller than the linear one. The coefficient of the harmonic term, determining the frequency of vibrations $\Omega_{e(g)}(I,\omega,\xi)$ in a lattice trap, depends on a combined E1-E2-M1 polarizability $\alpha^{dqm}_{e(g)}(\omega) = \alpha^{E1}_{e(g)}(\omega) - \alpha^{qm}_{e(g)}(\omega)$ and on the hyperpolarizability, as follows

$$U^{harm}_{g(e)}(I,\omega,\xi) = \frac{\mathcal{M}\Omega^2_{g(e)}(I,\omega,\xi)}{2} = \left[\alpha^{dqm}_{g(e)}(\omega)I + 2\beta_{g(e)}(\omega,\xi)I^2\right]k^2. \quad (8)$$

From this equation the eigenfrequency of the atomic vibrations inside the potential well (6) may be presented as

$$\Omega_{g(e)}(I,\omega,\xi) = 2\sqrt{\mathcal{E}^{rec}\left[D_{g(e)}(I,\omega,\xi) - \alpha^{qm}_{g(e)}(\omega)I + \beta_{g(e)}(\omega,\xi)I^2\right]}, \quad (9)$$

where $\mathcal{E}^{rec} = k^2/(2\mathcal{M}) = \omega^2/(2\mathcal{M}c^2)$ is the recoil energy of a lattice photon, $\mathcal{M} \approx 1823\mathcal{A}$ is the mass of atom in the atomic (electron mass) units, $\mathcal{A}$ is the mass of atom in the atomic mass units (1/12 of the carbon atom $^{12}C_6$ mass), presented for each atom in periodic tables. It becomes evident, that together with linear and quadratic in intensity terms, the dependence of the energy-level shifts in the field of the lattice wave includes also the terms of half-integer powers of I.

The coefficient of the anharmonic term is

$$U^{anh}_{g(e)}(I,\omega,\xi) = \left[\alpha^{dqm}_{g(e)}(\omega)I + 5\beta_{g(e)}(\omega,\xi)I^2\right]\frac{k^4}{3} = \frac{k^2\mathcal{M}\Omega^2_{g(e)}(I,\omega,\xi)}{6} + k^4\beta_{g(e)}(\omega,\xi)I^2. \quad (10)$$

Thus, the atom trapped in the potential energy well (6) effectuates vibrations in a bound state of an oscillator quantum number *n* and of the vibration energy

$$\mathrm{E}^{vib}_{g(e)}(I,\omega,\xi,n) = -D_{g(e)}(I,\omega,\xi) + \Omega_{g(e)}(I,\omega,\xi)\left(n+\frac{1}{2}\right) - \mathrm{E}^{anh}_{g(e)}(I,\omega,\xi)\left(n^2+n+\frac{1}{2}\right), \quad (11)$$

added to the energy of atom in its ground (excited) state. Here the anharmonic constant is determined by the recoil energy, and the fraction of hyperpolarizability over the E1-E2-M1 polarizability $\alpha^{dqm}_{e(g)}(\omega)$, as follows

$$\mathrm{E}_{g(e)}^{anh}(I,\omega,\xi)=\frac{\varepsilon^{rec}}{2}\left[1+\frac{3\beta_{g(e)}(\omega,\xi)I}{\alpha_{g(e)}^{dqm}(\omega)}\right]. \qquad (12)$$

Therefore, the hyperpolarizability contributes, via the anharmonic effect, to the linear in I term of the vibration energy (11), which is however essentially smaller than the linear term of the depth (7), since the fraction in the brackets at the operational intensity is on the order of 10^{-7}-10^{-6}.

3. Lattice-induced clock-frequency shift

Thus, the shift of the clock-transition frequency in a lattice-trapped atom is determined by the difference of vibration energies (11) of atom in clock states:

$$\begin{aligned}\Delta\nu_{cl}^{latt}(I,\omega,\xi,n)&=\mathrm{E}_e^{vib}(I,\omega,\xi,n)-\mathrm{E}_g^{vib}(I,\omega,\xi,n)\\&=-\Delta D(I,\omega,\xi)+\Delta\Omega(I,\omega,\xi)\left(n+\frac{1}{2}\right)-\Delta\mathrm{E}^{anh}(I,\omega,\xi)\left(n^2+n+\frac{1}{2}\right),\end{aligned} \qquad (13)$$

where the oscillation quantum number n is one and the same for the excited and ground-state atom, independent of transition between its clock states (the Lamb-Dicke regime). Therefore the dependence of the clock-frequency shift for an atom in the field of the lattice standing wave differs significantly from that of the traveling wave (1) and may be presented, as follows:

$$\Delta\nu_{cl}^{latt}(n,I,\omega,\xi)=c_{1/2}(n,\omega)I^{1/2}+c_1(n,\omega,\xi)I+c_{3/2}(n,\omega,\xi)I^{3/2}+c_2(\omega,\xi)I^2. \qquad (14)$$

The half-integer powers of the laser intensity I appear in the right-hand side of this equation from the eigenfrequency (9) of atomic oscillations in lattice traps. The basic contributions to this shift comes from the difference of electric-dipole polarizabilities $\Delta\alpha^{E1}(\omega)=\alpha_e^{E1}(\omega)-\alpha_g^{E1}(\omega)$, which should vanish at the "magic" frequency ω_m, determined as the root of equation $\alpha_e^{E1}(\omega_m)-\alpha_g^{E1}(\omega_m)=0$. Equalization of the E1 polarizabilities means the equivalence of only the linear terms of the depths (7) of the Stark-effect energy wells. However, an inevitable in practice uncertainty of the magic frequency $\Delta\omega_m$ may introduce corresponding uncertainties into the differences of electric dipole polarizabilities, transferred into the differences ΔD and $\Delta\Omega$. The latter appear in the coefficients $c_{1/2}^{E1}(n)\equiv c_{1/2}(n,\omega_m)$ and $c_1^{E1}(n,\xi)\equiv c_1(n,\omega_m,\xi)$ of the square-root and linear in I terms of the resolution (14) taken at the magic frequency ω_m. The multipole (E2-M1) effects, transferred into the difference $\Delta\Omega=\Omega_e-\Omega_g$, appear only in the square-root term of (14), whereas effects of the hyperpolarizability appear in all three terms ΔD, $\Delta\Omega$ and $\Delta\mathrm{E}^{anh}$ of equation (13) and finally contribute to the values of all the rest coefficients $c_j^{E1}(n,\xi)=c_j(n,\omega_m,\xi)$ $(j=1,\ 3/2,\ 2)$. So, the coefficients in the right-hand side of (14) may be presented in terms of the magic frequency shift $\Delta\omega_m$, the differences of multipole polarizabilities $\Delta\alpha_m^{qm}=\alpha_e^{qm}(\omega_m)-\alpha_g^{qm}(\omega_m)$ and the differences of hyperpolarizabilities $\Delta\beta_m(\xi)=\beta_e(\omega_m,\xi)-\beta_g(\omega_m,\xi)$, as follows:

$$\begin{aligned}c_{1/2}^{E1}(n)&=\left(\frac{\partial\Delta\alpha_m^{E1}}{\partial\omega}\Delta\omega_m-\Delta\alpha_m^{qm}\left(n+\frac{1}{2}\right)\right)\sqrt{\frac{\varepsilon_{E1}^{rec}}{\alpha_m^{E1}}},\\c_1^{E1}(n,\xi)&=-\frac{\partial\Delta\alpha_m^{E1}}{\partial\omega}\Delta\omega_m-\frac{3\varepsilon_{E1}^{rec}}{2\alpha_m^{E1}}\Delta\beta_m(\xi)\left(n^2+n+\frac{1}{2}\right),\\c_{3/2}^{E1}(n,\xi)&=2\Delta\beta_m(\xi)\sqrt{\frac{\varepsilon_{E1}^{rec}}{\alpha_m^{E1}}}\left(n+\frac{1}{2}\right),\qquad c_2^{E1}(\xi)=-\Delta\beta_m(\xi).\end{aligned} \qquad (15)$$

The values of all frequency-dependent quantities here are taken at the magic frequency of equal E1 polarizabilities (denoted by indexes "m" and "E1").

As is seen from equations (14) and (15), the possibility appears to introduce an "operational magic" frequency [7], which could minimize the lattice-induced clock-frequency shifts (13) within the operational distribution of the laser intensity over the lattice sites occupied by trapped atoms. To this end, the magic frequency shift $\Delta\omega_m$ should be controllable value, together with the circular polarization degree ξ and the intensity of the lattice laser I. In this case, the contribution of $\Delta\alpha_m^{qm}$ to the coefficient $c_{1/2}^{E1}(n)$ and the hyperpolarizability contribution to $c_1^{E1}(n,\xi)$ may be compensated by the E1-polarizability derivative term. Evidently, for the most efficient compensation, the atoms should be cooled down to the lowest possible oscillator quantum number $n=0$. In addition, the polarization dependence of the hyperpolarizability (2) may be used. In the case of opposite signs of the components $\Delta\beta_m^l$ and $\Delta\beta_m^c$ the hyperpolarizability-dependent terms may be removed at the "magic ellipticity" [6] of the lattice wave $\xi_m = \pm 1/\sqrt{1-\Delta\beta_m^c/\Delta\beta_m^l}$, for which the hyperpolarizability will vanish $\Delta\beta_m(\xi_m)=0$. As is seen from equations (15), the coefficients $c_{3/2}^{E1}(n,\xi_m)$ and $c_2^{E1}(\xi_m)$ also vanish at the magic ellipticity.

For evaluation of the shift (13), the values of the frequency-dependent E1,E2-M1 polarizabilities and hyperpolarizabilities at the magic frequency should be known. We have used the Fues' model potential method for describing wavefunctions in the single-electron approximation. The details of calculations in the FMP approach and numerical data for susceptibilities of Ca, Sr, Yb, Zn, Cd and Hg atoms were presented in [3]. In this paper we report the results of calculations of the data for susceptibilities of Mg atoms. First of all, the frequency-dependent polarizabilities of the clock states were calculated in the region of the magic wavelength. The results are presented on figure 1 for the wavelength dependence of excited- (full curve) and ground-state (dashed curve) polarizabilities. The position of the curves intersection at 471.6 nm differs from the result of experimental measurements of the magic wavelength [1] by less than 1%., thus confirming satisfactory precision of calculations in the FMP approach.

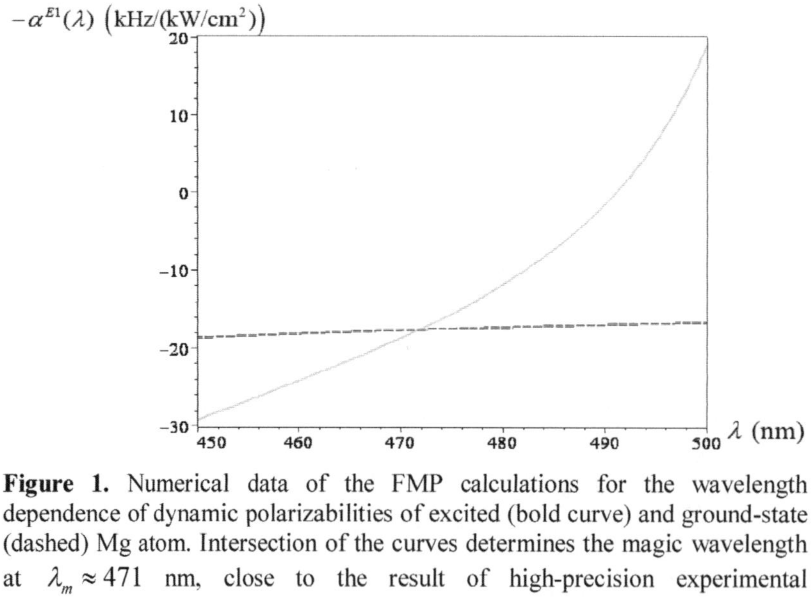

Figure 1. Numerical data of the FMP calculations for the wavelength dependence of dynamic polarizabilities of excited (bold curve) and ground-state (dashed) Mg atom. Intersection of the curves determines the magic wavelength at $\lambda_m \approx 471$ nm, close to the result of high-precision experimental measurement $\lambda_m^{\exp} \approx 468.46$ nm [1].

The magic frequency $\omega_m / 2\pi = c / \lambda_m = 639.95$ THz locates in the violet region of the light spectrum, rather close to the clock frequency (see table 1). The energy of two photons $2\omega_m$ exceeds the ionization potential of Mg atom. Therefore the hyperpolarizability is a complex value with the imaginary part determining the two-photon ionization rate, similar to the case of the group IIb atoms [3] and to the case of the blue-detuned magic wavelength of Sr atoms [9]. In table 1, the numerical data for the magic wavelength, dynamic polarizabilities and hyperpolarizabilities at the magic frequencies, the recoil energies and the blackbody radiation-induced clock-frequency shifts are presented for Mg atoms together with corresponding data for the group IIb atoms. The real parts of the hyperpolarizability components $\Delta\beta_m^l$ and $\Delta\beta_m^c$ of Zn, Cd and Hg atoms have opposite signs, providing the existence of the magic circular polarization degree ξ_m for which $\mathrm{Re}\{\Delta\beta_m(\xi_m)\} = 0$.

As is seen from equations (14)-(15), there are four basic parameters which determine the lattice-induced shift of the clock frequency: (i) the laser intensity I, which should ensure the trapping condition $D > 100\,\mathcal{E}^{rec}$, (ii) the vibration quantum number n, which should be reduced to its minimal value $n = 0$ by means of the side-band cooling, (iii) the degree of circular polarization ξ, which can eliminate or reduce to minimum the hyperpolarizability effects, (iv) the magic frequency shift $\Delta\omega_m$, which may reduce to minimal values the coefficients $c_{1/2}^{E1}(n)$ of the square-root and $c_1^{E1}(n,\xi)$ of linear in I terms. The data of table 1 allows to determine the most suitable values of ξ and $\Delta\omega_m$ to ensure the best trapping conditions simultaneously with the lowest lattice-induced clock-frequency uncertainties, on the other hand. In figure 2 the corresponding numerical data for the shifts (14) are presented for Mg (a) and Hg (b) atoms in the regions of intensities, sufficient to ensure trapping conditions.

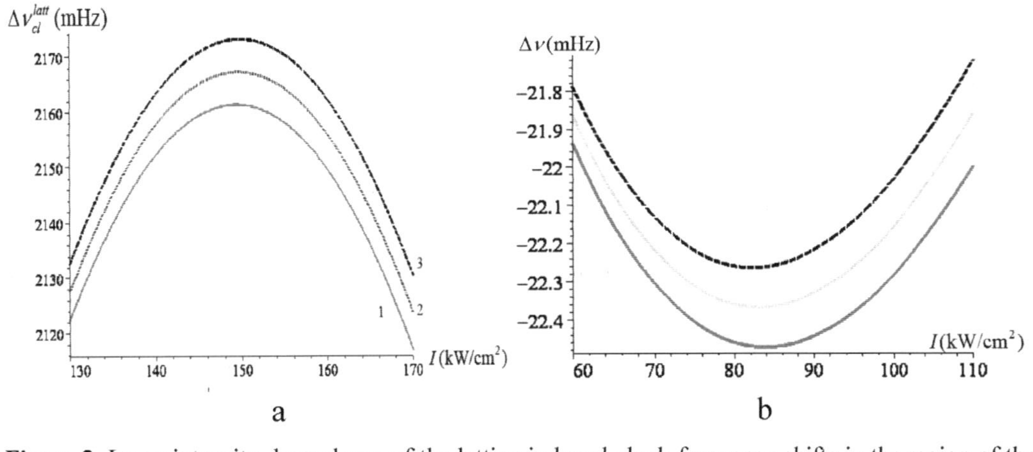

Figure 2. Laser-intensity dependence of the lattice-induced clock-frequency shifts in the region of the most suitable intensities and polarizations (linear for Mg and magic circular polarization degree $\xi_m = \pm 0.86$ for Hg): (a) in Mg atoms at the magic frequency shifts $\Delta\omega_m = -74.9$ MHz (curve 1), -75.0 MHz (curve 2) and -75.1 MHz (curve 3); (b) in Hg atoms at $\Delta\omega_m = -2.0$ MHz (lower solid curve), -2.01 MHz (medium dotted) and -2.02 MHz (upper dash-dotted curve).

Table 1. Characteristics of atoms in optical lattices of magic wavelengths. Bold-faced are experimentally determined data on the magic wavelengths for Mg [1] and Hg [10]. The data for the MWL of Zn and Cd are evaluated theoretically. The coefficient ν_0^{BBR} of the blackbody radiation-induced clock-frequency shift dependence $\Delta\nu_{cl}^{\mathrm{BBR}}(T) = \nu_0^{\mathrm{BBR}}(T/300)^4$ on the temperature T of environment is given in the last line of the table.

Atom	Mg	Zn	Cd	Hg
λ_m (nm)	**468.46**	406.5	414.4	**362.57**
$\nu_{clock}\,(\mathrm{THz})$	655	969	903	1129
$\alpha_m^{E1}\left(\frac{\mathrm{kHz}}{\mathrm{kW/cm^2}}\right)$	17.5	8.11	9.76	5.70
$\Delta\alpha_m^{qm}\left(\frac{\mathrm{mHz}}{\mathrm{kW/cm^2}}\right)$	5.48	15.3	5.86	8.25
$\Delta\beta_m^{l}\left[\frac{\mu\mathrm{Hz}}{(\mathrm{kW/cm^2})^2}\right]$	$111+5.88i$	$-4.3+1.64i$	$-5.47+2.02i$	$-2.67+0.82i$
$\Delta\beta_m^{c}\left[\frac{\mu\mathrm{Hz}}{(\mathrm{kW/cm^2})^2}\right]$	$1735+8.69i$	$42.6+2.45i$	$19.5+3.01i$	$0.94+1.21i$
$\frac{\Omega_m}{\sqrt{I}}\left[\frac{\mathrm{kHz}}{\sqrt{\mathrm{kW/cm^2}}}\right]$	51.5	24.1	19.9	13.1
$\frac{\partial(\Delta\alpha_m^{E1})}{\partial\omega}\left(\frac{10^{-9}}{\mathrm{kW/cm^2}}\right)$	0.420	0.187	0.200	0.134
$\varepsilon^{rec}\,(\mathrm{kHz})$	37.9	17.9	10.14	7.57
$\nu_0^{BBR}\,(\mathrm{Hz})$	-0.424	-0.23	-0.22	-0.188

Comparison between the lattice-wave induced shifts presented in figure 2 for Mg (plot a) and Hg (plot b) demonstrates significant advantage of the heavy atom: despite smaller magic frequency ($\omega_m^{Mg}/2\pi \approx 640$ THz against $\omega_m^{Hg}/2\pi \approx 827$ THz) the recoil energy for Mg exceeds almost 5 times that for Hg atoms, therefore the trapping laser intensity required for Mg is almost twice that for Hg, the latter should locate in the region of intensities near 90 kW/cm^2. The uncertainty of the lattice-induced shift due to the uncertainty of the intensity distribution over occupied by Mg atoms lattice sites between 140 and 160 kW/cm^2 achieves 10 mHz (between $\Delta\nu_{cl}^{latt}$=2157 and 2167 mHz) at the magic frequency shift $\Delta\omega_m = -75$ MHz (see figure 2a). For Hg atoms this kind of uncertainty does not exceed 0.5 mHz (around $\Delta\nu_{cl}^{latt} = -22$ mHz) for the intensities between 60 and 110 kW/cm^2 of the lattice wave of the magic circular polarization at the magic frequency shift $\Delta\omega_m = -2.01$ MHz (see figure 2b). To reduce the $\Delta\nu_{cl}^{latt}$ uncertainty to the level of 0.5 mHz, the fractional uncertainty of the operational intensity distribution over lattice sites, occupied by Mg atoms, should not exceed 2% around $I = 152$ kW/cm^2.

Acknowledgments
This work was supported in part by the Ministry of Education and Science of Russia in the framework of the state order on research activities in 2014-2016 (Project No 1226) and by the Russian Foundation for Basic Research (RFBR grant No 14-02-00516-a).

References

[1] Kulosa A P *et al.* 2015 *Phys. Rev. Lett.* **115** 240801
[2] Manakov N L, Ovsiannikov V D and Rapoport L P 1986 *Phys. Rep.* **141** 319
[3] Ovsiannikov V D, Marmo S I, Pal'chikov V G and Katori H 2016 *Phys. Rev.* A **93** 043420
[4] Taichenachev A V Yudin V I , Ovsiannikov V D, Pal'chikov V G and Oates C W 2008 *Phys. Rev. Lett.* **101** 193601
[5] Katori H, Hashiguchi K, Il'inova E Yu and Ovsiannikov V D 2009 *Phys. Rev. Lett.* **103** 153004
[6] Taichenachev A V, Yudin V I, Ovsiannikov V D and Pal'chikov V G 2006 *Phys. Rev. Lett.* **97** 173601
[7] Katori H, Ovsiannikov V D, Marmo S I and Palchikov V G 2015 *Phys. Rev.* A **91** 052503
[8] Ovsiannikov V D, Palchikov V G, Taichenachev A V, Yudin V I and Katori H 2013 *Phys. Rev.* A **88** 013405
[9] Takamoto M, Katori H, Marmo S I, Ovsiannikov V D and Pal'chikov V G 2009 *Phys. Rev. Lett.* **102** 063002
[10] Takamoto M. *et al.* 2015 *Comptes Rendus Physique* **16** 489

Deep sub-Doppler cooling of Mg in MOT formed by light waves with elliptical polarization

O N Prudnikov[1], D V Brazhnikov[1,2], A V Taichenachev[1,2], V I Yudin[1,2,3] and A N Goncharov[1,2,3]

[1] Novosibirsk State University, 2 Pirogova str., Novosibirsk 630090, Russia
[2] Institute of Laser Physics SB RAS, 13/3 Acad. Lavrentyev ave., Novosibirsk 630090, Russia
[3] Novosibirsk State Technical University, 20 K. Marx ave., Novosibirsk 630073, Russia

E-mail: oleg.nsu@gmail.com

Abstract. We study magneto-optical trap (MOT) of Mg atoms operating on the closed triplet $^3P_2 \rightarrow {}^3D_3$ (λ = 383.3 nm) transition formed by the light waves with elliptical polarization ($\varepsilon - \theta - \bar{\varepsilon}$ configuration). Compare to well-known trap formed by light waves with circular polarization ($\sigma_+ - \sigma_-$ configuration) the suggested $\varepsilon - \theta - \bar{\varepsilon}$ configuration offer the lower sub-Doppler temperature for trapped Mg atoms, that can't be reached in conventional $\sigma_+ - \sigma_-$ MOT.

Mg atoms one of perspective candidates for realization of new-generation of atomic clock, based on optical lattices. These atoms have the narrow spectroscopic lines due to forbidden optical transition from the ground state $^1S_0$ to the lowest excited states $^3P_{0,1,2}$ (see figure 1). The closed singlet optical transition $^1S_0 \rightarrow {}^1P_1$ can be used for Doppler cooling of Mg atoms down to the temperature T = 1.2 mK. The ground state of this optical transition has angular momentum J_g=0 that forbids the possibility of sub-Doppler cooling here. In order to get the lower cooling temperature, the cooling on closed triplet $^3P_2 \rightarrow {}^3D_3$ (λ =383.3 nm) optical transition was suggested in [1, 2]. In principle, the ground state here $^3P_2$ is degenerated over angular momentum (J_g=2) that should allow sub-Doppler cooling [3]. However, the experimental realization of laser cooling operating on $^3P_2 \rightarrow {}^3D_3$ in $\sigma_+ - \sigma_-$ MOT did not bring significant progress to reaching lower temperature. In particular, Mg atoms were cooled by laser radiation only down to the temperature **1 mK** that much higher than the Doppler limit for this optical transition $\boldsymbol{T_D}$ **= 425 μK**. The recent analysis [2] has revealed the limitation of temperature of laser cooling for ^{24}Mg atoms in $\sigma_+ - \sigma_-$ light field configuration and explained the results of previous experiments [1].

The evolution of atoms in the resonant light waves in presence of a static magnetic field is described by equation

$$\frac{\partial}{\partial t}\hat{\rho} = -\frac{i}{\hbar}\left[\hat{H}, \hat{\rho}\right] - \hat{\Gamma}\{\hat{\rho}\} \tag{1}$$

for atomic density matrix $\hat{\rho}$. Here $\hat{H}$ is the Hamiltonian and term $\hat{\Gamma}\{\hat{\rho}\}$ describes the relaxation of atomic levels in the process of spontaneous decay (e.g., see for details [4, 5]). Nowadays a lots of methods have been developed for solving this problem, and they can be separated to semiclassical [4, 6–9] and quantum [10–17] methods. The semiclassical methods require some conditions to be fulfilled [4, 6–9] that allow reducing the quantum equation for atomic density matrix to the Fokker-Plank type equation for atomic density function $W(r,p) = Tr\{\hat{\rho}(r,p)\}$:

$$\left(\frac{\partial}{\partial t}+\sum_i \frac{p_i}{m}\nabla_i\right)W = -\sum_i \frac{\partial}{\partial p_i}F_i(\vec{r},\vec{p})W + \sum_{ij}\frac{\partial^2}{\partial p_i\,\partial p_j}D_{ij}(\vec{r},\vec{p})\,W \tag{2}$$

where *F(r,p)* and *D(r,p)* are light force on moving atom and diffusion coefficient in light field.

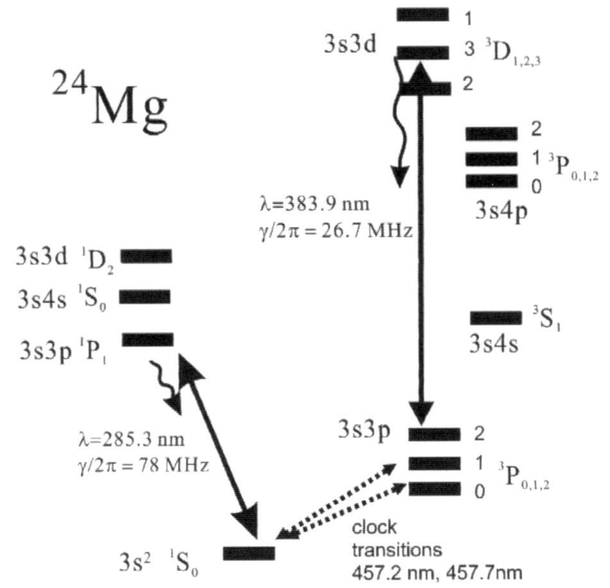

Figure 1. Diagram of low energy levels of ^{24}Mg. Solid lines denote cooling transitions, and dashed lines denote possible "clock" transitions.

The quantum methods as well utilize different approaches and simplifications. In particular, the method described in [16, 17], is so called band theory, use secular approximation that assumes enough large detuning $\delta = \omega - \omega_0$ of laser light with frequency ω from atomic resonance frequency ω_0, $U_0/E_R << 36\ \delta^2/\gamma^2$ with U_0 is depth of optical potential and $E_R=\hbar^2k^2/2M$ is recoil energy (see for details [16]). Additionally developed approach is based on simplified equation for atomic density matrix of ground state only $\hat{\rho}^{gg}$ that also limits applications of the results [18].

Recently we suggest a new method that allows to find solution of equation (1) for 1D problem with taken into account quantum recoil effects without limitations to field parameters and configuration [18, 19]. Our method allows one to compare different field configurations to find a minimum temperature of laser cooling. Strictly speaking, the distribution of cold atoms in electromagnetic field is significantly non-Maxwellian one. However, we define the temperature here as the average kinetic energy that for 1D case results $k_B T = \langle p^2 \rangle / M$ that is measured in $\hbar\gamma$ units ($k_B T/\hbar\gamma = 1$ corresponds T ≈ 1.28 mK for $^3P_2 \rightarrow {}^3D_3$ optical transition linewidth γ).

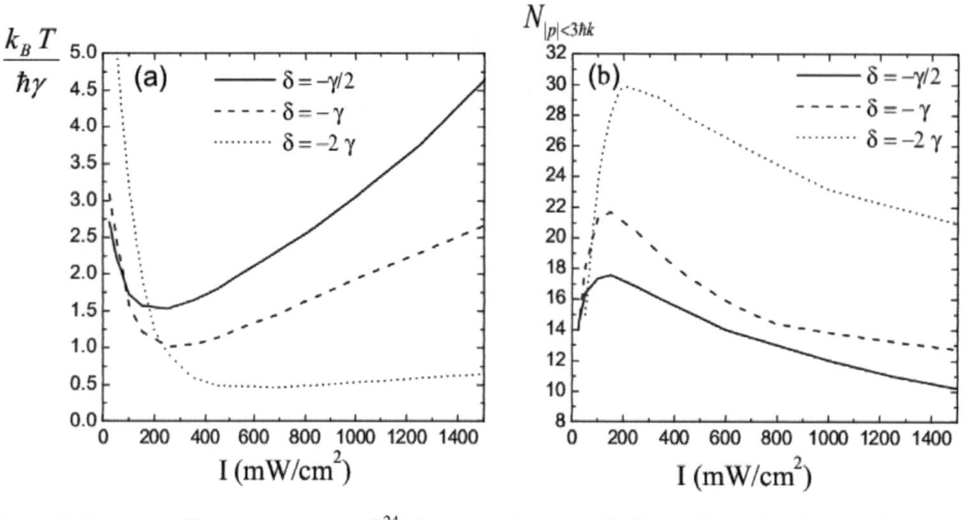

Figure 2. Laser cooling temperature of ^{24}Mg atoms in σ_+–σ_- field configuration (a) and fraction of cold atoms (b) as function of intensity of light waves. $k_B T/\hbar\gamma=1$ corresponds to $T \approx 1.28$ mK.

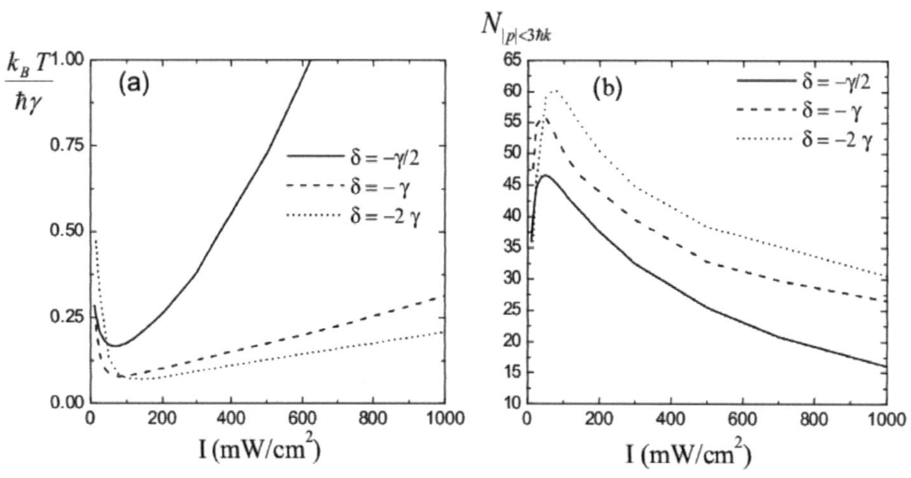

Figure 3. Laser cooling temperature of ^{24}Mg atoms in $lin \perp lin$ field configuration (a) and fraction of cold atoms (b) as function of intensity of light waves. $k_B T/\hbar\gamma=1$ corresponds to $T \approx 1.28$ mK.

Additionally, we introduce the fraction of cold atoms $N_{|p|<3\hbar k}$, i.e. the atoms with momentum $|p|<3\hbar k$, that for ^{24}Mg atoms cooled by the light field resonance to closed triplet transition $^3P_2 \rightarrow {}^3D_3$ (λ = 383.3 nm) corresponds to an effective temperature $(3\hbar k)^2/M \approx 48.4$ μK.

First we find the temperature of laser cooling ^{24}Mg atoms on $^3P_2 \rightarrow {}^3D_3$ optical transitions in conventional σ_+–σ_- field configuration commonly used in MOT do not reach deep sub-Doppler values (see figure 2a). The $lin \perp lin$ configuration of light (formed by counterpropagating waves with orthogonal linear polarization) offers much dipper laser cooling temperatures (figure 3a) and much large fraction of cold atoms $N_{|p|<3\hbar k}$, but cannot be used for MOT, because of zero magneto-optical trapping force in $lin \perp lin$ field.

Here to get the deep sub-Doppler laser cooling temperature for ^{24}Mg atoms in MOT we suggest $\varepsilon-\theta-\overline{\varepsilon}$ light field configuration to use in MOT for Mg atoms (figure 4). The first study of MOT in $\varepsilon-\theta-\overline{\varepsilon}$ light field was done by us in [5], where we discovered some peculiarities of laser cooling dealing with an elliptical polarization of the light waves, differ from linear $\varepsilon_0 = 0$ and circular polarization $\varepsilon_0 = \pm\pi/4$ [20].

Figure 5 shows the temperature of laser cooling in $\varepsilon - \theta - \bar{\varepsilon}$ field configuration as function of light intensity for particular case of $\varepsilon_0 = 0$ (linear polarization of counterpropagating waves). Figure 6 shows the temperature of laser cooling and the fraction of cold atoms (i.e. the atoms with momentum $|p|<3\hbar k$) in $\varepsilon - \theta - \bar{\varepsilon}$ field configuration as function of light ellipticity parameter ε_0. As one can see, the temperature dependence is not symmetrical function of light wave ellipticity parameter ε_0.

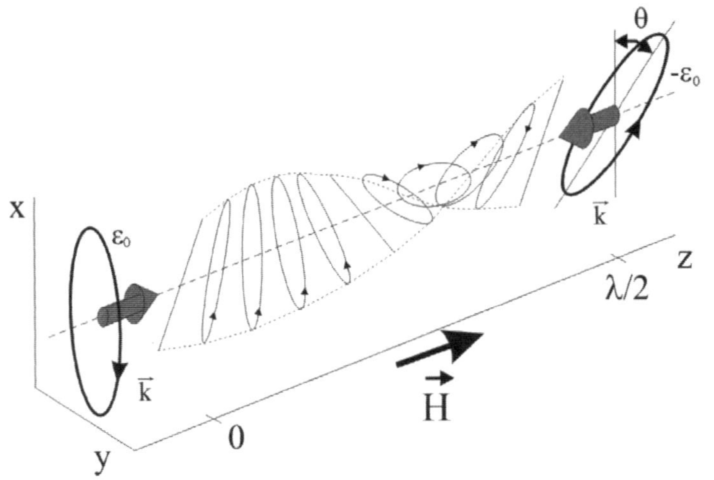

Figure 4. $\varepsilon - \theta - \bar{\varepsilon}$ field configuration formed by counterpropagating light waves with opposite elliptical polarization with parameters ε_0 and $-\varepsilon_0$ ($-\pi/4<\varepsilon_0<\pi/4$ and $\varepsilon_0=0$ corresponds to linear polarization and $\varepsilon_0=\pm\pi/4$ corresponds to right and left circular polarizations). The θ is the mutual orientation angle between main polarization axis of light waves.

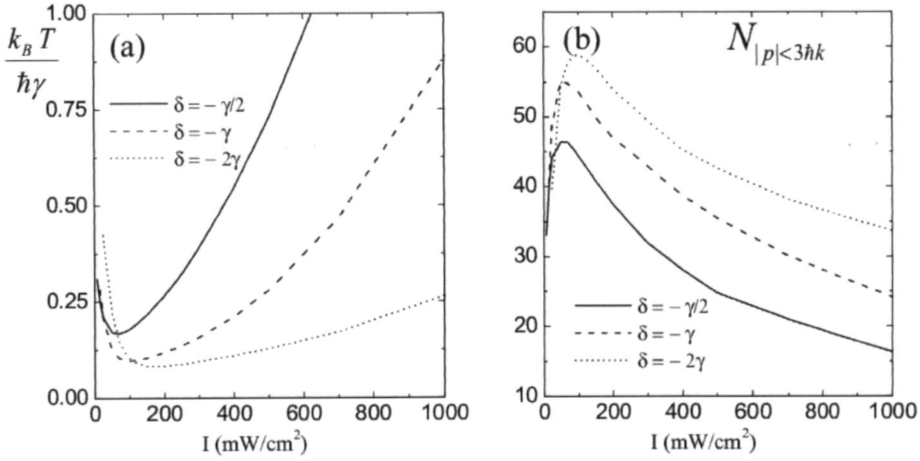

Figure 5. Laser cooling temperature of ^{24}Mg atoms in $lin - \theta - lin$ (θ=$-\pi/4$) field configuration (a) and fraction of cold atoms (b) as function of intensity of light waves. $k_B T/\hbar\gamma = 1$ corresponds to $T \approx 1.28$ mK.

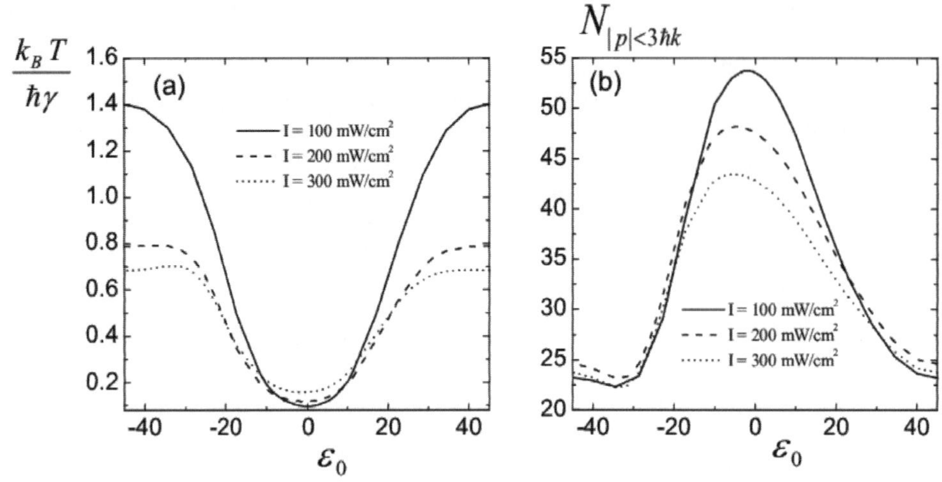

Figure 6. Laser cooling temperature of ^{24}Mg atoms in $\varepsilon-\theta-\bar{\varepsilon}$ (θ=–π/4) field configuration (a) and fraction of cold atoms (b) as function of intensity of parameter ellipticity of light waves ε_0.

The anomaly terms in the friction force results to minimum temperature is reached for an elliptical polarization. However, the anomaly force terms dominant for small detuning $|\delta|<<\gamma$ only [20], thus for the considered case of $\delta=-\gamma$ the anomaly force effect is small and results to temperature of laser cooling get the minimum values for ellipticities are close to linear polarization: for I = 100 mW/cm^2 minimum of temperature is reached at $\varepsilon_0\approx-0.28^o$, for I = 200 mW/cm^2 at $\varepsilon_0\approx-0.86^o$, and for I = 300 mW/cm^2 at $\varepsilon_0\approx-1.15^o$. The maximum fraction of cold atoms $N_{|p|<3\hbar k}$ for I = 100 mW/cm^2 is reached at $\varepsilon_0\approx-2^o$, for I = 200 mW/cm^2 at $\varepsilon_0\approx-5.16^o$, and for I = 300 mW/cm^2 at $\varepsilon_0\approx-5.26^o$.

Nonlinear dependence of magneto-optical trapping force as function of Zeeman shift Ω_H of the ground energy sublevel $|J_g=2,\mu_g=1\rangle$ is shown on figure 7a for conventional σ_+–σ_- field configuration and figure 8a for $\varepsilon-\theta-\bar{\varepsilon}$ field configuration (θ=–π/4) at different waves ellipticity parameters ε_0. The Zeeman shift $\Omega_H=\gamma$ corresponds to magnetic field H = 12.7 G.

Assuming the linear dependence of magnetic field on position near the center of the trap $H(z)\approx\partial_z H\,z$ for $|z|<R_W$ the depth of magneto-optical trap can be estimated as:

$$U^{(H)}=-\frac{R_W}{\Omega_H(R_W)}\int_0^{\Omega_H(R_W)}\left\langle F^{(H)}(v=0,\Omega_H)\right\rangle d\Omega_H\,, \tag{3}$$

where R_W is the radius of the light beams forming the trap and Ω_H (R_W) is the Zeeman shift on the trap boarder (entrance into the trap). The depth of magneto-optical trap is determined by the gradient of magnetic field $\partial_z H$ and can be expressed as function of magnetic field at the trap boarder (figure 7b and figure 8b) for the area of linear grow of magnetic field (3).

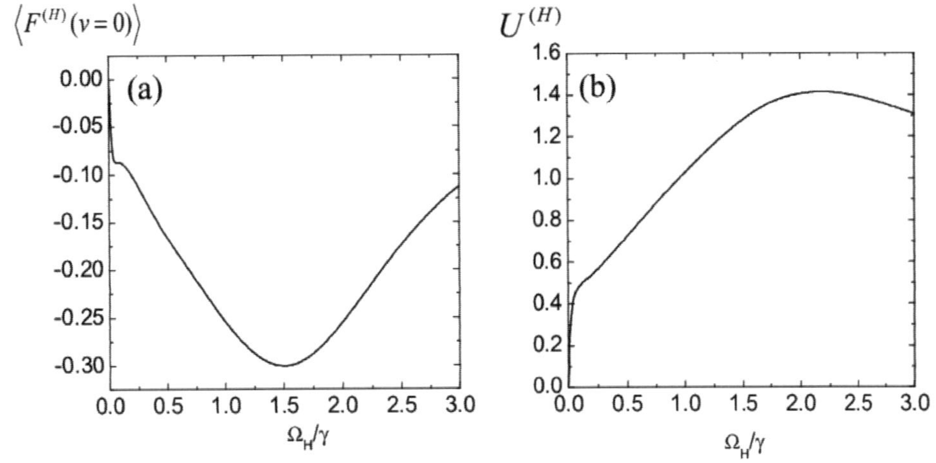

Figure 7. Magneto-optical trapping force (in $\hbar k\gamma$ units) on ^{24}Mg atoms (a) and depth of magneto-optical potential in $\hbar\gamma R_W/\lambda$ units (b) as function of Zeeman shift Ω_H in σ_+–σ_- field configuration (I = 100 mW/cm^2 and δ=$-\gamma$).

The depth of magneto-optical trap, formed by $\varepsilon-\theta-\bar{\varepsilon}$ field configuration, is significantly less the depth of magneto-optical trap formed by conventional σ_+–σ_- field configuration, but still much deep to capture cold atoms. As an example, the depth of the trap, formed by field with $\varepsilon-\theta-\bar{\varepsilon}$ field configuration for R_W = 0.5 cm and gradient of magnetic field of $\partial_z H = 12.7$ G/cm (that corresponds to $\Omega_H/\gamma \approx 0.5$ at the trap boarder, figure 8b) reaches $U^{(H)} = 0.094\ \hbar\gamma R_W/\lambda \approx 1.56$ K that much exceed the temperature of cold atoms (T = 124 μK for considered parameters).

The elliptical polarization of the light waves that could results to smaller temperature due to anomaly parts in the friction force (i.e. for $\varepsilon_0 < 0$ at considered θ=$-\pi/4$) unfortunately leads to decrement of the depth of magneto-optical potential (figure 8b). As an example for $\varepsilon_0 = -3^{\circ}$ the magneto-optical force reverse when magnetic field exceeds 9 G ($\Omega_H/\gamma \approx 0.71$).

For estimation of number of trapped atoms in MOT one should analysis of nonlinear dependence of magneto-optical force for slow atoms in magnetic field. The number N_c of trapped atoms $N_c \propto v_c^4$ is determined by the critical velocity v_c, i.e. the maximum velocity of slow atom can have and be captured in MOT [21]. By using the methods we developed in [5], we study magneto-optical force on slow ^{24}Mg atoms as function of magnetic field. For atoms in σ_+–σ_- MOT the critical velocity $v_c \approx 3.5\gamma/k$ [21] that for magneto-optical trap formed by waves with R_W = 0.5 cm results to $N_c \approx 7\times10^7$ atoms. For atoms in $\varepsilon-\theta-\bar{\varepsilon}$ MOT at some large values of magnetic field the magneto-optical force is reversed (see figure 9) for some range of velocity of slow atoms. Thus the atoms in this range of parameters (v, Ω_H) cannot be captured by the trap, that define v_c as a function of magnetic field. Here for $lin-\theta-lin$ field configuration (θ=$-\pi/4$) at large magnetic field (Ω_H/γ>1.5) the critical velocity decreased to $v_c \approx 0.014\gamma/k$ (0.14 m/s) that almost nullify the number of trapped atoms.

Thus for stable work of $\varepsilon-\theta-\bar{\varepsilon}$ MOT the low gradients of magnetic field is required, i.e the magnetic field within the trap area $r<R_W$ should never reach the critical values H_c when the force on the atoms moving with some velocity range is reversed (figure 9) . As an example, for stable work of $\varepsilon-\theta-\bar{\varepsilon}$ MOT for $\varepsilon_0 = -1^{\circ}$ magnetic field should not exceed H_c= 9.3 G ($\Omega_H/\gamma \approx 0.73$).

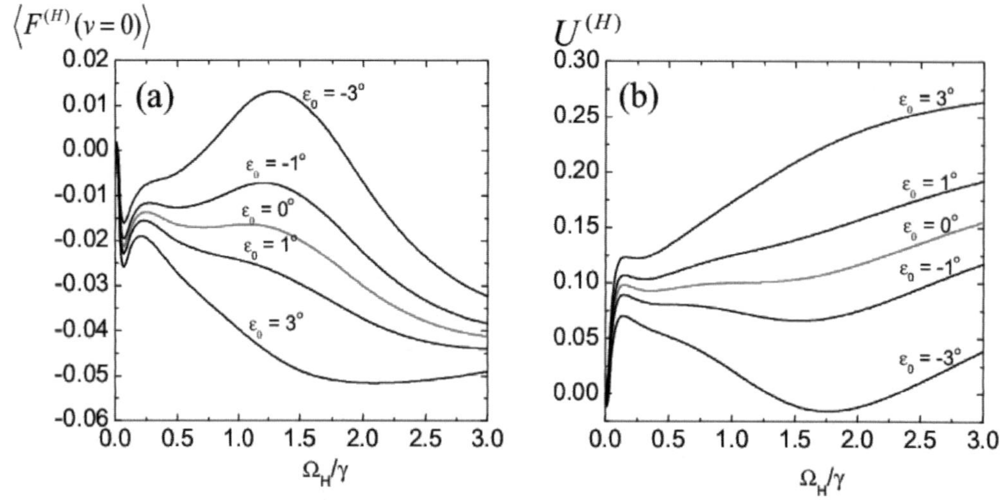

Figure 8. Magneto-optical trapping force (in $\hbar k\gamma$ units) on ^{24}Mg atoms (a) and depth of magnet-optical potential in $\hbar\gamma R_W/\lambda$ units (b) as function of Zeeman shift Ω_H in $\varepsilon-\theta-\bar{\varepsilon}$ field configuration for different parameters ellipticity of light waves ε_0 ($\theta = -\pi/4$, I=100 mW/cm^2 and $\delta = -\gamma$).

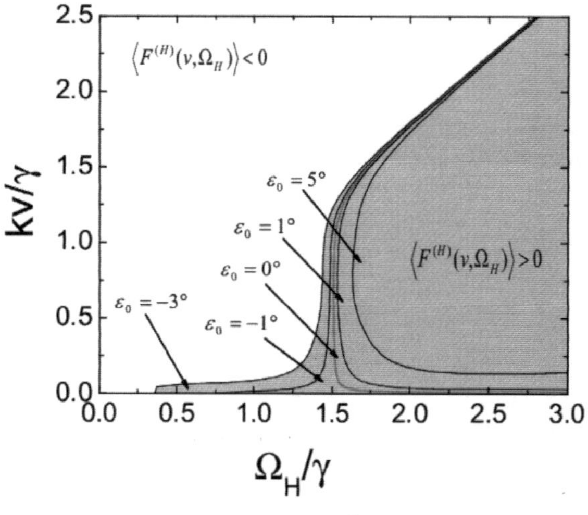

Figure 9. The zones of parameters (v, Ω_H) where magneto-optical force on ^{24}Mg atoms is repulsive force in $\varepsilon-\theta-\bar{\varepsilon}$ field configuration ($\theta = -\pi/4$) for different parameters ellipticity of light waves ε_0 (I=100 mW/cm^2 and $\delta = -\gamma$).

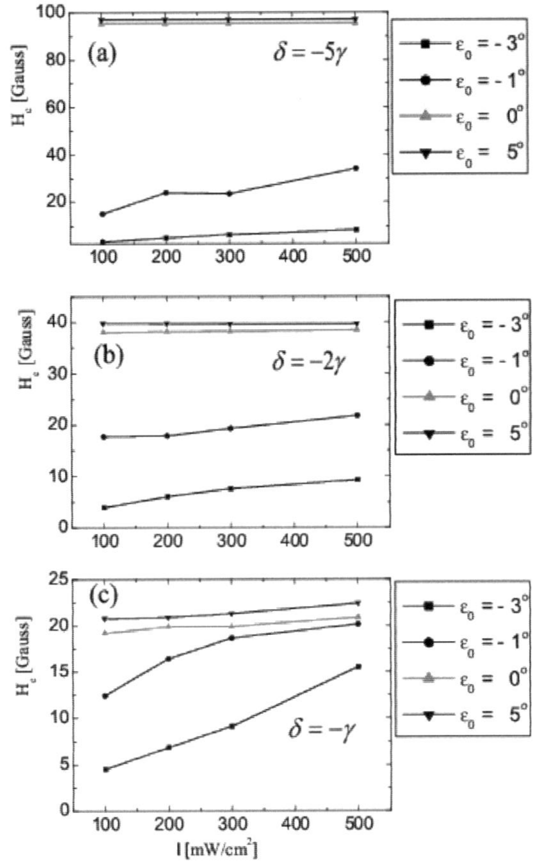

Figure 10. Critical values of magnetic field as function of intensity of light waves forming magneto-optical trap for different detunings and ellipticity of light waves for ^{24}Mg atoms in $\varepsilon-\theta-\bar{\varepsilon}$ MOT ($\theta = -\pi/4$).

Figure 10 shows the critical values of magnetic field as function of intensity of light waves forming the trap for different detunings and ellipticity of light waves for ^{24}Mg atoms in $\varepsilon-\theta-\bar{\varepsilon}$ MOT ($\theta = -\pi/4$).

Conclusion

In this paper we have considered a magneto-optical trap for ^{24}Mg atoms operating on the closed triplet $^3P_2 \rightarrow {}^3D_3$ transition, formed by the light waves with elliptical polarizations ($\varepsilon-\theta-\bar{\varepsilon}$ configuration, see figure 2). In the limit of a 1D model we have studied the magneto-optical potential, temperature and fraction of atoms extremely cooled bellow the momentum $p=3\hbar k$ (12.9 cm/s) as the functions of intensity, frequency detuning and polarizations of light waves that form the MOT. For our simulations we have used recently suggested method [18, 19] that allows taking into account the quantum recoil effects of interaction of atoms with a light field and correctly taking into consideration the slow atoms localized in the optical potential wells as well as the atoms moving above the potential wells.

We have found that the $\varepsilon-\theta-\bar{\varepsilon}$ light field configuration formed by the waves with elliptical polarizations and orientation angle $\theta=-\pi/4$ with parameters ellipticity are close to linear polarization can offer the lowest cooling temperatures on the level $\boldsymbol{T \approx 100}$ **μK** together with enough depth of the magneto-optical potential. In comparison with the conventional MOT, formed by waves with $\sigma_+ - \sigma_-$ polarizations, the suggested $\varepsilon-\theta-\bar{\varepsilon}$ MOT should operate with lower gradient of a static magnetic field. Indeed, in comparison with the $\sigma_+ - \sigma_-$ MOT, the proposed $\varepsilon-\theta-\bar{\varepsilon}$ MOT is more exacting to the field parameters, because of strong magnetic field may reverse magneto-optical trap force for slow atoms in $\varepsilon-\theta-\bar{\varepsilon}$ configuration that limit the number of atoms captured in the trap. Parameters of the critical magnetic fields have been also figured out.

Acknowledgments
The work was partially supported by the Russian Science Foundation (project no. 16-12-00054). A.N. Goncharov thanks the Russian Foundation for Basic Research for the support (grant no. 15-02-06087).

References

[1] Riedmann M *et al.* 2012 *Physical Review* A **86** 043416
[2] Brazhnikov D V *et al.* 2015 *Phys. Rev.* A **92** 063413
[3] Dalibard J and Cohen-Tannoudji C 1989 *J. Opt. Soc. Am.* B **6** 2023
[4] Prudnikov O N *et al.* 2004 *JETP* **98** 438
[5] Prudnikov O N *et al.* 2008 *Phys. Rev.* A **77** 033420
[6] Kazantsev A P, Surdutovich G I and Yakovlev V P 1990 *Mechanical Action of Light on Atoms* (Singapore: World Scientific) p 392
[7] Dalibard J and Cohen-Tannoudji C 1985 *J. Phys. B: At. Mol. Phys.* **18** 1661
[8] Javanainen J 1991 *Phys. Rev.* A **44** 5857
[9] Yoo S M and Javanainen J 1992 *Phys. Rev.* A **45** 3071
[10] Dum R, Zoller P and Ritsch H 1992 *Phys. Rev.* A **45** 4879
[11] Dalibard J, Castin Y and Mølmer K 1992 *Phys. Rev. Lett.* **65** 580
[12] Mølmer K, Castin Y and Dalibard J 1992 *J. Opt. Soc. Am.* B **10** 524
[13] Castin Y and Dalibard J 1991 *Europhys. Lett.* **14** 761
[14] Guo J and Berman P 1993 *Phys. Rev.* A **48** 3225
[15] Deutsch I H, Grondalski J and Alsing P M 1997 *Phys. Rev.* A **56** R1705
[16] Berg-Sorensen K, Castin Y, Mølmer K and Dalibard J 1993 *Europhys. Lett.* **22** 663
[17] Castin Y, Berg-Sorensen K, Dalibard J and Mølmer K 1994 *Phys. Rev.* A **50** 5092
[18] Prudnikov O N *et al.* 2011 *JETP* **112** 939
[19] Prudnikov O N *et al.* 2007 *JETP* **104** 839
[20] Prudnikov O N *et al.* 1999 *JETP Lett.* **70** 443
[21] Monroe C, Swann W, Robinson H and Wieman C 1990 *Phys. Rev. Lett.* **65** 1571

Pulsed UV laser technologies for ophthalmic surgery

A M Razhev[1, 2], V V Chernykh[4], S N Bagayev[1], D S Churkin[1, 3], E S Kargapol'tsev[1], I A Iskakov[4] and O V Ermakova[4]

[1]Institute of Laser Physics SB RAS, 13/3 Acad. Lavrentyev ave., Novosibirsk 630090, Russia
[2]Novosibirsk State University, 2 Pirogova str., Novosibirsk 630090, Russia
[3] Novosibirsk State Technical University, 20 K. Marx ave., Novosibirsk 630073, Russia
[4] Novosibirsk Branch of the Federal State Institution "Intersectoral Research and Technology Complex "Eye Microsurgery" named after academician S.N. Fyodorov, Ministry of Health of the Russian Federation", 10 St. Colchis, Novosibirsk 630071, Russia.

Abstract. The paper provides an overview of the results of multiyear joint researches of team of collaborators of Institute of Laser Physics SB RAS together with NF IRTC "Eye Microsurgery" for the period from 1988 to the present, in which were first proposed and experimentally realized laser medical technologies for correction of refractive errors of known today as LASIK, the treatment of ophthalmic herpes and open-angle glaucoma. It is proposed to carry out operations for the correction of refractive errors the use of UV excimer KrCl laser with a wavelength of 222 nm. The same laser emission is the most suitable for the treatment of ophthalmic herpes, because it has a high clinical effect, combined with many years of absence of recrudescence. A minimally invasive technique of glaucoma operations using excimer XeCl laser (λ=308 nm) is developed. Its wavelength allows perform all stages of glaucoma operations, while the laser head itself has high stability and lifetime, will significantly reduce operating costs, compared with other types of lasers.

1. Introduction

Today, the use of UV lasers for eye surgery has become an integral part of ophthalmic surgery. This area is one of the most important for the Institute of Laser Physics SB RAS and SF IRTC "Eye Microsurgery", the experience of successful cooperation which has more than 30 years. Joint studies on the possibility of using high-power short-pulse UV laser radiation in medicine and development of laser medical technologies began by collaborators of these organizations in the 1980s. By the time the team has had a complete set of pulsed excimer laser sources, most of which were created at the Institute for the first time in the world [1, 2]. Then in 1986 the world's first operation of laser correction of refractive errors were made. In order to keep the Bowman's membrane with epithelial layer, which would inevitably evaporate when exposed to and to hold direct ablation of the corneal stroma, the authors have been proposed and practiced separation of the corneal stroma, thus, the corneal flap is folded to one side. Results of clinical trials for the correction of high myopia in a number of patients were reported in [3]. The famous ophthalmologist professor at Columbia University (New York, USA) S. Trokel, noting a significant contribution to Russian researchers called the first operation "proto-LASIK" [4]. Later, with the introduction of mechanical microkeratome for corneal bundle, this method of correction was finally formed as a Laser-Assisted in Situ Keratomileusis - LASIK, which to date remains the most popular method of laser correction of refractive errors.

Then, in the late 1980s one began researches on the possibility of laser treatment of viral and other eye diseases, based on the bactericidal effect of UV short-pulse laser radiation. On the first positive results of the use of short-pulse UV laser radiation for the treatment of various eye diseases: glaucoma, herpes simplex virus, corneal ulceration was reported in [3].

The results of the researches were first developed UV laser ophthalmic medical devices for the treatment of refractive error correction eye surgery and the treatment of ophthalmic otrkrytougolnoy glaucoma. At the beginning of 2000 in the Institute of Laser Physics SB RAS is created ophthalmic system under trademark *Medilex*^{TM} at a wavelength of 193 nm and 222 nm.

In the course of further joint researches ILP SB RAS and NF IRTC "Eye Microsurgery" continues the continuous improvement of existing and development of new laser medical technologies for ophthalmic surgery, creating new laser medical devices.

2. The perspectives of using the wavelength of 222 nm of the KrCl excimer laser for refractive surgeryand for the treatment of some eye's diseases

For years there has been a growing interest to the refractive surgery of the human eye by the emission of the excimer lasers surgical means rather than the use of spectacles or contact lenses [5].

Excimer lasers are capable of emitting different wavelengths in the ultraviolet region, including 193, 222, 248, 308 and 351 nm. For corneal surgery, 193 nm is the wavelength of choice because it produces the most precise cuts with the minimum amount of thermal distortion of the surface of the tissue [6]. Therefore surgeons usually use the excimer ArF laser with radiation wavelength of 193 nm [7-9].

However a different research groups study the possibility of the application another laser wavelengths for the refractive surgery [10-12].

This paper describes a new UV ophthalmic laser system *Medilex*^{TM} based on excimer ArF (193 nm) or KrCl (222 nm) lasers. We have shown that the use of the wavelength of 222 nm for photorefractive surgery has several medical and technical advantages compared with typical wavelength of 193 nm and we propose that the emission of the wavelength of 222 nm of the KrCl excimer laser is preferable for refractive surgery than traditionally used wavelength of 193 nm [13, 14].

2.1. Features of excimer laser system

UV ophthalmic laser system *Medilex*^{TM} can be used for the correction of the myopia, hyperopia, astigmatism and myopic or hyperopic astigmatism by the Photorefractive Keratectomy (PRK) or Laser in Situ Keratomileusis (LASIK) procedures and for Phototerapeutic Keratectomy (PTK) to treat corneal pathologies. All procedures mentioned above can be realized using either the wavelength of 193 nm of ArF excimer laser or the wavelength of 222 nm of KrCl excimer laser

The UV ophthalmic laser system *Medilex*^{TM} consists of a laser subsystem, an optical subsystem, an alignment subsystem, a measurement and a control subsystem. The laser subsystem generates the pulsed coherent UV radiation of suitable wavelength and energy level. The optical subsystem provides a means for conditioning the beam and delivering it to the patient's eye. The alignment subsystem provides a microscope coaxial with the laser beam alignment to the center of patient's eye. The measurement and control subsystem provides the computer and electronics circuitry requared for monitoring and controlling to all systems of the set-up via a microprocessor device.

The laser system is built as a cabinet. In the cabinet, the excimer laser, the gas subsystem, the high voltage power supply, the vacuum pump, the control units and other auxiliary aggregates are included. The unit measures 1500×800×1300 mm and the weight is about 400 kg. Electric power lines needed for the excimer laser systems are: single phase 220V, 50Hz, 5 A.

In addition the system is supplied by XYZ patient's bed with fluently changing speed of movement and video camera with monitor to observe the patient's eye during process of treatment. A video recorder or computer can be connected to the video camera for archiving. All surgical proceduress can be documented using the computer system connected to a printer.

2.2. The excimer laser

For UV ophthalmic system we use the simple, reliable and efficient excimer ArF and KrCl lasers which were designed in Institute of Laser Physics SD RAS [15]. The technical parameters of these lasers using in UV ophthalmic system are shown in the Table 1.

Table 1. The technical parameters of the excimer lasers.

Laser medium	ArF or KrCl
Wavelength	193 nm or 222 nm
Fluence	up to 250 mJ/cm^2
Pulse durations	12±1 ns (FWHM)
Pulse repetition rate	up to 20 Hz

2.2. Beam delivery system

The beam delivery system has been developed to permit precise thickness of tissue to be ablated at exact locations on the anterior corneal surface without involving normal adjaced or deeper corneal tissue. Optical scheme allowed the laser beam to be projected onto the cornea either as a fairly uniform beam of 0,2-5 mm diameter or as a radially symmetric nonuniform pattern for the correction of the refraction anomalies over 4-7 mm diameter areas. Beam delivery system operates under total computer control.

Figure 1 is a functional schematic diagram of the major parts in the beam delivery system *Medilex*^{TM}. The entering laser beam is rectangular with dimensions of approximately 10×22 mm. The intensity profiles in the short and long directions are essentially Gaussian and ''top-hat''. Since there are significant ''hot spots'' at the edges of this beam, firstly it is passed through a 7×20 mm aperture mask *2* to ''scrape-off'' these edges. Then it turns by mirror *3* and passes through cylindrical telescope *4*. As the compressed by cylindrical telescope *4* the laser beam still has different intensity profiles in the principal directions it next passes through a beam homogenizer *5*. In this article we describe the optical delivery system with beam homogenizer – the Dove prism. In other our optical delivery systems we used the special waveguide or diffractive homogenizers.

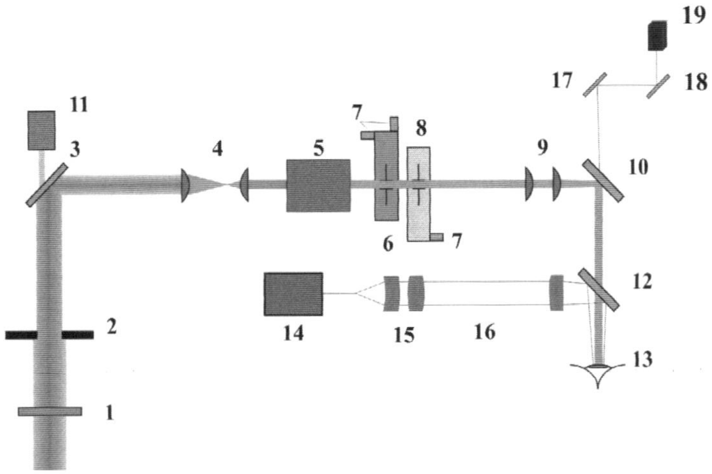

Figure 1. Functional schematic diagram of the major parts in the beam delivery system.

The rotating beam having time-averaged axially-symmetric intensity profiles is next passed through an expandable rotating slit *6* and patented special masks *8* [16].Expandable rotating slit *6* is used for treatment of astigmatism and has two computer-controlled motors *7*, with one of them being controlled the expansion of slit and the second one controlled the angle of inclination. When the PRK, LASIK or PTK procedures are realized the slit is opened at a maximum dimension. The using of the special masks *8* for refractive surgery and PTK procedures enables us to achieve very smooth surface of the cornea and good quality of central reprofiling form figures 2 and 3 illustrate schematically the form of masks (a), distribution of laser emission intensity during process of ablation (b) and the ablation profiles achieved with the use of masks in the myopia (figure 2) and hyperopia (figure 3) cases (c), respectively. The diameter of masks depends on the degree of anomaly of refraction and it is changed within 4 - 7 mm. The rotation of the masks with a variable velocity under the special law allows to correct the myopic or hyperopic astigmatism.

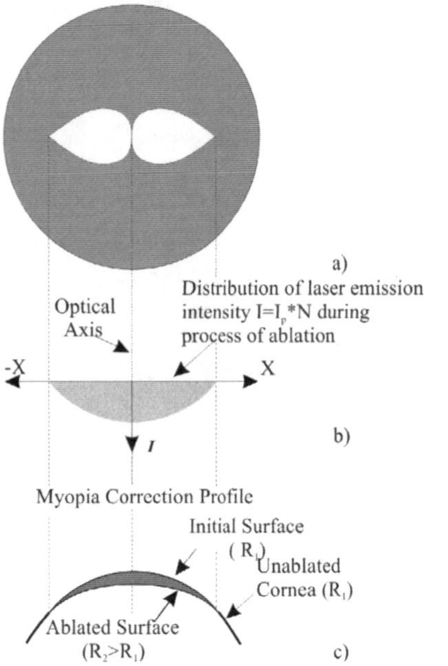

Figure 2. Myopia rotating mask.

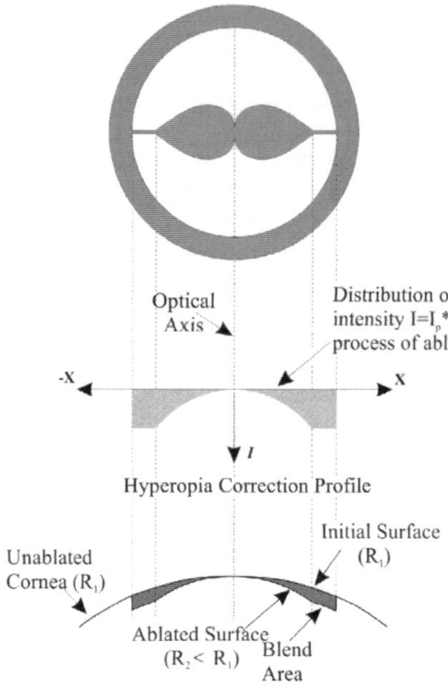

Figure 3. Hyperopia rotating mask.

After the mask the laser beam is delivered to the cornea of patient's eye *13* via spherical lenses *9* and mirror *10*. The photodiode *11* provides the measurements of radiation energy before and during process of the operation allowing to make the ablation with great accuracy. Fixation green light diode 16 provides the alignment of laser beam with the center of the patient's eye. The observation of the patient's eye during process of operation by a video system is possible using beam splitter *14*, beam-bender *16* and video camera *17*.

2.3. Discussion

2.3.1. The comparative analysis of the interaction of excimer lasers radiation with the tissue of the human eye cornea. For determination of the possibility of using different radiation wavelengths of the excimer lasers for refractive surgery we have performed the comparative analysis of the interaction of those ones with the tissue of human eye cornea. The first stage of our investigations was the deriving of the transmission spectrum of the human cornea in ultraviolet region of the spectra. The results are given in[13]. The curve of transmission of the cornea illustrates that three wavelength 193 nm, 222 nm and 248 nm of ArF, KrCl and KrF lasers is found in the region of strong absorption (from 2700 up to 210 cm^{-1}) of cornea and only two wavelength 193 nm and 222 nm have the total absorption in thin layers of cornea [14]. The laser radiation with a longer wavelength cannot be used for microsurgery of the cornea as it will penetrate inside on eye.

The determination of the ablation rate of the cornea using wavelength 193, 222 and 248 nm was the next stage of our investigations. We have obtained that the highest rate of ablation process is achieved at 248 nm up to 3 um/p, and the lowest rate is achieved at 193 nm-up to 0,8 um/p for maximal energy density. When 222 nm is used the ablation of corneal tissue is twice as quick as that (2 um/p) at 193 nm and only by 30% slower that at 248 nm [13].

2.3.2. The effect of the UV laser radiation wavelength on the cornea tissue damage. The experiments on determination of thickness of the zone of damage at the cut surface of corneal stroma following laser ablation at three different wavelengths were performed [14].

In exposing the cornea of an isolated rabbit or human eye to laser radiation a tissue defect was formed, its depth being dependent on the number of pulses. Along the defect margin, an opaque eosinophilic (damaged) layer was observed, its thickness being 0.3 μm after radiation at wavelength of 193 nm, about 0.6 μm at wavelength of 222 nm and about 10 μm at wavelength of 248 nm. The examination of the eosinophilic layer on the floor of the lesion demonstrated its homogeneity and the eosinophilic layer thickness did not change depending on the number of pulses. In all cases the epithelium along the wound margin was practically not damaged. At the power density approaching the threshold values which depend on the radiation wavelength the epithelium may be damaged, collagen fibres remaining intact. It should be emphasized that when non-penetrating cornes defects were formed close to the Descemets membrane and endothelium the latter remained intact.

Experiment in vivo. The laser beam on the rabbit cornea produced lesions of a given configuration. No visible thermal coagulation was observed following laser radiation (λ =193; 222; 248 nm). After laser radiation a normal response in the form of moderate pericorneal injection, pupil narrowing and mild irritation was observed during 7 days. Epithelization of the lesions took three days. Inflammatory phenomena disappeared completely in 10 days. Histologic examination in the areas exposed to laser radiation revealed thickening of the epithelial layer, vacuolization of the epithelium, appearance of compact and contracted nuclei and formation of keraticsquamas. In the connective tissue underlying the new epithelium bands of collagen fiberes were located chaotically, there appeared accumulations of fibroblasts and fibrocytes. The extent of such changes depended on focusing the laser beam. The histologic examination of the areas of the corneaexposed with the energy density lower than the threshold levels irrespective of the wavelength revealed as a rule no changes of the connective tissue. The epithelium was normal opacification of the cornea after the repair of the tissue lesions caused by radiation (energy density 0.2-0.5 J/(cm^2.pulse). No damage of the deep eye media was noted.

2.3.3. The study the chromosome rearrangement in the eye cornea. The study of the chromosome rearrangement was made upon the cornea and the bone marrow of 1-1.5 months old female nonline mice. The animals were divided into two groups. In the first group, one eye of the animal was exposed to laser radiation, the other being used as a control. The mice of the second group were not subjected to any special effect and were used as an intact control group. In the course of the experiments the energy density of the radiation was changed in the range of 0.015-1.5 J/(cm^2.pulse) for the wavelength of 248 nm, 0.027-0.04 J/(cm^2.pulse) for 222 nm and 0.3-0.9 J/(cm^2.pulse) for 193 nm. The area of the spot with the maximum energy density was 0.001 cm^2. To characterize the process of mutation, the dose of exposure equal to the product of the energy density in one pulse and the number of pulses was determined. The cornea were fixed in ethanol-acetous mixture (3:1) 10 and 72 hours after exposure, stained according to Erlich and stable specimens were made.

It was shown that in this case the concentration of the cells with mutation does not differ from the control over the whole range of doses [20]. Thus, the exposure of the corneal tissue to powerful UV short-pulse radiation does not present any danger from the point of view of the mutation appearance in the epithelium cells at the parameters being used.

All these results mentioned above allowed us to turn to the clinical tests of ophthalmic devices. The correction of anomalies of refraction (myopia, hyperopia and astigmatism) and healing some diseases of the eye (herpetic keratities and glaucoma) were performed together with Eye Microsurgery Intersectoral Research and Technology Complex (Novosibirsk).

There were performed 82 refractive treatments. There were 18 men and 23 women of 18 to 42 years of the age old. Photorefractive keratectomy (PRK) was performed for 20 patients, LASIK for 21 patients. 15 patients had the myopia degree up to 6.0 D, 26 patients had the myopia degree over 6.0 D. Preoperative and postoperative examination involved visual acuity, refractometry, keratometry, pachimetry and computerized topography. The analysis of PRK efficiency in-group of patients with the myopia degree up to 6.0 D revealed that desired effect (visual acuity 0.8-1.0) was achieved in

72%. Refractive effect was stable throughout 5-months follow-up period. The result, close to emmetropy, was achieved in 92% in the case of the myopia degree over 6.0 D. 21 patients with myopia over 6.0 D, were operated by LASIK technology. There were no intraoperative complications. 94% of patients, 7 days after surgery, had refractive closed to emmetropy. Refractive effect was stable throughout 3-months follow-up period.

3. The choice of the laser wavelength for a herpetic keratitis treatment

Recently the utilizing the 193 nm radiation wavelength of an ArF excimer laser took a world-wide extension as more simple and efficient method of the refraction anomalies correction such as myopia, hyperopia, astigmatism and as an instrument for therapeutic procedures on the surface of a cornea [5, 18]. However there was a question unexplored on an opportunity of use UV radiation of excimer lasers for treatment of other superficial diseases of an eye. There are communications on use of ArF laser at treatment of the open-angle glaucoma [19]. But laser treatment method is not known to have some advantages over the existing traditional non-laser ones. At the same time the question about the possibility of using the UV excimer lasers radiation for the treating the virus eye diseases such as herpetic keratitis and also ulcers of a cornea is interesting. This task arises from that fact, that one of the most widespread virus infections on today is herpetic one. The main aim of our work was the investigation of the possibility of a herpetic keratitis laser treatment.

3.1.Materials and methods

The three ophthalmic laser system based on ArF (193 nm), KrCl (222 nm) and KrF (248 nm) excimer lasers were developed and created for making the investigations. The constructions of these systems were almost the same as for the ophthalmic system Medilex were described in detail in [20]. To implement the purpose of the investigation the trial of 75 patients with herpetic keratitis manifests by a surface arborescent keratitis before and after treatment has been carried out. All patients undergoing surgery were volunteer and only one eye was operated on in each individual. The clinical diagnosis was determined on the base of a complex of ophthalmologic methods. Depending upon a laser radiation wavelength used (193 nm, 222 nm or 248 nm) all patients were allocated to 3 groups (25 patients in each group), where lacrymal fluid was obtained on day 7-10 after laser exposure and testing a changes of the parameters studied was performed. Additionally a group of 28 healthy individuals was used for intact control. The choice of a time point for investigation was determined by literature findings that indicated a necessity of such time period for antibody production and its pronounced utilisation or formation of circulating immune complexes (CIC).

Certainly, it is interesting to determine a herpes simplex virus (HSV) effect and laser treatment of HSV on some integral immunobiochemical parameters of lacrimal fluid. We have chosen testing of antibodies to native DNA-antigens CIC and lactoferrin in the capacity of the parameters. The choice of these parameters was determined by the following factors:

1. At present one may ascertain that inflammation of any etiology and localization as well as some pronounced degenerative-destructive processes follow herpetic keratitis, a destruction of cell membranes, DNA-antigens elimination.
2. A pathogenic significance of CIC in the arising and forming of the immune response, an interrelation of this parameter with inflammation and immunologic homeostasis maintenance is beyond doubt.

3.2. Results and discussion

3.2.1. The 193nm laser radiation treatment. The 1st group of the patients (25 patients) was treated by 193 nm laser radiation wavelength. During laboratory investigations in this group of patients with herpetic keratitis the autoantibody (AAB) levels to native DNA-antigens in lacrimal fluid was 0.46±0.03 a.u. (arbitrary units) and was not significantly different from normal value.

The level of CIC was 42.6±6.3 a.u. that is significantly ($p<0.02$) higher than normal value (25.1±3.9 a.u.) and is at the level of the parameter studied in patients with herpetic keratitis before treatment (42.9±6.9 a.u.). A positive clinical effect of the laser treatment with 193nm radiation wavelength was obtained in 9 of 25 patients treated that was equal 36%. A clinical effect appears as a decrease of the pain characterised for patients with herpetic keratitis, reduction of photophobia, epiphora, and eyelid edema. However, a positive clinical effect was not obtained in 16 patients (64% of this patient group) after laser ablation of the herpes inflammation area by 193 nm radiation wavelength.

As a result one may ascertain that laser therapy by 193 nm laser radiation wavelength does not produce a destructive effect on cornea tissues in an irradiated area that is confirmed by a reduction of antibody levels to native DNA-antigens. However, the absence of such effect can indicate inefficiency in treatment as to an effect upon HSV. The laser with the given wavelength has an inefficient energy for coagulation of HSV on the surface of the cornea and does not cease an inflammation formation. The further formation of an inflammation is confirmed by CIC high levels that indicate a continued humorial response to antigen stimulation.Thus, the use of the 193-nm radiation wavelength is low effective to treat herpetic keratitis manifested with a surface arborescent keratitis.

3.2.2. The 222nm laser radiation treatment. The 2nd group of the patients (25 patients) was treated by 222 nm laser radiation wavelength. In the result of treatment by 222 nm radiation wavelength there is a significant increasing the lacrimal fluid antibodies level to native DNA-antigens that are equal 0.76±0.11 a.u. compared with the results obtained in control group 0.47±0.03 a.u. ($p<0.008$) as well with the results obtained before treatment - 0.54±0.02 a.u. ($p<0.04$). Comparing these results with those obtained by 193 nm wavelength treatment a reliable increase ($p<0.006$) of a parameter studied was also revealed. After exposure by 222 nm wavelength the level of CIC (figure 8) was 35.6±3.5 a.u. that reliable exceeded ($p<0.05$) a normal value (25.1±3.9 a.u.). An unreliable tendency to the decrease of a studied parameter compared with 1st group and this group before treatment (35.6±3.5 a.u. versus 42.6±6.3 a.u. and 42.9±6.9 a.u. respectively) is also revealed.

As a result of clinical trials the data were obtained that allowed to evaluate the efficiency of the treatment by 222 nm radiation wavelength. 22 patients of 25 treated ones (88%) showed a positive clinical results. But it is necessary to note that the treatment of a herpetic inflammation area by 222 nm radiation wavelength results to the absence of a complete clinical effect in 12% patients (3 persons) though the area damaged was entirely ablated.

3.2.3 The 248nm laser radiation treatment. The 3rd group of the patients (25 patients) was treated by 248nm radiation wavelength laser therapy. In this group of the patients autoantibody levels to native DNA-antigens (figure 7) was equal 0.95±0.12 a.u. that reliable exceeded ($p<0.004$) the normal values 0.47±0.03 a.u. and as well parameter levels obtained in the group of patients with herpetic keratitis before the treatment 0.54±0.02 a.u. ($p<0.001$). When compared the results obtained in this group with ones of the groups of patients treated by 193nm and 222nm radiation wavelengths the reliable tendency to increasing the autoantibody levels to normal DNA-antigens ($p<0.002$ and $p<0.04$ respectively) has been revealed.

CIC level after exposure by 248nm radiation wavelength (figure 8) was equal 27.8±8.5 a.u. that was not significantly differed from normal value 25.1±3.9 a.u. The tendency to the decrease of a CIC levels in comparing with those of the 1st, 2nd groups and the data obtained in 3rd groups before treatment (42.6±6.3 a.u., 35.6±3.5 a.u. and 42.9±6.9 a.u. respectively) has been also revealed.

The clinical results obtained in this group of patient were the same as for the patients of the 2nd group treated by 222 nm radiation wavelength. A positive clinical effect was obtained in 22 patients of 25 ones (88%). However, in patients with a positive clinical dynamics in given group a number of negative accessory effects after laser exposure were found. When determined a visual acuity level did not exceed 0.5 and was in the range of 0.2-0.5. This fact is explained by a high level of a thermal effect of 248 nm radiation wavelength upon affected cornea area that manifested as a semitransparent epithelium formation and in further formation of cloud-like opacity in the exposure area.

The results obtained indicate that with increasing the wavelength from 193 nm to 248 nm the autoantibody levels to native DNA-antigens are raising and the CIC levels decrease that indicated by reduction of antigen-antibody intensity reaction and antigen stimulation.

The data obtained indicate that 248 nm radiation wavelength possesses the maximal thermal effect from all wavelengths used in our experiments. It is confirmed by higher autiantibody levels to native DNA-antigens in lacrimal fluid. This wavelength has a high level of a destructive effect upon the eye cornea, which results to cornea opacity in some cases.

4. Ab-externo glaucoma surgery with UV laser radiation

Today glaucoma called chronic eye disease characterized by constant or periodic increase in intraocular pressure (IOP) with the development of trophic disorders in the outflow tracts, in the retina and the optic nerve, causing the appearance of typical defects in the field of view and the development of the edge of excavation (groove punching) of the optic nerve.Glaucoma mainly affects people over 40 years, but is also found youth glaucoma and even congenital glaucoma (newborns). Thus, the frequency of congenital glaucoma was 1 per 10-20 thousand infants aged 40-45 years. Primary glaucoma occurs in about 0.1% of the population. In the age group 50-60 years have glaucoma occurs in 1.5% of cases and in persons over 75 years - more than 3%.

The aim of our work was to development of a safe, clinically effective and convenient way to practice laser treatment of glaucoma and implement the methodology of the operation on the treatment of patients with open-angle glaucoma (ab-externo) using a short-pulse excimer laser radiation.
Main tasks of our investigationwere:

- to determine the transmission spectrum of the sclera in the range of 190-400 nm;
- to investigate the effects of KrF, XeCl, XeF excimer lasers to the tissue of the sclera, which are capable of producing ablation of sclera amid filtration intraocular fluid, without causing damage of the scleral tissue;
- to determine the threshold energy for ablation and ablation rate for these wavelengths;
- find out a convenient, effective way to deliver UV radiation to the area of operation;
- to carry out of a clinical testing.

Studies were carried out using two UV laser systems. The typical laser chamber design with high-voltage excitation system and view of *Medilex*^{TM} excimer lasers are presented in figure 4.

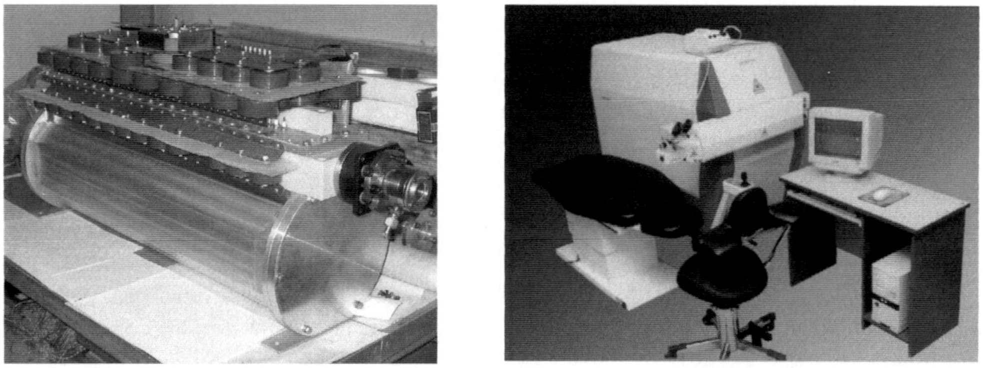

Figure 4. The laser chamber design and HV excitation system(left) and view of excimer laser system *Medilex*^{TM} design (right), based on the gas-discharge pumped ArF /KrCl excimer laser.

4.1. Experiment and discussion

Experimental part was carried out with scleral tissue of isolated human eye using 3 wavelengths: 248, 308 and 353 nm - specially designed excimer laser systems were created.Ablation threshold and ablation speed for studied wavelengths were investigated.Dependence of thickness of an isolated human eye's sclera layer removed per pulse on the energy density of radiation for different wavelengths is presented on the figure6.It was shown that in terms of the maximum rate of ablation, the XeCl (308 nm) and XeF (353 nm) laser radiation is the most acceptable for practical use, as it allows to provide higher speed of operations.

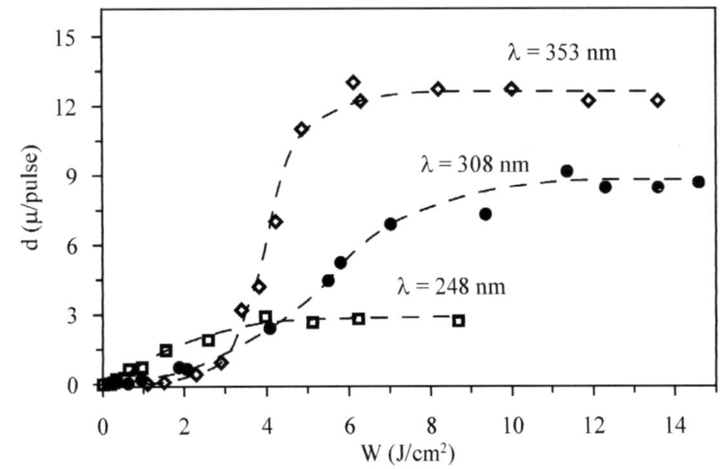

Figure 6. The dependence of thickness of an isolated human eye's sclera layer removed per pulse on the energy density of radiation for different wavelengths.

Histological examination of the sclera preparations subjected to radiation with a wavelength of 308 and 353 nm, edge grooves have been presented structured collagen fibers without evidence of damage, and coagulation of visible lesions (figure 7). The walls of the arteries and veins around the zone of ablation and at the edges of the flap all over were not damaged; foci of hemorrhage and hemosiderosis were absent. At the same time, research of the transmission spectrum of UV radiation of human scleral tissue in the 190 - 500 nm showed that the maximum absorption of the UV radiation in tissue is shorter than 300 nm (figure 8). At longer wavelengths, the absorption becomes less, and the laser light starts to penetrate deep into the scleral tissue.

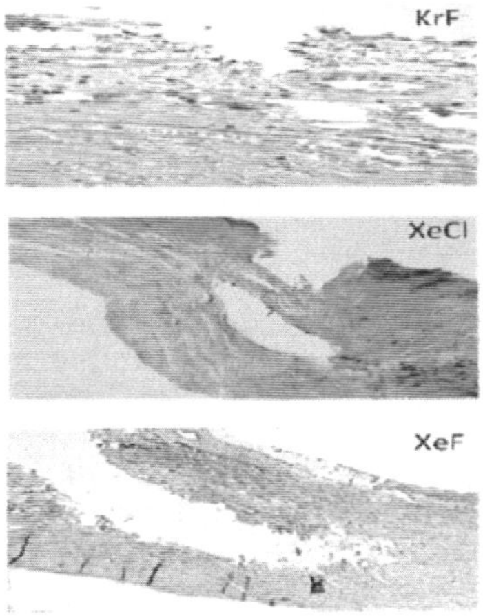

Figure 7. The histological test of sclera tissue.

At wavelengths over 310 nm radiation can penetrate through the entire thickness of the sclera, reaching the underlying tissues. This could lead to possible damage to the internal parts of the eye with unpredictable consequences, therefore safer when exposed sclera is radiation having a wavelength less than 310 nm. In this case, such is a XeCl laser with a wavelength of 308 nm. As a result, the basis for the establishment of a medical device for the treatment of open-angle glaucoma was chosen XeCl excimer laser.

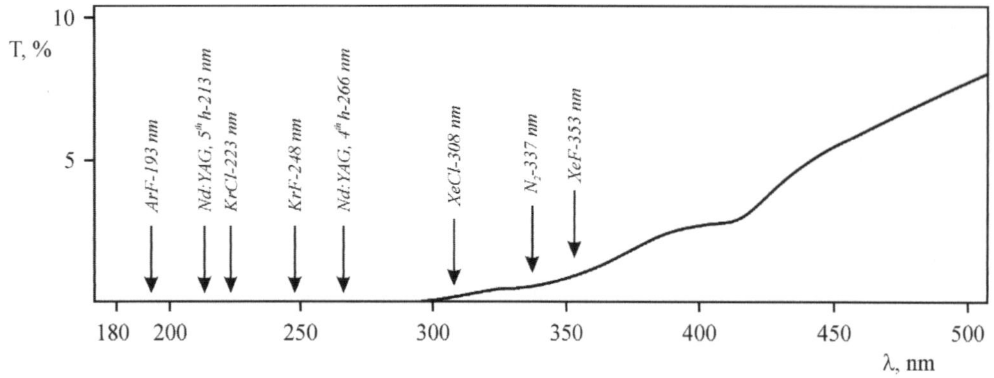

Figure 8. The transmission of the human sclera in the UV range of spectra.

Based on experimental results new method for the ab-externo open-angle glaucoma surgery with excimer laser radiation (λ = 308 nm) is suggested and clinically realized.The UV laser ophthalmological medical system *Medilex*^{TM} was developed on the XeCl excimer laser (see on the figure 17). The basic *ab-externo*method of laser irradiation is the maximum thinning of trabeculae and Descemet's membrane while maintaining the integrity of these tissues in order to achieve the maximum antihypertensive effect by restoring the ocular fluid filtration.The procedure for the intensification of filtering properties is layered evaporation (ablation) trabeculae and Descemet membrane. Duration of exposure is determined by the ophthalmologist in each individual case as the current intensification of aqueous humor. The main advantage of using the excimer laser surgery for this is the high precision tissue removal and surface quality without thermal injury.

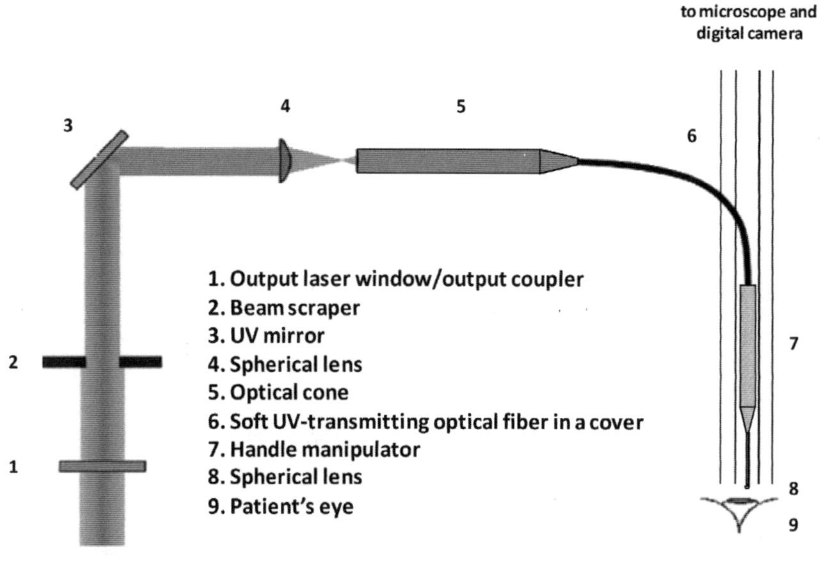

Figure 9. The optical beam delivery system of UV ophthalmic excimer laser system *Medilex*^{TM}using UV-transmitting optical fiber laser.

To thinning was necessary to develop an optical system that allows the delivery form and focus the UV laser radiation on the surface of the biological tissue.To deliver 308 nm to the surgical field, flexible fiber light guide with easy-to-use handle was made (figure 9). The optical system of the laser system is designed for the transport of laser radiation, transforming geometrical dimensions of the laser beam is focused to a diameter of about 0.05 cm and, subsequently, the transmission of a focused radiation on the surface of Descemet's membrane.

The basis of the future medical system has been taken previously developed UV excimer laser system*Medilex*^{TM} based on the gas-discharge pumped excimer laser (see figure 10).Energy characteristics of the laser system are provided by gas mixtures which do not involve the usage of expensive buffer gases, such as helium grade "60" and neon (see figure 15).

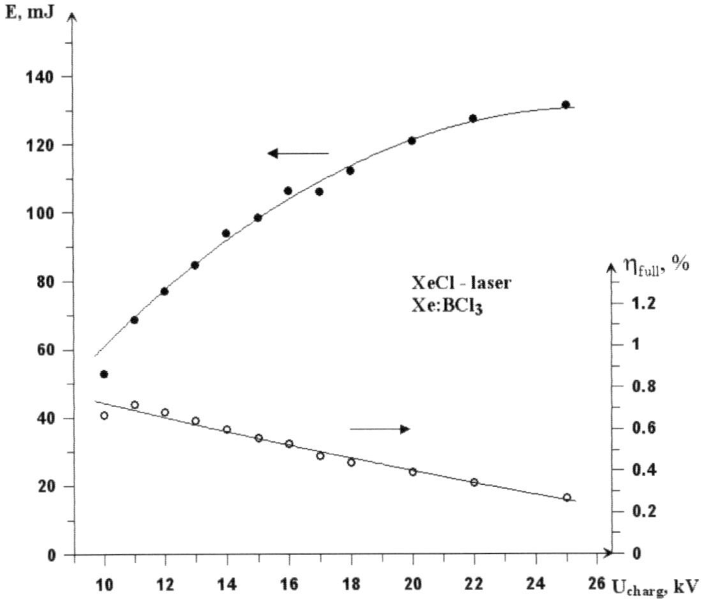

Figure 10. The pulse energy E and total efficiency h of XeCl (308 nm) excimer laser as the function of charging voltage U_{charg} (left) and the XeCl-laser radiation print on the thermofilling paper (right).

The usage of a pulsed gas-discharge XeCl excimer laser, the active medium which is a mixture of gaseous components without the buffer gas is first proposed. The source of intense UV laser radiation is able to compete effectively on a combination of factors with an excimer laser, the active medium which is a mixture of gaseous components with helium buffer gas (grade "B").

The long standing studies were carried out in cooperation with the Novosibirsk ophthalmologists.

5. Conclusion

On the base of investigations carried out and results obtained we have choose the 222 nm KrCl laser for the photorefractive keratectomy. The use of the KrCl laser has several advantages over the ArF (193 nm) laser as the ablation rate is about one-and-a-half times higher than that of the ArF laser, allowing shorter laser treatment times and less chance of eye movement during the process of treatment. Tissue damage adjacent to the area of ablation is histologically about the same as and clinically indistinguishable from that of the ArF laser and little or no mutagenic potential has been attributed to the 193 nm wavelength, it is possible that mutagenesis is even less with the 222 nm wavelength, as it does not penetrate the nuclear membrane as well as the 193 nm wavelength. There are technical advantages as well. Importantly, there is no toxic fluoride gas to deal with because in gas active media of KrCl laser the HCl is used as halogen gas. The lifetime of the KrCl gas mixture is higher than that of ArF gas mixture and the wavelength of 222 nm is better transmitted through the optical system, allowing increased efficiency and prolonged life of the optics.

In ophthalmic devices for photorefractive keratectomy the using of the radiation of the KrCl excimer laser was suggested and realized. Using the results of investigations the new surgical UV ophthalmic laser systems *Medilex*^{TM} based on ArF and KrCl excimer lasers for the photorefractive keratectomy were created and the first trial utilizing the KrCl laser to produce a photorefractive keratectomy in human eyes has been performed. The process of KrCl laser superficial keratectomy has proved to be one of the promising areas of surgical intervention for photorefractive keratectomy in the future. Intensive investigations need to be undertaken on the corneal wound healing process following laser ablation as well as the nature, and long-term stability of the corneal excisions or induced refractive corrections.

For the first time the effect of the UV laser radiation to human eye cornea with herpetic keratitis was experimentally investigated. On the base of results obtained we consider that the use of 222nm laser radiation wavelength is most efficient to treat the herpetic keratitis that manifests by a surface arborescent keratitis. As it allows treating the herpetic keratitis with sufficient curative and low traumatic effects.

A new method for ab-externo excimer laser surgical treatment of open angle glaucoma is developed. For the first time, the technology allows for the operation incisions and scleral ablation trabeculae and Descemet's membrane UV laser radiation with a small footprint and traumatic surgery. As a consequence, informed choice of laser wavelength is noted that the procedure for thinning of trabeculae does not stop even when a filtering aqueous humor. The result is a higher clinical effect – a significant reduction in postoperative complications related to inflammation and scarring. Established installation reduces trauma and increase the hypotensive effect of the operation due to the possibility to carry out cuts and dosed separation sclera ablation of trabeculae and Descemet's membrane.

References

[1] Ishchenko V N, Lisitsyn V N and Razhev A M 1976 *Pisma ZhTF* **18** 839–42
[2] Ishchenko VN, Lisitsyn VN and Razhev A M 1977 *Optics Communications* **21** 30–3
[3] Razhev AM 1988 *Digest Conf. on Lasers and Electrooptics CLEO-88* vol.7 (Anaheim, CA,USA) p 334
[4] Trokel S L 2010 *EuroTimes* **15** is. 7/8
[5] Wheeler M D 1999 *Photonics Spectra* **8** 76–8
[6] Trokel S L, Srinivasan R and Braren B 1983 *American J. of Ophthalmology* **96** 710–5
[7] Marshal J, Trokel S L, Rothery S and Schubert H 1985 *Ophthalmology* **92** 749–58
[8] Munnerlyn C R, Koons S J and Marshall J 1988 *J. Cataract. Refract. Surg.* **14** 46–52
[9] Krauss J M and Puliafito C A 1995 *Lasers in Surgery and Medicine* **17** 102–59
[10] Loertscher H, Mandelbaum S, Parrish R K and Parel J-M 1986 *Am. J. of Ophthalmology* **102** 217–21
[11] Berns M W and Gaster R N 1985 *Lasers in Surgery and Medicine* **5** 371–75
[12] Ren Q, Gailitis R P, Thomson K P and Lin J T 1990 *IEEE J. of Quant. Elect.* **26** 2284–8
[13] Lantukh V V *et al.* 1988 *Lasers in the Life Sciences* **2** 271–84
[14] Razhev A M, Lantukh V V and Pyatin M M 1991 *Journal De Physicue IV* **1** C7-235–7
[15] Razhev A M and Zhupikov A A 1997 *Quantum Electronics* **27** 665–9
[16] Fedorov S N, Lantukh V V, Pyatin M M, Ischenko V N, Kochubei S A, Razhev A M, Rydannykh O V, Tsibizov A V and Chebotaev V P 1990 *Un dispositivo de Correccion de anomalias de refraccion del ojo Patent № P 8902371* (Spain)
[17] Lebedeva L I, Akhmametyeva E M, Razhev A M, Kochubei C A and Ridannich O V 1990 *Radiobiology* **30** 821–6 [in Russian]
[18] Robinson K 2001 *Biophotonics* **5** 54–5
[19] Berlin M S, Martinez M, Papaioannou T, Grundfest W, Goldenberg T and Laudenslager J 1988 *Lasers and Light in Ophthalmology* **2** 17–24
[20] Razhev A M, Bagayev S N, Gelikonov V M, Kargapoltsev E S, Kuranov R V, Turchin I V and Zhupikov A A 2001 *Proc. SPIE Ophthalmic Technologies XI* **4245** 1–6

The scissors oscillation of a quasi two-dimensional Bose gas as a local signature of superfluidity

Camilla De Rossi[1,2], Romain Dubessy[1,2], Karina Merloti[1,2], Mathieu de Goër de Herve[1,2], Thomas Badr[2,1], Aurélien Perrin[2,1], Laurent Longchambon[1,2] and Hélène Perrin[2,1]

[1] Université Paris 13, Sorbonne Paris Cité, Laboratoire de physique des lasers, F-93430, Villetaneuse, France
[2] CNRS, UMR 7538, F-93430, Villetaneuse, France

E-mail: helene.perrin@univ-paris13.fr

Abstract. We test the superfluid character of a two-dimensional Bose gas confined in an anisotropic harmonic trap through the frequency of the scissors mode. We show that a local analysis of this frequency allows to evidence the boundary between a superfluid phase at the centre of the cloud and a thermal phase at the edge. The location of this boundary agrees well with the prediction of the Berezinskii-Kosterlitz-Thouless theory within local density approximation.

Superfluidity is an intriguing property of certain quantum fluids, which is characterized by a few manifestations in its dynamics: absence of viscosity, existence of a critical velocity for the appearance of excitations, vanishing moment of inertia, hydrodynamic behavior including irrotational flow, quantum vortices and collective oscillations. Widely studied in the context of liquid helium, it has been extended to three-dimensional quantum degenerate weakly interacting Bose gases which appear to present a superfluid character as Bose-Einstein condensation (BEC) is reached.

The case of two-dimensional quantum gases is very different in this respect. In homogeneous gases, BEC is absent while a superfluid transition still occurs at low temperature when local phase fluctuations are reduced by vortex-antivortex pairing, as described by Berezinskii, Kosterlitz and Thouless (BKT) [1, 2]. Superfluidity then exists even in the absence of long range order. In trapped Bose gases, BEC is recovered but the BKT superfluidity mechanism still holds and is responsible for the superfluid character of the sample. As the density is inhomogeneous in a trapped gas, the central core is expected to present a superfluid dynamics while the outer region of the sample is still normal. At equilibrium at a given temperature, the gas properties (density, chemical potential) can be defined locally, which is known as the local density approximation (LDA). The criterion for the BKT transition could then be applied locally, giving rise to the existence of a boundary in the sample between a normal component and a superfluid component.

However, as explained above, superfluidity is a dynamical property and should be tested in a dynamics experiment. In particular, the scissors mode, which is a collective mode describing the oscillation of an anisotropic superfluid around one of the trap axes [3], is characteristic of the superfluid behavior of a dilute gas, and has already been used in the past to identify the superfluid character of a three-dimensional Bose gas [4]. In a recent experiment we have used the

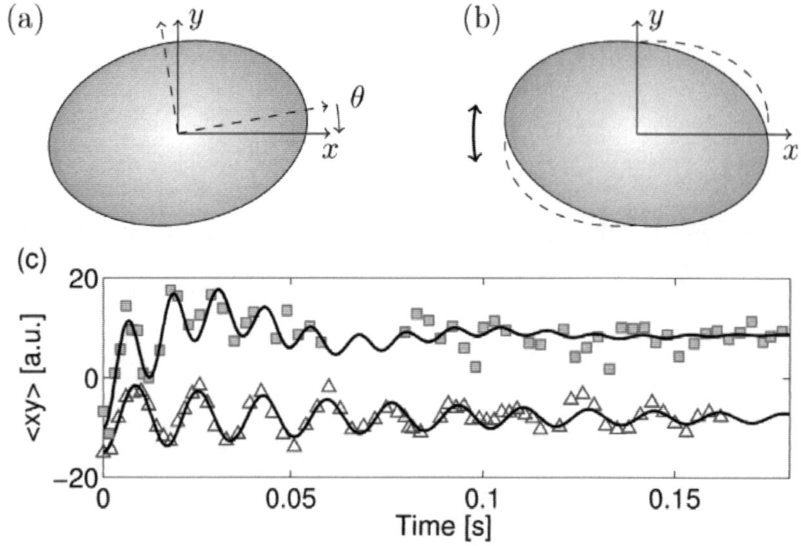

Figure 1. (color online) Excitation of the scissors mode: (a) the trap axes are suddenly rotated by an angle θ and (b) the gas starts oscillating around the new long trap axis. (c) Typical evolution of the observable $\langle xy \rangle$ (average value of xy), which is characteristic of the scissors mode. Red squares: thermal gas, two damped frequencies are present but no superfluid mode; blue triangles: superfluid oscillation of a well defined scissors mode.

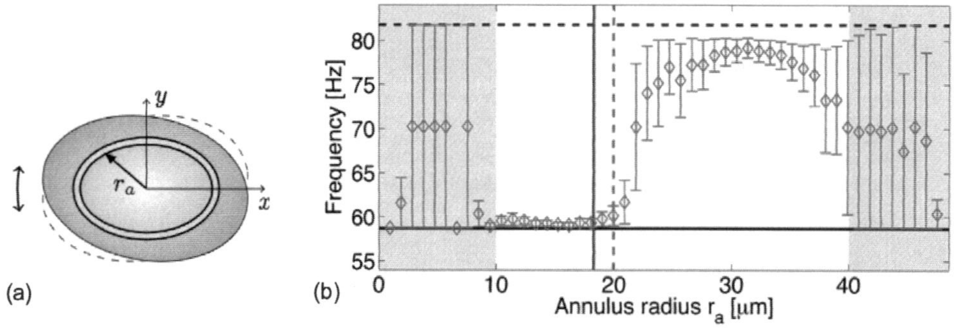

Figure 2. (color online) (a) Region of local average analysis: the average $\langle xy \rangle$ is computed in a thin annulus of radius r_a. (b) Frequency of the scissors mode as a function of the analysis annulus radius r_a. A transition from the superfluid frequency near the center (small r_a) to a normal frequency near the edges (large r_a) is clearly visible. The blue vertical line corresponds to the expected BKT threshold computed for a homogeneous gas within the LDA.

scissors mode to characterize the superfluid nature of a trapped two-dimensional Bose gas, for various values of the temperature and chemical potential at the center [5]. Some of the samples are purely superfluid and evolve at the scissors frequency, others are purely normal and present beat notes between the two trap frequencies (figure 1). On the other hand, we also observe samples which present a clear bimodal behavior. Thanks to a local average analysis of the gas oscillation frequency (figure 2(a)), we are able to isolate the superfluid phase from the normal

IOP Conf. Series: Journal of Physics: Conf. Series **793** (2017) 012023 doi:10.1088/1742-6596/793/1/012023

phase in these samples. It is evidenced by a sudden change in the scissors mode frequency as a function of the distance to its center, located at a well-determined boundary corresponding to the normal to superfluid threshold (see figure 2(b)).

Acknowledgments
We thank Markus Holzmann for very helpful discussions.

References

[1] Berezinskii V L 1971 *J. Exp. Theor. Phys.* **34** 610 URL `http://www.jetp.ac.ru/cgi-bin/e/index/e/34/3/p610?a=list`
[2] Kosterlitz J M and Thouless D J 1973 *J. Phys. C Solid State Phys.* **6** 1181 1203 ISSN 0022-3719 URL `http://iopscience.iop.org/article/10.1088/0022-3719/6/7/010`
[3] Guéry-Odelin D and Stringari S 1999 *Phys. Rev. Lett.* **83** 4452–4455 ISSN 0031-9007 URL `http://link.aps.org/doi/10.1103/PhysRevLett.83.4452`
[4] Maragò O M, Hopkins S A, Arlt J, Hodby E, Hechenblaikner G and Foot C J 2000 *Phys. Rev. Lett.* **84** 2056–2059 ISSN 0031-9007 URL `http://link.aps.org/doi/10.1103/PhysRevLett.84.2056`
[5] De Rossi C, Dubessy R, Merloti K, de Goër de Herve M, Badr T, Perrin A, Longchambon L and Perrin H 2016 *New Journal of Physics* **18** 062001 URL `http://stacks.iop.org/1367-2630/18/i=6/a=062001`

Controlling the interactions between cold Rydberg atoms by a time-varying electric field

I I Ryabtsev[1,2], D B Tretyakov[1,2], V M Entin[1,2], I I Beterov[1,2], E A Yakshina[1,2] and C Andreeva[3]

[1]Rzhanov Institute of Semiconductor Physics SB RAS, 630090 Novosibirsk, Russia
[2]Novosibirsk State University, 630090 Novosibirsk, Russia
[3]Institute of Electronics, Bulgarian Academy of Sciences, 1784 Sofia, Bulgaria

E-mail: ryabtsev@isp.nsc.ru

Abstract. Long-range interactions between cold Rydberg atoms are being investigated for neutral-atom quantum computing, quantum simulations, phase transitions in cold Rydberg gases and other applications. Fine tuning of the interaction strength can be implemented using Förster resonances between Rydberg atoms controlled by an electric field. The observation of the Stark-tuned Förster resonances between Rydberg atoms excited by narrowband cw laser radiation requires usage of a Stark-switching technique in order to excite the atoms first in a fixed electric field and then to induce the interactions in a varied electric field, which is scanned across the Förster resonance. The application of the radio-frequency field causes additional Förster resonances between collective states, whose line shape depends on the interaction strengths and time. Spatial averaging over the atom positions in a single interaction volume yields a cusped line shape of the Förster resonance. We present a detailed experimental and theoretical analysis of the line shape and time dynamics of the Stark-tuned Förster resonances $\mathrm{Rb}(nP_{3/2}) + \mathrm{Rb}(nP_{3/2}) \rightarrow \mathrm{Rb}(nS_{1/2}) + \mathrm{Rb}([n+1]S_{1/2})$ for two Rb Rydberg atoms interacting in a time-varying electric field.

1. Introduction

Long-range interactions between highly excited Rydberg atoms are of interest for quantum information processing with single neutral atoms [1, 2]. Resonant dipole-dipole interaction between atoms in an identical nL Rydberg state can be implemented via Stark-tuned Förster resonances [3-7]. A Rydberg state should be exactly midway between two other Rydberg states of the opposite parity to induce a Förster resonance (figure 1a).

In our recent experiments [2, 7-9] we used three-photon excitation of the $\mathrm{Rb}(nP)$ Rydberg atoms by narrowband cw lasers. Scanning the electric field by just 50 mV/cm shifts the Rydberg level out of resonance with the laser radiation, as the cw lasers have line widths below 1 MHz [7]. In order to avoid this problem, we applied a Stark-switching technique [10, 11] to excite the atoms first in a fixed electric field and then to induce the interactions in a lower electric field, which is scanned across the Förster resonance, as shown in figure 1b.

In this report, we present a detailed experimental and theoretical analysis of the Förster resonance line shapes and time dynamics in a time-varying electric field used for the Stark switching. The resonance under study is the Förster resonant energy transfer $\mathrm{Rb}(nP_{3/2}) + \mathrm{Rb}(nP_{3/2}) \rightarrow \mathrm{Rb}(nS_{1/2}) + \mathrm{Rb}([n+1]S_{1/2})$ due to dipole-dipole interaction of two Rb Rydberg

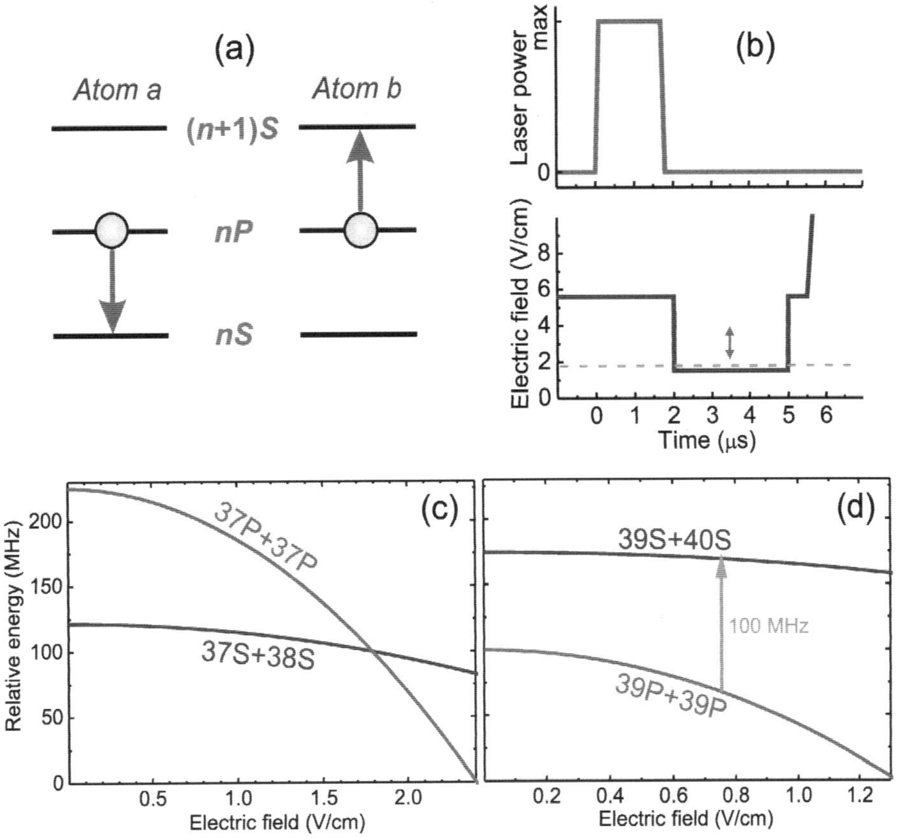

Figure 1. (a) Förster resonance Rb(nP)+Rb(nP)→Rb(nS)+Rb($[n+1]S$) for two Rb Rydberg atoms. (b) Timing diagram of the laser and electric-field pulses. (c) Collective states of the Förster resonance Rb(37P)+Rb(37P)→Rb(37S)+Rb(38S) in the dc electric field. Intersection at 1.79 V/cm corresponds to a Stark-tuned Förster resonance. (d) Collective states of the Förster resonance Rb(39P)+Rb(39P)→Rb(39S)+Rb(40S) in the dc electric field. A dc Stark-tuned Förster resonance is impossible, but it can be induced by adding a ~100-MHz radio-frequency field (green arrow) that binds the collective states [8, 9].

atoms (figure 1a) in a small single laser excitation volume of a frozen Rydberg gas. The energy detuning of this resonance, $\hbar\Delta = E(nS_{1/2}) + E([n+1]S_{1/2}) - 2E(nP_{3/2})$, is controlled by a weak dc electric field F. The energy shift of a Rydberg level nL with nonzero quantum defect is quadratic and is defined by its polarizability, $\delta E_{nL}/\hbar = -\alpha_{nL}F^2/2$. The detuning is then given by

$$\Delta = \Delta_0 + (\alpha_{nP} - \frac{1}{2}\alpha_{nS} - \frac{1}{2}\alpha_{[n+1]S})F^2. \qquad (1)$$

Here, Δ_0 is the detuning in a zero electric field (for example, -103 MHz for n=37 and +74 MHz for n=39). The detuning can be tuned to zero for Rydberg states with $n \leq 38$, as shown in figure 1c for the $37P_{3/2}(|M_J| = 1/2)$ state, while for states with $n \geq 39$, the dc electric field only increases Δ, as shown in figure 1d for the $39P_{3/2}(|M_J| = 1/2)$ state. For the latter, a Förster resonance can be obtained by adding a 90–100 MHz rf field that induces rf transitions between collective states and compensates for the Förster energy defect, as demonstrated in our recent papers [8, 9].

Our previous numerical Monte Carlo simulations using the Schrödinger's equation [5, 12] and our experimental data [5, 8, 9] have shown that spatial averaging over the random positions of $N = 2 - 5$ Rydberg atoms interacting in a single laser excitation volume results in cusp-shaped Förster resonances as soon as the resonances saturate and the interaction time is long enough. Cusp-shaped Förster resonances were also observed recently at high Rb Rydberg atom density in Ref. [13] and an analytical model has been proposed. Three-body Förster resonances in Cs Rydberg atoms were observed in Ref. [14]. The line-shape analysis of the resonant dipole-dipole interaction thus remains a hot topic in the study of cold Rydberg atom samples. The present paper complements our recent paper [9] in the part of comparison between exact numerical and approximate analytical calculations of the time dynamics and line shapes of the Stark-tuned Förster resonances in Rydberg atoms.

2. Time dynamics and line shapes of the Förster resonances

2.1. Theory with density-matrix equations

We consider an example of the Stark-tuned Förster resonance for two Rb Rydberg atoms randomly placed in a single excitation volume. Our Förster resonance induces transitions between Rydberg states with $\Delta M_J = 0$. This corresponds to the z-oriented dipoles, and the operator of the dipole-dipole interaction is

$$\hat{V} = \frac{\hat{d}_a \hat{d}_b}{4\pi\varepsilon_0}\left[\frac{1}{R_{ab}^3} - \frac{3\,Z_{ab}^2}{R_{ab}^5}\right], \tag{2}$$

where $\hat{d}_{a,b}$ are the z components of the dipole-moment operators of the two interacting atoms a and b, Z_{ab} is the z component of the vector connecting the two atoms $\mathbf{R}_{ab}$ (z axis is chosen along the dc electric field), and ε_0 is the dielectric constant.

In order to calculate the time evolution of the populations in the two Rydberg atoms interacting via a Förster resonance, we elaborate a density-matrix model, which takes into account the additional broadening due to unresolved hyperfine structure and fluctuations of the controlling dc electric field. Our considerations are limited to the case of N=2 interacting Rydberg atoms, when approximate analytical solutions can be found and compared with our experimental data.

Let us denote the lower energy state $37S$ as state 1, the middle state $37P_{3/2}$ as state 2, and the upper state $38S$ as state 3, as shown in figure 2a. Then, for two interacting Rydberg atoms with dipole-dipole matrix element V given by equation (2), $|22\rangle$ is the initial collective state populated by a short laser pulse at $t = 0$, and $|13\rangle$ and $|31\rangle$ are the two equally populated final states having a small energy detuning Δ from state $|22\rangle$ (figure 2b). A simpler effective two-level system, shown in figure 2c, with a reduced dipole-dipole matrix element $\sqrt{2}\,V$ can replace the three-level system of figure 2b in the theoretical calculations, because collective states $|13\rangle$ and $|31\rangle$ are equivalent and behave identically. We now denote state $|22\rangle$ as state a in the effective two-level system and the symmetric composite state $(|13\rangle+|31\rangle)/\sqrt{2}$ as state b in figure 2c. The other, antisymmetric composite state $(|13\rangle - |31\rangle)/\sqrt{2}$ is not considered, as it is unaffected by the operator of the dipole-dipole interaction $\hat{V}$ and can be excluded from the analysis.

The most convenient form of equations for the density-matrix elements ρ_{ij} is the optical Bloch equations [15]:

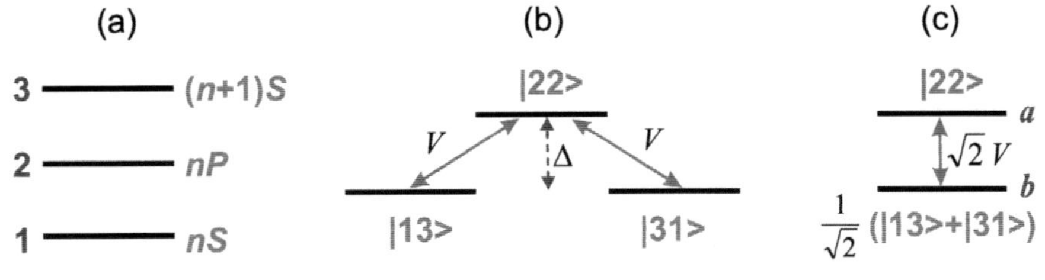

Figure 2. (a) Rydberg states related to the Förster resonance Rb(nP)+Rb(nP)→ Rb(nS)+Rb($[n+1]S$). (b) Collective states of two Rydberg atoms interacting via Förster resonance with dipole-dipole matrix element V. The energy defect of the Förster resonance Δ is controlled by the electric field. (c) Effective two-level system can replace the three-level system in the theoretical calculations, as collective states |13> and |31> are equivalent and behave identically. The reduced dipole-dipole matrix element is now $\sqrt{2}\,V$.

$$\begin{aligned}
\dot{\rho}_{aa} &= i\sqrt{2}\,\Omega\,\left(\rho_{ab}-\rho_{ba}\right),\\
\dot{\rho}_{bb} &= i\sqrt{2}\,\Omega\,\left(\rho_{ba}-\rho_{ab}\right),\\
\dot{\rho}_{ab} &= -(i\Delta+\Gamma/2)\rho_{ab}+i\sqrt{2}\,\Omega\,\left(\rho_{aa}-\rho_{bb}\right),\\
\dot{\rho}_{ba} &= (i\Delta-\Gamma/2)\rho_{ba}+i\sqrt{2}\,\Omega\,\left(\rho_{bb}-\rho_{aa}\right).
\end{aligned} \tag{3}$$

Here, $\Omega = V/\hbar$ is the matrix element of the dipole-dipole interaction in circular frequency units. In order to take into account the additional broadening Γ phenomenologically, equations (3) are modified using the method we have applied previously in Ref.[7] to account for the finite laser linewidths in a four-level theoretical model of the three-photon laser excitation of Rydberg states. In the equations for the coherences, we add the terms with $\Gamma/2$ in order to introduce additional coherence decay.

The two-atom signal S_2 measured in our experiments is a fraction of Rydberg atoms in the final state $37S$ or a population of the final state $37S$ per atom, which is calculated for the interaction time t as

$$S_2 = \frac{1}{2}\rho_{bb}(t). \tag{4}$$

In order to find $\rho_{bb}(t)$, equations (3) can be solved analytically with the initial conditions $\rho_{aa}(0) = 1;\ \rho_{bb}(0) = \rho_{ab}(0) = \rho_{ba}(0) = 0$. Their solution reduces to finding the roots of a cubic equation, if Ω, Δ, and Γ are independent of time [9]. The resulting general analytical expressions are rather complicated, however, and in what follows we consider the analytical solutions only for some particular cases.

2.2. Theory for the Förster resonance amplitude

The exact analytical solution of equations (3) for the time evolution of the Förster resonance amplitude $S_2(\Delta = 0)$ is given by

$$S_2(\Delta=0) = \frac{1}{4} - \frac{1}{4}\mathrm{e}^{-\Gamma t/4}\left[\mathrm{ch}\left(\sqrt{\Gamma^2/16-8\Omega^2 t}\right) + \frac{\Gamma/4}{\sqrt{\Gamma^2/16-8\Omega^2}}\,\mathrm{sh}\left(\sqrt{\Gamma^2/16-8\Omega^2 t}\right)\right]. \tag{5}$$

At the weak dipole-dipole interaction $8\Omega^2 \ll \Gamma^2/16$, equation (5) reduces to

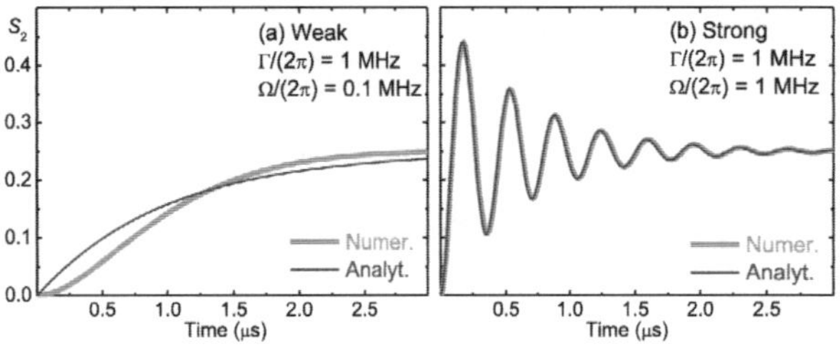

Figure 3. (a) Comparison between approximate (equation (6), blue line) and exact (equation (5), green line) analytical calculations of the time dependence for Förster resonance amplitude at the weak dipole-dipole interaction of two frozen Rydberg atoms. (b) The same for the strong interaction and equations (5) and (7).

$$S_2^{weak}(\Delta = 0) \approx \frac{1}{4}\left[1 - \mathrm{e}^{-16\Omega^2 t/\Gamma}\right]. \tag{6}$$

The amplitude slowly goes to its steady-state value 1/4 as t increases. For the strong dipole-dipole interaction $8\Omega^2 \gg \Gamma^2/16$, the damped Rabi-like oscillations appear, while the steady-state value is also 1/4:

$$S_2^{strong}(\Delta = 0) \approx \frac{1}{4}\left[1 - \mathrm{e}^{-\Gamma t/4}\cos\left(2\sqrt{2}\,\Omega t\right)\right]. \tag{7}$$

Figure 3a demonstrates the comparison between the approximate (equation (6), blue line) and exact (equation (5), green line) analytical calculations of the time dependence for the Förster resonance amplitude at the weak dipole-dipole interaction of two spatially frozen Rydberg atoms. Although there is some discrepancy at the short interaction time due to unaccounted coherence in equation (6), the overall agreement is rather good. Figure 3b shows the same for the strong interaction and demonstrates almost perfect agreement between equations (5) and (7). The calculated time dependences in figure 3 is what can be observed with two Rydberg atoms, interacting in two spatially separated optical dipole traps, as in the experiment of Ref. [16].

In our experiments, the two atom positions are not fixed. The interaction term Ω in equations (6) and (7) has a fluctuating value that results in decoherence and washing out the Rabi-like oscillations in equation (7), even at $\Gamma = 0$. To calculate the resonance amplitude measured in our experiments, equations (6) and (7) should be averaged over all possible spatial positions of the two interacting atoms in the excitation volume, which is formed by the two intersecting laser beams. In Ref. [12] we have shown that the averaging can be done using the nearest-neighbor probability distribution [17] with the average distance between nearest-neighbor atoms $R_0 \approx [3/(4\pi n_0)]^{1/3}$ at volume density n_0. Using this method, we find the approximate analytical solutions to the averaged amplitudes of equations (6) and (7) [9]:

$$< S_2^{weak}(\Delta = 0) > \approx \frac{1}{4}\left(1 - \mathrm{e}^{-\left[0.44\Omega_0^2\, t/\Gamma\right]^{1/3}}\right), \tag{8}$$

$$< S_2^{strong}(\Delta = 0) > \approx \frac{1}{4}\left(1 - \mathrm{e}^{-0.55\Omega_0\, t - \Gamma t/4}\right), \tag{9}$$

where

$$\Omega_0 = \frac{\sqrt{2} d_1 d_2}{4\pi\varepsilon_0 \hbar R_0^3}$$

is the orientation-averaged reduced interaction matrix element at the average distance R_0. Here, d_1 and d_2 are the dipole moments of transitions $nP_{3/2}\,(\mathrm{M}_J = 1/2) \to nS_{1/2}\,(\mathrm{M}_J = 1/2)$ and

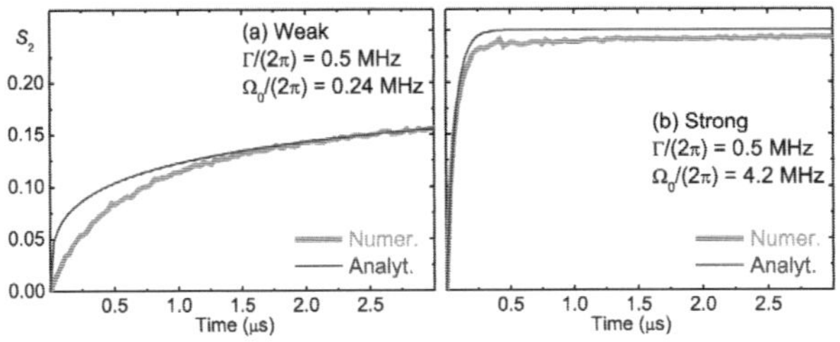

Figure 4. (a) Comparison between analytical (equation (8), blue line) and numerical (green line) calculations of the time dependence for Förster resonance amplitude at the weak dipole-dipole interaction of two spatially averaged Rydberg atoms. (b) The same for the strong interaction and equation (9).

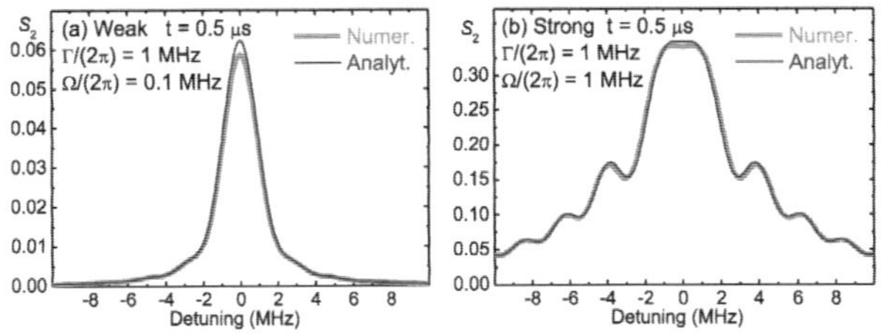

Figure 5. (a) Comparison between analytical (equation (10), blue line) and numerical (green line) calculations of the Förster resonance line shape at the weak dipole-dipole interaction of two frozen Rydberg atoms. (b) The same for the strong interaction and equation (13).

$nP_{3/2}\,(\mathrm{M}_J = 1/2) \to (n+1)S_{1/2}\,(\mathrm{M}_J = 1/2)$. The fitting coefficients 0.44 and 0.55 have been found from the numerical simulations.

Figure 4a demonstrates the comparison between the analytical (equation (8), blue line) and numerical (green line) calculations of the time dependence for the Förster resonance amplitude at the weak dipole-dipole interaction of two spatially averaged Rydberg atoms. Again, there is some discrepancy at the short interaction time, but at the longer times the agreement is good. Figure 4b shows the same for the strong interaction and demonstrates good agreement between the analytical calculations with equation (9) and numerical results. We note that the time dependences of the averaged amplitudes in equations (8) and (9) significantly differ from those in equations (6) and (7) for two spatially fixed atoms. This should be taken into account when theory is compared with experiment.

2.3. Theory for the Förster resonance line shape

The analytical expressions for the line shape of the two-atom Förster resonances at $\Delta \neq 0$ are much more complicated than equations (5)–(7). In the case when $\rho_{bb} \ll \rho_{aa}$ (weak interaction or short interaction time), the line shape for two frozen Rydberg atoms is approximately given by

$$S_2^{weak} \approx \frac{2\Omega^2}{\Delta^2+\Gamma^2/4}\left(\frac{\Gamma t}{2} + \frac{\Delta^2-\Gamma^2/4}{\Delta^2+\Gamma^2/4}\left[1-\mathrm{e}^{-\Gamma t/2}\cos\left(\Delta t\right)\right] - \frac{\Delta\Gamma}{\Delta^2+\Gamma^2/4}\mathrm{e}^{-\Gamma t/2}\sin\left(\Delta t\right)\right) \quad (10)$$

Our numerical simulations have shown that this formula is quite precise (error less than 10%) for arbitrary Ω, Γ, Δ, and t as far as $S_2 < 0.1$. In fact, this is a general formula for the line shape of weak Förster resonances in two frozen Rydberg atoms. Figure 5a presents the comparison between analytical (equation (10), blue line) and numerical (green line) calculations of the Förster resonance line shape at the weak dipole-dipole interaction of two frozen Rydberg atoms and the interaction time $t \sim 0.5$ μs.

In the opposite case of strong interaction $8\Omega^2 \gg \Gamma^2/16$, the line shape is approximately given by

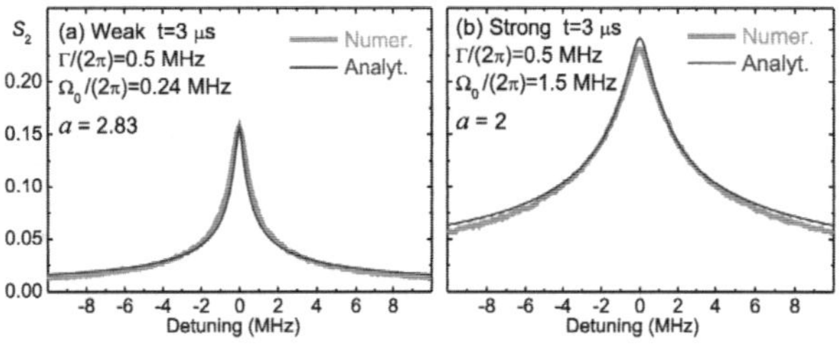

Figure 6. (a) Comparison between analytical (equation (12), blue line) and numerical (green line) calculations of the Förster resonance line shape at the weak dipole-dipole interaction of two spatially averaged Rydberg atoms. (b) The same for the strong interaction.

$$S_2^{strong} \approx \frac{1}{4} - \frac{\Delta^2/4}{8\Omega^2+\Delta^2} e^{-\frac{4\Omega^2}{8\Omega^2+\Delta^2}\Gamma t} - \frac{2\Omega^2}{8\Omega^2+\Delta^2} e^{-\frac{4\Omega^2+\Delta^2}{8\Omega^2+\Delta^2}\Gamma t/2} \cos\left(\sqrt{8\Omega^2+\Delta^2}\, t\right). \quad (11)$$

This formula describes the saturation of the Förster resonance accompanied by the damped Rabi-like oscillations. As t or Ω grow, there appears a flat-top contour with the width of the flat-top part $4\sqrt{2}\Omega$, while the resonance wings are close to a Lorentzian. At $t \to \infty$, equation (11) gives $S_2^{strong} \to 1/4$ independently of Δ. This is because the steady-state solutions $\rho_{aa} = \rho_{bb} = 1/2$; $\rho_{ab} = \rho_{ba} = 0$ of equations (3) are independent of Δ if $\Gamma \neq 0$. For large Δ, however, the time evolution is very slow, as described by the second term in equation (11). Figure 5b presents the comparison between analytical (equation (11), blue line) and numerical (green line) calculations of the Förster resonance line shape at the strong dipole-dipole interaction of two frozen Rydberg atoms and the interaction time $t \sim 0.5$ μs.

The position-averaged line shape of the Förster resonance for two atoms in a single interaction volume can be obtained analytically only for equation (11), because it correctly describes the saturation of resonances, otherwise the integral in the averaging diverges at short interaction distances. In our averaging here, we use not the Lorentz line shape but equation (11) with $\Gamma \neq 0$, as discussed in Ref. [9]. The averaged line shape turns out to be a cusp, which is approximately described by the following formula:

$$\left\langle S_2^{strong} \right\rangle \approx \frac{1}{4}\left[1 - \exp\left(-\left\{\frac{0.44\,\Omega_0^2 \Gamma t}{a^2\Delta^2+\Gamma^2}\right\}^{1/3}\right)\right]. \quad (12)$$

The fitting coefficients 0.44 and a have been obtained by comparing equation (12) with the numerical simulations of equations (3) averaged over a cubic interaction volume for the Förster resonance Rb(37P)+Rb(37P)→Rb(37S)+Rb(38S). At $\Gamma/(2\pi) \sim 1$ MHz, we have found that $a \approx 2$ for $\Omega_0/(2\pi) > 1$ MHz and $t > 10$ μs; $a \approx 2\sqrt{2}$ for $\Omega_0/(2\pi) \sim 0.1 - 0.5$ MHz and $t \sim 1 - 5$ μs; and $a \approx 4$ for $\Omega_0/(2\pi) < 0.1$ MHz and $t \sim 1 - 5$ μs. In all intermediate cases, a should be varied to fit experimental or numerical data. Figure 6 presents the comparison between analytical (equation (12), blue lines) and numerical (green lines) calculations of the Förster resonance line shape at the weak (a) and strong (b) dipole-dipole interactions of two spatially averaged Rydberg atoms and the interaction time $t \sim 3$ μs. It demonstrates almost perfect agreement between the analytical and numerical calculations.

Equation (12) is thus a general formula that describes the cusped line shape of the Förster resonance for two Rydberg atoms randomly positioned in a single interaction volume. It is valid for a broad range of all parameters, and in addition it describes the time dynamics of the Förster resonance for the interaction times $t > 1$ μs.

The formation of the cusp-shaped resonance upon spatial averaging can be understood from the fact that at zero detuning, the interaction is long range (resonant dipole-dipole) and it is effective even for the distant atoms, while for nonzero detuning the interaction is short range (van der Waals) and it is much weaker for the distant atoms [8]. Therefore, upon spatial averaging, the resonance wings are reduced stronger than the resonance center, and this finally results in the cusped line shape.

The only drawback of equation (12) is that it is invalid for the short interaction times ($t < 1$ μs), when the Fourier broadening dominates. For the analytical estimates, the averaged Fourier broadening is described with a certain accuracy by equations (10) or (11) if $2\Omega^2$ is replaced by Ω_0^2.

2.4. Experimental Förster resonance line shape

The experiments were performed with cold ^{85}Rb atoms in a magneto-optical trap (MOT). Our experiments feature atom-number-resolved measurement of the signals obtained from $N = 1-5$ detected Rydberg atoms with a detection efficiency of 65% [5, 18]. It is based on a selective field ionization (SFI) detector with channel electron multiplier (CEM) and post-selection technique. We can record Förster resonances for up to five of the detected Rydberg atoms and compare these with the theoretical simulations.

The excitation of Rb atoms to the $nP_{3/2}(|M_J|{=}1/2)$ Rydberg states is realized via three-photon transition $5S_{1/2} \to 5P_{3/2} \to 6S_{1/2} \to nP_{3/2}$ (figure 2a) by means of three cw lasers modulated to form 2 μs exciting pulses at a repetition rate of 5 kHz [7]. Small Rydberg excitation volume of 20–40 μm in size is formed using crossed laser-beam geometry [5]. The laser intensities are adjusted to obtain about one Rydberg atom excited per laser pulse on average.

We use a Stark-switching technique [10, 11] to switch the Rydberg interactions on and off, as depicted in figure 1b. Laser excitation occurs during 2 μs at a fixed electric field of 5.6 V/cm. Then the field decreases to a lower value near the resonant electric field (1.79 V/cm for the $37P_{3/2}$ state), which acts for 3 μs or less until the field increases back to 5.6 V/cm. Then, 0.5 μs later, a ramp of the strong field-ionizing electric pulse of 200 V/cm is applied. The lower electric field is slowly scanned across the Förster resonance and the SFI signals are accumulated for $10^3 - 10^4$ laser pulses. A pulse of the rf field with variable amplitude (0 – 0.5 V/cm) and frequency (10 – 100 MHz) can be admixed to the lower dc field.

Figure 7 presents the line shapes of the Förster resonance Rb($37P$)+Rb($37P$)$\to$ Rb($37S$)+Rb($38S$) recorded in our experiment for N=2 detected Rydberg atoms at various t. The interaction time t is set by the length of the square-shaped controlling electric-field pulse.

At the long interaction times (1.975 and 2.975 μs), the resonance shape and width in figures 7g and 7h almost do not change with t . This means that the resonance takes its stationary form, and only its amplitude slowly grows with t according to equation (8). The resonance is cusp shaped, in agreement with equation (12). The narrowest resonance appears at t=2.975 μs. Its width in the electric-field scale is 16 mV/cm, corresponding to 1.9 MHz in the detuning scale. The detuning scale is obtained using the calculated polarizabilities of the collective Rydberg states in figure 1c. Our attempts to obtain a narrower resonance by increasing t were not successful. Since the Fourier width of the interaction pulse is $t^{-1} \approx$ 0.34 MHz and the estimated average interaction energy is $\Omega_0/(2\pi) \sim 0.25$ MHz, they cannot be fully responsible for the 1.9 MHz resonance width. Therefore, there is an additional broadening $\Gamma/(2\pi) \sim 1$ MHz, apparently due to the unresolved hyperfine structure and parasitic ac electric fields. This observation agrees with what we have observed in our previous experiments [5, 8].

The Fourier broadening of the resonances is demonstrated in figures 7a–7e at the shorter interaction times (< 1 μs). The Fourier width of the interaction pulse is significant at short t and the Förster resonance broadens, while its amplitude decreases.

In figures 7a–7h, we also compare the experimental two-atom spectra at various t with the line shapes numerically calculated with equations (3) and averaged over the random positions of two atoms in a cubic interaction volume. The fitting parameters in the theory were the volume size

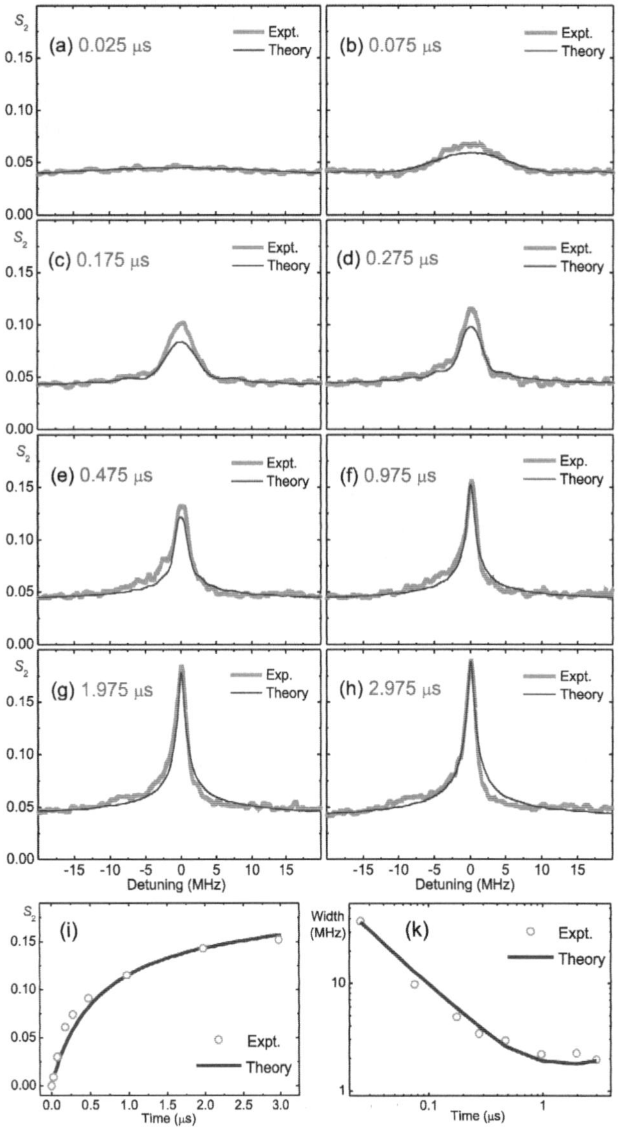

Figure 7. (a)-(h) Comparison between experimental and theoretical line shapes of the Förster resonance Rb($37P$) + Rb($37P$) → Rb($37S$) + Rb($38S$) for two Rydberg atoms at various interaction times. (i) Comparison between experiment and theory for the resonance amplitude at various interaction times. (k) Comparison between experiment and theory for the resonance width at various interaction times.

$26\times26\times26$ μm^3 and the additional broadening $\Gamma/(2\pi) = 0.5$ MHz. We also added a background signal $S_2 \approx 0.037$ that appears in the experiments due to parasitic transitions between Rydberg states induced by a 300 K blackbody radiation (BBR). These parameters allowed us to fit the time dependences of the experimental spectra, both for the amplitude (figure 7i) and width (figure 7k) of the two-atom Förster resonance. The amplitude and width are measured with respect to the BBR-induced background signal level.

The time dependence of the amplitude in figure 7i for $t > 1$ μs is also well fit by equation (8) at $\Omega_0/(2\pi) = 0.25$ MHz and $\Gamma/(2\pi) = 0.5$ MHz. The time dependence of the width in figure 7k for $t < 1$ μs is mainly represented by the Fourier transform width $FWHM/(2\pi) \approx 1/t$. For $t > 1\mu$s, the width becomes nearly constant.

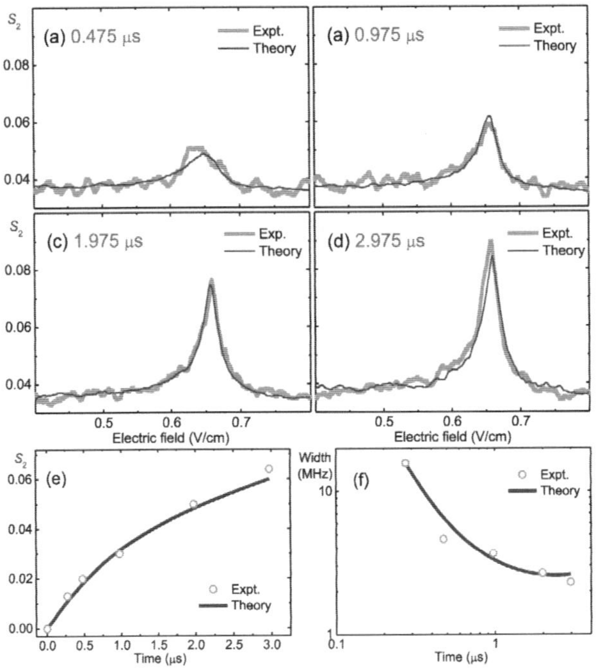

Figure 8. (a)-(d) Comparison between experimental and theoretical line shapes of the inaccessible Förster resonance Rb(39P) + Rb(39P) → Rb(39S) + Rb(40S) induced by the 95 MHz and 100 mV rf field for two Rydberg atoms at various interaction times. (e) Comparison between experiment and theory for the resonance amplitude at various interaction times. (f) Comparison between experiment and theory for the resonance width at various interaction times.

Now we will turn to the "inaccessible" Förster resonance Rb($39P_{3/2}$) + Rb($39P_{3/2}$) → Rb($39S_{1/2}$) + Rb($40S_{1/2}$) which cannot be tuned by the dc electric field. Its collective energy levels in the dc electric field are shown in figure 1d. The dc field alone increases the energy detuning Δ and makes the interaction between Rb(39P) atoms less efficient. However, as we have shown in Refs. [8, 9], the rf field can induce transitions between collective states, so the population transfer at the Förster resonance occurs for the Floquet sidebands irrespective of the possibility to tune it by the dc field. In the present experiment, we applied rf field with 95 MHz frequency and 100 mV amplitude, which created the Floquet sidebands.

In figures 8a–8d, we compare the experimental rf-induced Förster resonances recorded for two atoms at various t to the theoretical line shapes numerically calculated with equations (3) and averaged over random positions of two atoms in a cubic interaction volume. A BBR-induced background signal $S_2 \approx 0.03$ is added to the theoretical spectra. The fitting parameters in the theory were the volume size 16×16×16 μm^3, the additional broadening $\Gamma/(2\pi) = 1$ MHz, and the rf-field strength 150 mV/cm (the latter is 50% larger than the experimentally measured one; this discrepancy remains unclear). These allowed us to fit the time dependences of the experimental spectra both for the amplitude (figure 8e) and width (figure 8f) of the two-atom Förster resonance. The amplitude and width are measured with respect to the BBR-induced background-signal level.

The narrowest resonance appears at t=2.975 μs. It has a cusped shape, in agreement with equation (12), and its width in the electric-field scale is 18 mV/cm, corresponding to 1.1 MHz in the detuning scale, so this resonance is nearly two times narrower than that for the $37P_{3/2}$ state. The detuning scale is obtained using the calculated polarizabilities of the collective Rydberg states in figure 1d. The main contribution to Γ thus comes from the parasitic ac electric fields. That is why in the theory we have to take $\Gamma/(2\pi) = 1$ MHz (i.e., two times larger than that for the $37P_{3/2}$ state), and a smaller interaction volume to fit the amplitude of the two-atom resonances in figure 8.

3. Conclusions

Our simple theoretical density-matrix model works well and provides correct analytical and numerical results for the time dynamics and cusped line shapes of the Förster resonances in two randomly positioned Rydberg atoms interacting in a single laser-excitation volume.

Good agreement between experiment and theory confirms that any unresolved (hyperfine or Zeeman) structure or parasitic ac Stark broadening of the Förster resonances can be accounted for theoretically by a single parameter Γ. This significantly reduces the number of equations and simplifies the calculations of the line shapes and time dynamics of the Förster resonances.

Acknowledgments

This work was supported by the RFBR Grants No. 14-02-00680 and No. 16-02-00383, the Russian Science Foundation Grant No. 16-12-00028 (for laser excitation of Rydberg states), the Siberian Branch of the Russian Academy of Sciences, the EU FP7-PEOPLE-2009-IRSES Project COLIMA, and the Novosibirsk State University.

References

[1] Saffman M, Walker T G and Mølmer K 2010 *Rev. Mod. Phys.* **82** 2313
[2] Ryabtsev I I, Beterov I I, Tretyakov D B, Entin V M and Yakshina E A 2016 *Physics – Uspekhi* **59** 196
[3] Safinya K A, Delpech J F, Gounand F, Sandner W and Gallagher T F 1981 *Phys. Rev. Lett.* **47** 405
[4] Tretyakov D B, Beterov I I, Entin V M, Ryabtsev I I and Chapovsky P L 2009 *J. Exper. Theor. Phys.* **108** 374
[5] Ryabtsev I I, Tretyakov D B, Beterov I I and Entin V M 2010 *Phys. Rev. Lett.* **104** 073003
[6] Tretyakov D B, Beterov I I, Entin V M, Ryabtsev I I, Bezuglov N N and Arimondo E 2012 *J. Exper. Theor. Phys.* **114** 14
[7] Entin V M, Yakshina E A, Tretyakov D B, Beterov I I and Ryabtsev I I 2013 *J. Exper. Theor. Phys.* **116** 721
[8] Tretyakov D B, Entin V M, Yakshina E A, Beterov I I, Andreeva C and Ryabtsev I I 2014 *Phys. Rev.* A **90** 041403(R)
[9] Yakshina E A, Tretyakov D B, Beterov I I, Entin V M, Andreeva C, Cinins A, Markovski A, Iftikhar Z, Ekers A and Ryabtsev I I 2016 *Phys. Rev.* A **94** 043417
[10] Ryabtsev I I and Beterov I M 2000 *Phys. Rev.* A **61** 063414
[11] Ryabtsev I I, Tretyakov D B and Beterov I I 2003 *J. Phys.* B **36** 297
[12] Ryabtsev I I, Tretyakov D B, Beterov I I, Entin V M and Yakshina E A 2010 *Phys. Rev.* A **82** 053409
[13] Richards B G and Jones R R 2016 *Phys. Rev.* A **93** 042505
[14] Faoro R, Pelle B, Zuliani A, Cheinet P, Arimondo E and Pillet P 2015 *Nature Comm.* **6** 8173
[15] Stenholm S 2005 *Foundations of Laser Spectroscopy* (Dover, NY: Mineoloa).
[16] Ravets S, Labuhn H, Barredo D, Beguin L, Lahaye T and Browaeys A 2014 *Nature Physics* **10** 914
[17] Chandrasekhar S 1943 *Rev. Mod. Phys.* **15** 1
[18] Ryabtsev I I, Tretyakov D B, Beterov I I and Entin V M 2007 *Phys. Rev.* A **76** 012722; Erratum: 2007 Phys. Rev. A **76** 049902(E)

IOP Conf. Series: Journal of Physics: Conf. Series **793** (2017) 012025 doi:10.1088/1742-6596/793/1/012025

Laboratory simulation of energetic flows of magnetospheric planetary plasma

I F Shaikhislamov, V G Posukh, A V Melekhov, E L Boyarintsev, Yu P Zakharov, P A Prokopov and A G Ponomarenko

Dep. of Laser Plasma, Institute of Laser Physics SB RAS, 13/3 Acad. Lavrentyev ave., Novosibirsk 630090, Russia

E-mail: ildars@ngs.ru

Abstract. Dynamic interaction of super-sonic counter-streaming plasmas moving in dipole magnetic dipole is studied in laboratory experiment. First, a quasi-stationary flow is produced by plasma gun which forms a magnetosphere around the magnetic dipole. Second, explosive plasma expanding from inner dipole region outward is launch by laser beams focused at the surface of the dipole cover. Laser plasma is energetic enough to disrupt magnetic field and to sweep through the background plasma for large distances. Probe measurements showed that far from the initially formed magnetosphere laser plasma carries within itself a magnetic field of the same direction but order of magnitude larger in value than the vacuum dipole field at considered distances. Because no compression of magnetic field at the front of laser plasma was observed, the realized interaction is different from previous experiments and theoretical models of laser plasma expansion into uniform magnetized background. It was deduced based on the obtained data that laser plasma while expanding through inner magnetosphere picks up a magnetized shell formed by background plasma and carries it for large distances beyond previously existing magnetosphere.

1. Introduction

Laboratory modelling of space plasma processes is an important method of study of basic physics. Despite of significant progress in spacecraft measurements and numerical simulations a laboratory experiment remains a source of unique data inaccessible by other means. One of the fields where namely laboratory experiments with lasers have pushed the advances in theory and numerical simulation is interaction of counter-streaming plasma flows in presence of magnetic field. In 1970-s and 80-s a number of works with laser-produced plasma expanding with super-Alfvenic velocity into magnetized background have been carried out with the aim to model active near-Earth releases AMPTE, CRRES, Argus, Starfish. The other field which was extensively studied by means of laboratory experiment is magnetosphere. At KI-1 simulation Facility such studies are based on two sources of plasma – induction Θ-pinch and Laser Plasma (LP) – which interact with compact magnetic dipoles. Combination of energetically and spatially different plasma flows allowed modelling of extreme compression of the Earth's magnetosphere by super powerful CME or by artificial near-Earth releases [1]. Such complex systems as field-aligned currents connecting boundary layer with ionosphere have been studied in detail [2]. Namely laboratory experiments supplied necessary data to formulate and verify a Hall model of mini-magnetospheres [3] which explains its unusual features.In the present experiment we investigate essentially new combination of interacting flows and magnetic field. Θ-pinch plasma fills the vacuum chamber and creates around magnetic dipole a magnetosphere with estimated size of about 30 cm. The novel feature is that laser plasma is generated inside of this magnetosphere at two targets symmetrically placed at dipole cover (figure 1). LP is directed opposite to the Θ-pinch flow and has kinetic energy large enough to sweep previously existing plasma and dipole magnetic field. We study the interaction at distances 40÷90 cm from the dipole beyond the previously existing magnetosphere. The specific case of plasma expansion from the inner region of magnetic dipole outward into the background flow has at least two possible applications. It directly relates to the concept of magnetosail. The other field is Hot Jupiters – close orbiting exoplanets heated by ionizing stellar radiation to a point of super-sonic expansion of upper atmosphere. The interaction of expanding planetary flow with counter-streaming stellar plasma in a case when such a planet possesses weak magnetic field [4, 5] was one of motivations of the present experiment.

2. Experimental set up and results

Before the laser plasma is produced on the surface of the dipole, the flow of background plasma creates around it a magnetosphere with size of about 30 cm. Thus, LP, when created, expands at first in the dipole magnetic field of inner magnetosphere filled from about X=15 to X=30 cm by background plasma and after that across magnetopause into the background plasma proper where magnetic field is zero. Such scenario principally differs from the previous experiments on LP expansion in magnetized uniform background in that the magnetized background in the present case is a compact localized shell. The energy density of LP (total energy about 16 J in initial volume of about 8 cm^3) is comparable to that of dipole magnetic field already at the laser target and with further LP expansion quickly becomes dominant. It was found that the LP flowing through background plasma partially expels it due to Coulomb collisions. However, collisional interaction is rather weak and counter-streaming flows deeply penetrate into each other, so LP expansion isn't significantly affected at distances up to 100 cm. The main dimension and dimensionless parameters are listed in the table 1 while figure 1 shows experimental set up.

Figure 1. Experimental set-up superimposed on snapshot of laser-produced plasma.

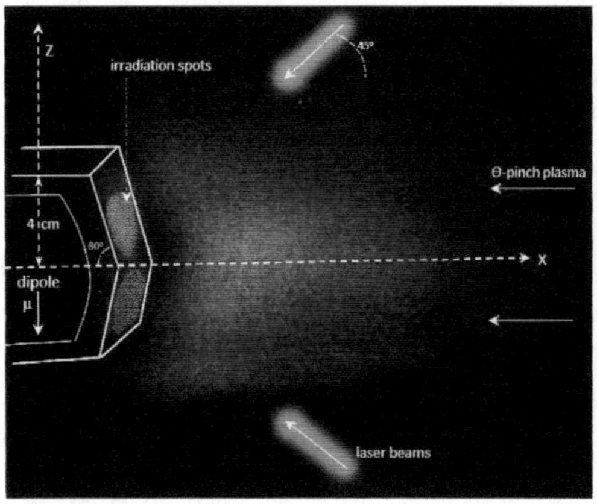

Table 1. Parameters of experiment.

Background plasma parameters		
Velocity	V_*, km/s	20-60
Density[1]	n_*, cm^{-3}	$2\cdot10^{12}$
Mach number	V_*/C_{S*}	~1
Magnetospheric parameters		
Magnetosphere size[2]	L_m, cm	30
Relative dipole radius	R_D/L_m	0.1
Knudsen number λ_*/L_m		~2
Reynolds number $4\pi\sigma L_m V_*/c^2$		~100
Hall parameter	$L_m\omega_{pi}/c$	2
Degree of ion magnetization[3]	L_m/R_{L*}	3
Laser plasma parameters		
LP energy to magnetic energy[4]	Q_{LP}/Q_d	~1
Velocity relative to background	V_o/V_*	≈ 3
Mean free path[5] λ_{i*}, cm		≈ 500

[1] Evaluated close to the dipole location at X=0.
[2] Calculated as pressure balance distance between Θ-pinch plasma and dipole field pressures.
[3] Ion gyroradius is calculated for the value of magnetic field of B_m=50 G.
[4] Q_{LP} is calculated as total energy of a single LP blob (16 J) divided by initial LP volume (~8 cm^3). Q_d is calculated as dipole field energy density at a distance of X=3.5 cm where the LP target is positioned.
[5] Coulomb scattering length of LP test ion in the Θ-pinch plasma at a density of 4 10^{12} cm^{-3}.

The main finding of the experiment is that when background plasma pre-fills the vacuum chamber LP carries with itself across distances 40÷90 cm the magnetic field order of magnitude larger than in case without ambient plasma. This field is also much larger than the vacuum dipole field at these distances. Typical measurements are demonstrated in figure 2. Spatial profile at the left panel shows that magnetic field carried by laser plasma decreases with distance much more weakly than the cubic fall off of the laser plasma density. Obtained results impose a question of origin of such magnetic field carried by LP. They suggest a novel and unexpected feature that LP captures magnetospheric field rather than simply stretching it and that effectiveness of this capture is directly related to the density of background plasma which creates magnetosphere.

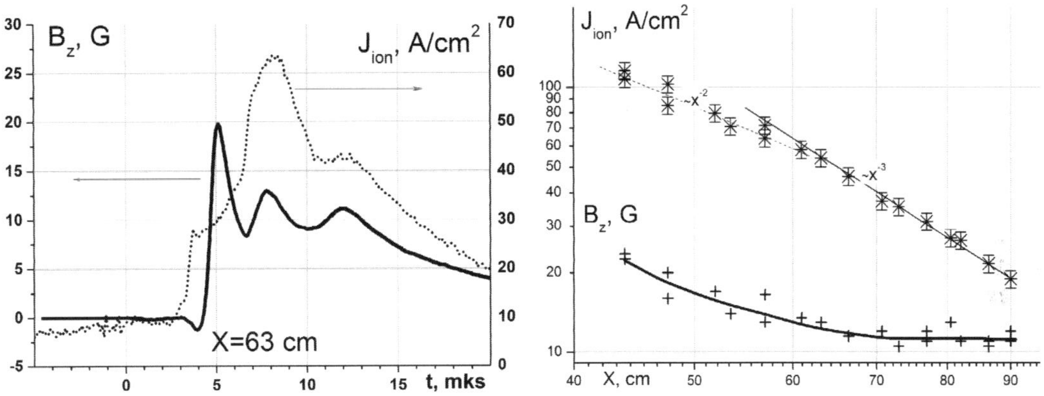

Figure 2. Right - typical dynamic signals of ion current and magnetic field measured by probes at a distance of 63 cm. Left - spatial profiles of ion current and magnetic field measured along the interaction axis. For the ion current amplitude the maximum of the second LP flow is taken. For the magnetic field the averaged value carried by the second and the third LP flows is calculated. Straight lines plotted over ion current data indicate square and cubic behavior.

In previous experiments with uniform background the compression and strong increase of magnetic field at the LP front and its total expulsion inside of LP proper has been observed [6]. Such behaviour is explained in the frame of displaced electrons model [7]. When the electron density of expanding plasma significantly exceeds that of background, which is true in our case at distances X<30 cm where magnetized shell exists, the electrons of background together with frozen-in magnetic field are displaced and strongly compressed at the front of LP. However, in present experiment no such compression is seen. Measured fields don't exceed the expected values of >50 G in the initial magnetized shell. Moreover, observed magnetic field is present in the whole LP flow.

On the base of the obtained experimental results a following conclusion was made, demonstrated in figure 3. Plasma expanding outward from the inner region of magnetic dipole can interact with it by catching and dragging the magnetic field lines. The effectiveness of such process of transfer of magnetic field far from the dipole is directly related to the density of background plasma prefilling the magnetic field lines close to the dipole. Without pre-made plasma magnetized into dipole field lines the impulsive energetic plasma doesn't carry any significant field after crossing the dipole region. There can be two reasons why the LP catches and carries within itself the magnetized shell formed by background plasma. First, the dipole field lines loaded with plasma can't move faster than with the Alfven speed, and sufficiently fast impulsive flow can overcome the magnetized shell instead of displacing it. Second, the curvature of dipole field lines makes it possible for electrons of LP to mix with electrons of magnetized shell. Only by such mixing the LP might pick up the magnetic field instead of displacing it. The last feature is a main difference of the present work from previous studies of LP interaction with uniform magnetized background.

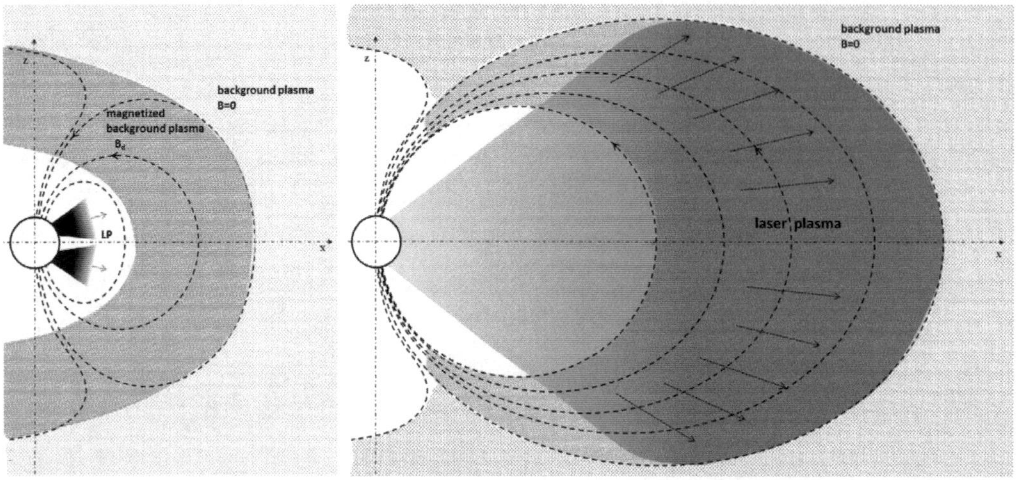

Figure 3. Schematic representation of the experimental results. Left picture – mini-magnetosphere including magnetized shell formed by Θ-pinch plasma around magnetic dipole. Right – laser plasma expanding flow which picks up and carries the magnetized shell. Dashed lines indicate magnetic field lines.

Acknowledgments
This work was supported by SB RAS Research Program (project II.10.1.4 N 01201374303), Presidium RAS program on fundamentals of double technologies and by the Russian Foundation for Basic Research (grant nos. 14-29-06036, 16-52-14006).

References

[1] Ponomarenko A G, Zakharov Yu P, Antonov V M, Boyarintsev E L, Melekhov A V, Posukh V G, Shaikhislamov I F, Vchivkov K V 2008 *Plasma Phys.Control. Fusion* **50** 074015
[2] Shaikhislamov I F, Antonov V M, Zakharov Yu P, Boyarintsev E L, Melekhov A V, Posukh V G, Ponomarenko A G 2009 *Plasma Phys. Control. Fusion* **51** 105005
[3] Shaikhislamov I F, Antonov V M, Zakharov Yu P,Boyarintsev E L, Melekhov A V, Posukh V G, Ponomarenko A G 2013 *Adv. Space Res.* **52** 422
[4] Antonov V M, Boyarintsev E L, Boyko A A, Zakharov Yu P, Melekhov A V, Ponomarenko A G, Posukh V G, Shaikhislamov I F, Khodachenko M L, Lammer H 2013 *The Astrophysical Journal* **769** 28
[5] Khodachenko M L, Shaikhislamov I F, Lammer H, Prokopov P A 2015 *The Astrophysical Journal* **813** 50
[6] Shaikhislamov I F, Zakharov Yu P, Posukh V G, Melekhov A V, Boyarintsev E L, Ponomarenko A G, Terekhin V A 2015 *Plasma Physics Reports* **41** 399
[7] Wright T P 1971 *Phys. Fluids* **14** 1905

IOP Conf. Series: Journal of Physics: Conf. Series **793** (2017) 012026 doi:10.1088/1742-6596/793/1/012026

Light localization in cold and dense atomic ensemble

Igor Sokolov

Peter the Great St.Petersburg Polytechnic University, Polytechnicheskaya st. 29, Saint-Petersburg, 194251, Russia

E-mail: ims@is12093.spb.edu

Abstract. We report on results of theoretical analysis of possibilities of light strong (Anderson) localization in a cold atomic ensemble. We predict appearance of localization in dense atomic systems in strong magnetic field. We prove that in absence of the field the light localization is impossible.

The analysis of possibilities of light strong (Anderson) localization in a cold atomic ensemble is based on the consistent quantum-posed theoretical approach developed previously in [1]. In the frame of this approach we solve nonstationary Schrodinger equation for the wave function of the joint system consisting of N motionless atoms and the weak electromagnetic field. We take into account vector nature of the field and do not introduce any model of continuous media. Restriction of the total number of states taken into account by the states with no more than one photon in the field allows us to obtain finite set of equation for Fourier component of amplitudes of states with one excited atom. This set of equation is solved numerically and amplitudes of the other states are calculated through found ones. The procedure gives us opportunity to find approximately the wave function of the system and consequently analyze both the properties of atomic system and the light. This approach was successful used in various applications [2]-[4].

To analyze the possibilities of strong localization at first we studied the fulfillment of Ioffe-Regel criterion [5]. For this purpose we calculated dispersion of dielectric susceptibility of cold atomic gases [6]. We showed that there was a spectral region where the photon mean free path was less than its wave length, i.e. where the criterion was satisfied. The fulfillment of the Ioffe-Regel criterion is likely a necessary, but not sufficient condition for strong localization. A clearer answer to the question about light localization in cold gases can be obtained by means of analysis of the incoherent radiation transfer in such gases. We studied the light transport and trapping in cold dense atomic clouds by several ways [7]-[9]. Particularly we have analyzed

(i) Afterglow dynamics of clouds excited by pulse radiation.
(ii) Spatial distribution of atomic excitation in quasi homogeneous ensemble caused by monochromatic coherent light.
(iii) Transmission coefficient of the cloud for different conditions.
(iv) Statistical properties of atomic excitation and light transmission.

This analysis showed no noticeable signs of light strong localization effects, even in those parameter regions where the Ioffe-Regel criterion of strong localization was satisfied. However, a comparative calculation performed in the framework of the often-used scalar approximation to the dipole-dipole interaction displayed explicit manifestation of strong localization for some conditions. This result was confirmed in the frame of scaling theory of localization [10]. Analysis of the Thouless number based on calculation of collective eigenstates of cold atomic ensemble, as well as analysis of their spatial localization based on calculation of inverse participation ratio showed that strong localization of light cannot be achieved in a random three-dimensional ensemble of atomic scattering [11]. Localization is reclaimed if the vector character of light is neglected. Performed calculations demonstrate the importance of the vector character of electromagnetic waves in the context of the Anderson localization problem and elucidate the role of resonant dipole-dipole interactions in multiple light scattering. At the same time they cause to anticipate that suppressing this interaction can open the way to localization. Our analysis shows that static magnetic field giving rise to Zeeman splitting changes the nature of atomic exchange of photons in the ensemble and essentially modifies resonant dipole-dipole interatomic interaction [12] (see also [13]). This effect influences strongly on light trapping. The efficiency of magnetic field depends on specific type of optically excited atomic transition, particularly on magnetic quantum numbers of quasi resonant states. We found out appearance of long lived polyatomic collective states which lifetimes essentially exceed those take place in the absence of magnetic field. Calculation of inverse participation ration shows that these states are localized. Scaling analysis of collective eigenstates distribution revealed that, in many aspects, this distribution exhibits the behavior expected for the Anderson transition driven by disorder [13]. On the basis of our approach we analyze also the possibility to observe manifestation of these long-lived localized states in experiment. We study the dynamics of fluorescence of atomic clouds initiated by pulse radiation. For an appropriate choice of frequency and polarization of the exciting pulse, the field is expected to speed up the fluorescence of a dilute atomic system. In a dense ensemble, the field does not affect the early-time superradiant signal but amplifies intensity fluctuations at intermediate times and induces a very slow, nonexponential long-time decay [14]. We analyze as well the influence of magnetic field on the steady-state transmission of plain layer of motionless atomic scatterers.

Acknowledgments

Acknowledgments The author acknowledges fruitful discussions with M.Havey and S. E. Skipetrov. This work was partially supported by the Russian Foundation for Basic Research (Grant No. RFBR-15-02-01013).

References

[1] Sokolov I M, Kupriyanov D V and Havey M D 2011 *J. Exp. Theor. Phys.* **112** 246
[2] Balik S, Win A L, Sukenik C I, M. D. Havey M D, Sokolov I M and Kupriyanov D V 2013 *Phys. Rev.* A **87** 053817
[3] Roof S, Kemp K J, Havey M D, Sokolov I M and Kupriyanov D V 2015 *Optics Letters* **40** 1137
[4] Roof S, Kemp K J , Havey M D and Sokolov I M 2016 *Phys. Rev. Lett.* **117** 073003
[5] Ioffe A F and Regel A R 1960 *Progress in Semiconductors* **4** 237
[6] Fofanov Ya A, Kuraptsev A S, Sokolov I M and Havey M D 2011 *Phys. Rev.* A **84** 053811
[7] Sokolov I M, Kupriyanov D V, Olave R G and Havey M D 2010 *Journal of Modern Optics* **57** 1833
[8] Fofanov Ya A, Kuraptsev A S, Sokolov I M and Havey M D 2013 *Phys. Rev.* A **87** 063839
[9] Sokolov I M 2015 *Laser Physics* **25** 065202
[10] Abrahams E, Anderson P W, Licciardello D C and Ramakrishnan T V 1979 *Phys. Rev. Lett.* **42** 673
[11] Skipetrov S E and Sokolov I M 2014 *Phys. Rev. Lett.* **112** 023905
[12] Skipetrov S E and Sokolov I M 2015 *Phys. Rev. Lett.* **114** 053902
[13] Afrousheh K, Bohlouli-Zanjani P, Carter J D, Mugford A and Martin J D D 2006 *Phys. Rev.* A **73** 063403
[14] Skipetrov S E, Sokolov I M and Havey M D 2016 *Phys. Rev.* A **94** 013825

Recent advances in precision spectroscopy of ultracold atoms and ions

A V Taichenachev[1,2], V I Yudin[1,2,3] and S N Bagayev[1,2]

[1] Institute of Laser Physics SB RAS, 13/3 Acad. Lavrentyev ave., Novosibirsk 630090, Russia
[2] Physics Department, Novosibirsk State University, 2 Pirogova str., Novosibirsk 630090, Russia
[3] Department of Physics and Technology, Novosibirsk State Technical University, 20 K. Marx ave., Novosibirsk 630073, Russia

E-mails: taichenachev@laser.nsc.ru, taichenachev@nsu.ru

Abstract. New methods and approaches in precision spectroscopy of ultracold atoms and ions are reviewed with an emphasis on contributions of Institute of Laser Physics SB RAS.

1. Introduction

Presently, laser spectroscopy and fundamental metrology are among the most important and actively developed directions in modern physics. Frequency and time are the most precisely measured physical quantities, which, apart from practical applications (in navigation and information systems), play critical roles in tests of fundamental physical theories (such as QED, QCD, unification theories, and cosmology) [1, 2]. Now, laser metrology is confronting the challenging task of creating an optical clock with fractional inaccuracy and instability at the level of 10^{-17} to 10^{-18}. Indeed, considerable progress has already been achieved along this path for both ion-trap- [3, 4] and atomic-lattice-based [5, 6] clocks.

Work in this direction has stimulated the development of novel spectroscopic methods such as spectroscopy using quantum logic [7], magnetically-induced spectroscopy [8], hyper-Ramsey spectroscopy [9], spectroscopy of "synthetic" frequency [10] and others [11]. Part of these methods was developed in order to excite and detect strongly forbidden optical transitions. The other part fights with frequency shifts of various origins. In the present paper we review both parts with a special emphasis on methods developed and studied in Institute of Laser Physics SB RAS, Novosibirsk.

2. Magnetically-induced spectroscopy of strongly forbidden optical transitions

Initially experimental work on optical lattice clocks [12] focused on the transition in the odd isotopes of Sr and Yb [13-16], which is weakly allowed due to the hyperfine mixing. In contrast to their even counterparts that have zero nuclear spin, however, the odd isotopes have an uncomfortably large ensitivity to magnetic fields (MHz/T) and lattice light polarization. They also have multiple ground state sublevels that considerably complicate spectroscopic line shapes. The method described in this section [8] uses a small magnetic field (~ 1 mT) to mix a small, controllable fraction of the nearby ${}^3P_1$ state into the ${}^3P_0$ state, thereby allowing single-photon excitation of the ${}^1S_0 \rightarrow {}^3P_0$ clock transition for the even isotopes. In this way we can replace the odd isotopes with the even in lattice-based atomic clocks, which should greatly accelerate their progress. Estimates for Yb using experimentally realistic linewidths (1 Hz) show that induced frequency shifts can be controlled at a level that could enable a fractional frequency uncertainty of below 10^{-17} for a lattice clock. Our method is equally effective for all alkaline-earth-like clock candidates (Yb, Sr, Ca, and Mg).

To excite the strongly forbidden $|1\rangle \rightarrow |2\rangle$ clock transition at frequency ω_{21} as shown in figure 1, we apply a static magnetic field **B** that couples the states $|2\rangle$ and $|3\rangle$. These states are split by a frequency Δ_{32} and have a coupling matrix element $\Omega_B = \langle 2|\mu \cdot \mathbf{B}|3\rangle$, where μ is the magnetic-dipole operator. According to first-order perturbation theory the state $|2\rangle$ acquires a small admixture of the state $|3\rangle$ due to the presence of the static magnetic field:

$$|2'\rangle = |2\rangle + \frac{\Omega_B}{\Delta_{32}}|3\rangle . \tag{1}$$

As a result the transition $|1\rangle \rightarrow |2\rangle$ becomes partially allowed. An optical field with amplitude **E** and frequency ω acts via the $|1\rangle \rightarrow |3\rangle$ transition (see figure 1). We assume that this transition is at least weakly allowed. The corresponding coupling matrix element is the Rabi frequency, $\Omega_L = \langle 3|\mathbf{d} \cdot \mathbf{E}|1\rangle / \hbar$. Because of the slight admixture of state $|3\rangle$ into the bare state $|2\rangle$, a resonance will be observed on the forbidden transition $|1\rangle \rightarrow |2\rangle$ when $\omega \approx \omega_{12}$.

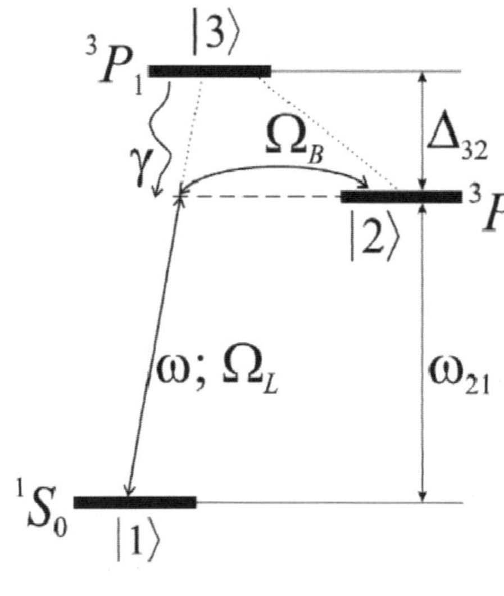

Figure 1. Magnetic field-induced excitation of a strongly forbidden transition in a generic three-level atomic system.

The effective Rabi frequency of the clock transition can be written in three different forms. First of all, taking into account the vector nature of the applied fields and Zeeman degeneracy of the intermediate state $|3\rangle$, we obtain

$$\Omega_{12} = \frac{d\mu(\mathbf{E} \cdot \mathbf{B})}{\hbar^2 \Delta_{32}} , \tag{2}$$

where d and μ are the reduced matrix elements of the dipole and magnetic dipole moments, respectively. From the other side the effective Rabi frequency is obviously proportional to the magnetic field strength and to square root from the light intensity:

$$\Omega_{12} = \alpha\sqrt{I} B \cos\theta . \tag{3}$$

Here θ stands for the angle between **E** and **B** directions. Finally, it is useful to express the effective Rabi frequency in terms of the light shift $\Delta_L = \kappa I$ and quadratic Zeeman shift $\Delta_B = \beta B^2$ produced by the probe fields:

$$\Omega_{12} = \xi\sqrt{\Delta_L \Delta_B} \cos\theta \, , \tag{4}$$

where $\xi \equiv \alpha / \sqrt{\beta\kappa}$ is a dimensionless quality factor for the clock transition in this scheme, since it relates the strength of the excitation to the magnitudes of the induced field shifts. A larger value for ξ implies that for a given spectroscopic linewidth, the induced shifts will be smaller. Table I summarizes the relevant parameters for four different alkaline-earth-like atoms that have been considered as excellent clock candidates.

Table 1. The calculated parameters for four different atoms.

	Δ_{32} [THz]	α [Hz/(T mW$^{1/2}$/cm)]	β [MHz/T^2]	κ [mHz/(mW/cm^2)]	ξ
Yb	21	186	-6.2	15	0.6
Sr	5.6	198	-23.3	-18	0.3
Ca	1.5	154	-83.5	-3.5	0.3
Mg	0.6	98	-217	-0.5	0.3

Using these values for Yb one can show that a magnetic field 1 mT with a probe light intensity 10 mW/cm^2 will produce the effective Rabi frequency about 1 Hz, accompanying by the quadratic Zeeman shift of 10 Hz and the probe light shift of 0.1 Hz. These shifts can be controlled at the level of a few mHz that results in a fractional frequency uncertainty well below 10^{-17} level. Further improvements can be achieved by the use a special method of hyper-Ramsey spectroscopy described in the next section.

Note that the method of magnetically-induced spectroscopy has been successfully realized first in experiments with Yb atoms [8], then with Sr atoms [17-19], and very recently with Mg atoms [20].

3. Hyper-Ramsey spectroscopy

3.1. Motivations

For some of the promising clock systems, one of the key unsolved problems is the frequency shift of the clock transition due to the excitation pulses themselves. For the magnetically-induced spectroscopy, these shifts (quadratic Zeeman and ac-Stark shifts) could ultimately limit the achievable performance. Moreover, for ultra-narrow transitions (e.g., electric octupole [21] and two-photon transitions [22, 23]) the ac-Stark shift can be large enough in some cases that it rules out high accuracy clock performance. A similar limitation exists for clocks based on direct frequency comb spectroscopy [24, 25] because ac-Stark shifts are induced by large numbers of off-resonant laser modes.

In this section, we describe a general solution to this important problem, which is based on the development and generalization of the Ramsey method [26]. We have found that in contrast with the single-pulse (Rabi) technique, multi-pulse Ramsey spectroscopy offers several ways (e.g., pulse durations, frequencies, and phases) to significantly manipulate (due to interference effects) the induced frequency shifts of the spectroscopy signals. In particular, for special excitation schemes (which we refer to as “hyper- Ramsey”) the resulting ac-Stark shift depends on the laser intensity in an essentially nonlinear way. Such unusual and unexpected behavior allows us to dramatically suppress these shifts and their uncertainties (most critical for clocks) by two to four orders of magnitude with strongly relaxed control requirements for the experimental parameters. Additionally, we have found that these schemes can have a greatly reduced sensitivity to the pulse areas, which makes the procedure robust and accessible experimentally. Thus, this method can be readily implemented in a variety of existing and proposed clock systems [3-7, 12-25, 27]. Our approach could lead to significant progress for atomic clocks: it will improve several key existing optical clock systems and could enable new systems that were not previously thought to be competitive.

3.2. The original proposal

The hyper-Ramsey spectroscopy schemes [9] (see in figure 2) are based on time-separated pulses that can have different durations, frequencies, and phases. The action of a single light pulse (with frequency ω_p, duration τ, and Rabi frequency Ω_0) on two-level atoms with ground $|g\rangle = \begin{pmatrix} 0 \\ 1 \end{pmatrix}$ and excited $|e\rangle = \begin{pmatrix} 1 \\ 0 \end{pmatrix}$ states, (separated by the unperturbed energy $\hbar\omega_0$), is described by the following matrix:

$$W(\tau,\Omega_0,\delta_p) = \begin{pmatrix} \cos\left(\frac{\Omega\tau}{2}\right) + i\frac{\delta_p}{\Omega}\sin\left(\frac{\Omega\tau}{2}\right) & i\frac{\Omega_0}{\Omega}\sin\left(\frac{\Omega\tau}{2}\right) \\ i\frac{\Omega_0}{\Omega}\sin\left(\frac{\Omega\tau}{2}\right) & \cos\left(\frac{\Omega\tau}{2}\right) - i\frac{\delta_p}{\Omega}\sin\left(\frac{\Omega\tau}{2}\right) \end{pmatrix}, \quad (5)$$

where $\Omega = \sqrt{\Omega_0^2 + \delta_p^2}$ is the generalized Rabi frequency. The detuning during the pulse $\delta_p = \omega_p - \omega_0 - \Delta_{sh}$ contains the excitation related shift Δ_{sh} (see figure 2, level scheme) due to the influence of other (far-off-resonant) transitions.

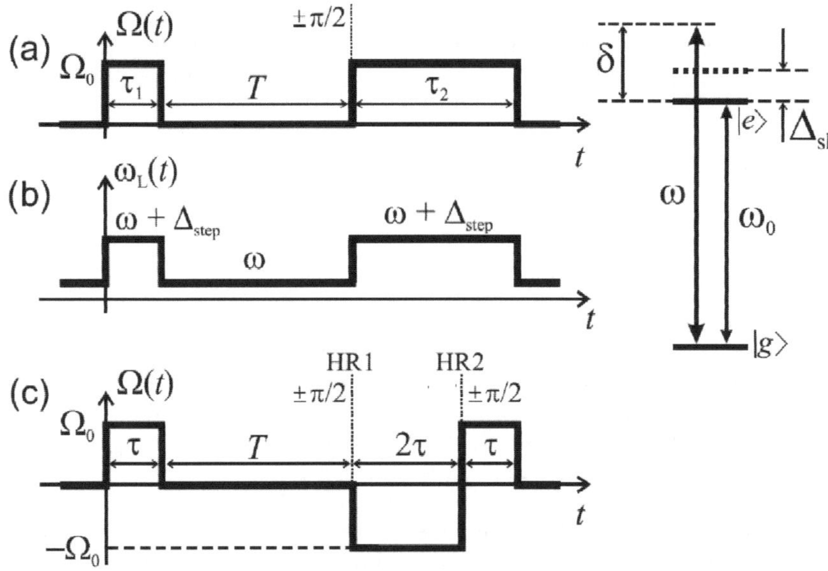

Figure 2. Ramsey pulses with Rabi frequency Ω_0 of different duration [τ_1 and τ_2 (a)]. During the pulses, we step the laser frequency ω by Δ_{step} [(b) and text]. Hyper-Ramsey scheme with composite second pulse 3τ [(c) and text]. Also shown is a two-level atom with splitting ω_0, detuning δ of the laser with frequency ω during dark time T , and excitation related shift Δ_{sh} during pulses.

During the dark period between the pulses, excitation-related shifts (which produce the total actual shift Δ_{sh}) are absent (e.g., the ac-Stark shift from the laser) or can be turned off (like the Zeeman shift). If during the dark period T the laser frequency is ω, then the free evolution is described by the matrix with detuning $\delta = \omega - \omega_0$:

$$V(T\delta) = \begin{pmatrix} \exp(iT\delta/2) & 0 \\ 0 & \exp(-iT\delta/2) \end{pmatrix} \tag{6}$$

In the general case, the laser frequency during the pulse does not have to be the same as the frequency during the dark time, i.e., $\omega_p \neq \omega$ [27]. As we will see, at times it can be useful to approximately offset the induced shift Δ_{sh} by stepping the laser frequency only during the pulses by a fixed Δ_{step}, i.e., $\omega_p = \omega + \Delta_{step}$ [see figure 2(b)]. Thus, in the general case the detuning during the pulses can be written as $\delta_p = \delta + \Delta$, where $\Delta = \Delta_{sh} - \Delta_{step}$ step is the effective frequency shift during the pulse. This manipulation allows us to stabilize the frequency ω under the controlled condition $|\Delta/\Omega_0| << 1$ (independently of the value Δ_{sh}). This condition is very important, because the shift of the central Ramsey fringe has a form of decreasing series in terms of powers of the small value Δ/Ω_0 (see [9]).

Formulas (5) and (6) are sufficient for description of the signal in Ramsey spectroscopy. For example, if at $t = 0$ atoms are in the lower level $|g\rangle$, then after the action of two pulses of duration τ_1 and τ_2 separated by dark period T [see figure 2(a)] the population $\pi^{(e)}$ of atoms in the excited state $|e\rangle$ is determined by

$$\pi^{(e)} = |\langle e|W(\tau_2,\Omega_0,\delta-\Delta)V(T\delta)W(\tau_1,\Omega_0,\delta-\Delta)|g\rangle|^2 \tag{7}$$

This formula describes Ramsey fringes as a function of variable detuning δ, but with fixed Δ. The presence of the additional shift Δ in the course of the pulse action leads to the shift of the central Ramsey fringe with respect to the unperturbed frequency ω_0.

Hereafter we assume that the position of the central fringe ω_T is determined by stepping the phase of one of the pulses by $\pm\pi/2$ in the way [28] and equalizing these signals. This approach is of greater relevance for clocks, because it directly generates an error signal with high sensitivity. In respect to the signal (7), this method is formulated as following. Let us introduce the phase steps φ after the dark time T, then the excited state population $\pi^{(e)}(\varphi)$ and the error signal $S^{(err)}$ can be expressed as:

$$\begin{aligned} \pi^{(e)}(\varphi) &= |\langle e|W(\tau_2,\Omega_0,\delta-\Delta)V(\varphi)V(T\delta)W(\tau_1,\Omega_0,\delta-\Delta)|g\rangle|^2 \\ S^{(err)} &= \pi^{(e)}(\pi/2) - \pi^{(e)}(-\pi/2). \end{aligned} \tag{8}$$

The shift δ_T of the stabilized frequency ω_T is determined from the solution of the equation $S^{(err)} = 0$ relative to the unknown δ.

In the standard Ramsey spectroscopy both exciting pulses have equal duration $\tau_1 = \tau_2 = \tau$. In the case of $\Omega_0\tau = \pi/2$ and $\tau << T$, the dominant contribution has the following linear dependencies on the small value $|\Delta/\Omega_0| << 1$:

$$\delta_T \approx \frac{2}{T}\frac{\Delta}{\Omega_0} \tag{9}$$

However, most interesting and important results can be obtained by the use of the hyper-Ramsey scheme [21,30] shown in figure 2(c). Here the main peculiarity is the composite pulse (with total duration 3τ), which consists of subpulse 2τ with inverted phase ($-\Omega_0$) and subpulse τ with initial phase (Ω_0). If for the error signal we apply additional phase $\pm\pi/2$ steps directly after dark time (as it was in [9]), then this method we will be referred to as HR1 [see in figure 1(c)]. In this case, we define the function $\pi^{(e)}_{HR1}(\varphi)$ and the error signal $S^{(err)}_{HR1}$:

$$\pi_{HR1}^{(e)}(\varphi) = \left| \langle e | W(\tau, \Omega_0, \delta - \Delta) W(2\tau, -\Omega_0, \delta - \Delta) V(\varphi) V(T\delta) W(\tau, \Omega_0, \delta - \Delta) | g \rangle \right|^2$$
$$S_{HR1}^{(err)} = \pi_{HR1}^{(e)}(\pi/2) - \pi_{HR1}^{(e)}(-\pi/2). \qquad (10)$$

As it was first shown in [21], for the hyper-Ramsey pulse sequence the dominant contribution in the shift δ_T has a cubic dependence on the small value $|\Delta/\Omega_0| << 1$. For instance, if $\Omega_0\tau = \pi/2$ and $\tau << T$, we have

$$\delta_T \approx \frac{4}{T}\left(\frac{\Delta}{\Omega_0}\right)^3 \qquad (11)$$

The suppression of the probe-field-induced shift in the hyper-Ramsey spectroscopy is illustrated by figure 3. As is clearly seen, a cubic nonlinearity appears. Moreover, this character is not changed under variations of Ω_0, τ, and T, i.e., we do not need the rigorous condition $\Omega_0\tau = \pi/2$. This circumstance is a key point to successfully realize our method in atomic clocks, because in real experiments the value of Ω_0 can not be controlled better than at the level of 1%–10%.

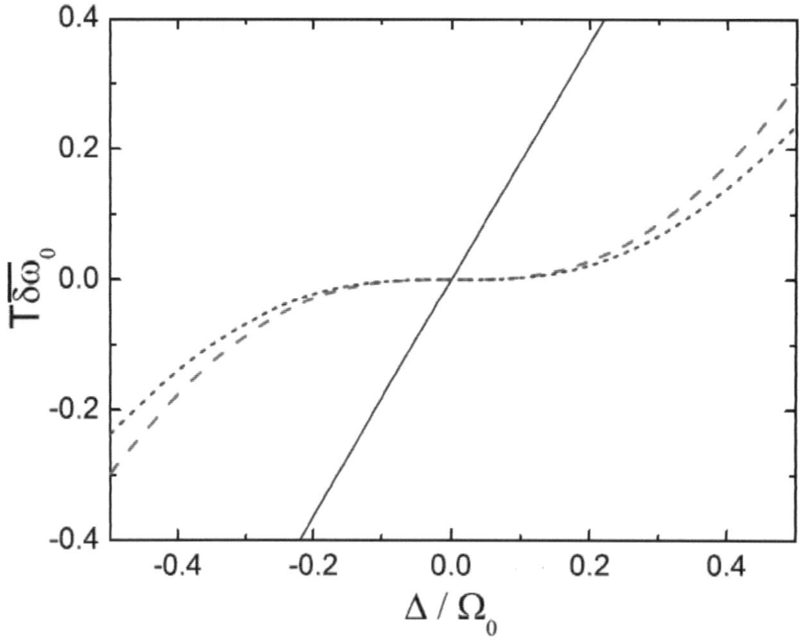

Figure 3. Numerically calculated shift of the central resonance $T\delta_T$ versus Δ/Ω_0 for the standard Ramsey spectroscopy (solid line) and the hyper-Ramsey method (dotted and dashed lines) at $\Omega_0\tau = \pi/2$; $\Omega_0 T = 20$. Dashed line: position of extremum of the excited-state population $\pi_{HR1}^{(e)}(0)$; dotted line: estimate of center from the error signal with $\pm\pi/2$ phase steps.

The hyper-Ramsey method was experimentally realized for the first time in Yb single-ion system [9]. The probe-field frequency shift was suppressed by more than three orders of magnitude. The use of the hyper-Ramsey scheme was crucial in development of the best single-ion clock [4]. Recently, a successful realization of the hyper-Ramsey excitation in magnetically-induced spectroscopy of neutral Sr atoms was reported in [19].

3.3. Further modifications

In the recent paper [19], the authors have proposed the use of $\pm\pi/2$ phase steps after inverted ($-\Omega_0$) subpulse 2τ [see in figure 2(c)] to form an error signal for the hyper-Ramsey approach (we will denote this modification as HR2). For its theoretical description, we introduce the following function $\pi^{(e)}_{HR2}(\varphi)$ and the error signal $S^{(err)}_{HR2}$:

$$\pi^{(e)}_{HR2}(\varphi) = \left|\langle e|W(\tau,\Omega_0,\delta-\Delta)V(\varphi)W(2\tau,-\Omega_0,\delta-\Delta)V(T\delta)W(\tau,\Omega_0,\delta-\Delta)|g\rangle\right|^2$$
$$S^{(err)}_{HR2} = \pi^{(e)}_{HR2}(\pi/2) - \pi^{(e)}_{HR2}(-\pi/2). \tag{12}$$

Then the shift δ_T of the stabilized frequency ω_T is determined from the solution of the equation $S^{(err)}_{HR2} = 0$ relative to the unknown δ. In this case, our theoretical estimations give us (for $\Omega_0\tau = \pi/2$ and $|\Delta/\Omega_0| << 1$):

$$\delta_T \approx -\frac{4}{T}\left(\frac{\Delta}{\Omega_0}\right)^3, \tag{13}$$

which almost coincide with equation (11), but with the opposite sign.

Besides HR2, Ref. [19] also describes the theory and successful experimental demonstration of a modified hyper-Ramsey (MHR) method, which is based on the combination of both protocols HR1 and HR2. In the case of MHR, the error signal is determined as

$$S^{(err)}_{MHR} = \pi^{(e)}_{HR1}(\varphi) - \pi^{(e)}_{HR2}(\varphi). \tag{14}$$

Again, the shift δ_T of the stabilized frequency ω_T is determined from the solution of the equation $S^{(err)}_{MHR} = 0$ relative to the unknown δ. This equation leads to an unexpected result: $\delta_T = 0$ for arbitrary $\varphi, \Delta, \Omega_0, \tau$, and T. In the other words : $\delta_T = 0$ in all orders with respect to Δ/Ω_0. An analogous result was also found for an alternative scheme (so-called generalized hyper-Ramsey (GHR)), presented in the theoretical paper [20].

At first glance, both MHR and GHR approaches [19, 20] are the best for frequency stabilization, because they allow us to totally eliminate probe-induced shifts. However, as it was shown in [29], MHR and GHR methods are unstable relative to the decoherence, which leads to an appearance of the shift $\delta_T \neq 0$. Moreover, this residual shift is very sensitive to variations of Rabi frequency Ω_0. A proper modification of the hyper-Ramsey method robust with respect to the decoherence (which is present in any real system due to the finite laser line width and/or the finite lifetime of the excited state, see in [30, 31]) was proposed by us recently [29] and it is based on the use of the synthetic frequency protocol briefly described in the next section.

4. Synthetic frequency method

4.1. General formulation

The essence of our approach consists of the following. Let us consider two clock frequencies $\omega_1^{(0)}$ and $\omega_2^{(0)}$ (different in the general case). Assume that due to a certain physical cause we have the stabilized frequencies ω_1 and ω_2, which are shifted relative to the unperturbed frequencies by the values Δ_1 and Δ_2:

$$\omega_1 = \omega_1^{(0)} + \Delta_1; \quad \omega_2 = \omega_2^{(0)} + \Delta_2. \tag{15}$$

Also assume that the ratio $\varepsilon_{12} = \Delta_1/\Delta_2 = const$ does not fluctuate, while the shifts $\Delta_{1,2}$ can vary during the experiment. In this case, we can construct the following superposition:

$$\omega_{syn} = \frac{\omega_1 - \varepsilon_{12}\omega_2}{1-\varepsilon_{12}} = \frac{\omega_1^{(0)} - \varepsilon_{12}\omega_2^{(0)}}{1-\varepsilon_{12}}, \tag{16}$$

which is insensitive to the perturbations $\Delta_{1,2}$ and their fluctuations. We will call this frequency the "synthetic frequency." A key advantage of this concept is that to construct the shift-free frequency ω_{syn}, we do not need to know the actual values of the shifts $\Delta_{1,2}$. We only need to know their ratio ε_{12}, which can be exactly calculated or measured for many cases.

4.2. Application to the problem of BBR shift

The largest effect that contributes to the systematic uncertainty of many atomic clocks is the interaction of the thermal blackbody radiation (BBR) with the atomic eigenstates. This effect was first considered in 1982 for cesium atomic clocks [32] but remains up to now a major problem for many modern atomic time and frequency standards. At present, there exist three approaches to tackle the BBR shift problem. The first one is the use of cryogenic techniques to suppress this shift to a negligible level. This approach is pursued for the mercury ion clock [33], for the Cs fountain clock [34], and for the Sr optical lattice clock [18]. The second approach is the precise temperature stabilization of the experimental setup in combination with theoretical and/or semiempirical numerical calculations of the shift at a given temperature [35, 36]. The third approach is based on the choice of an atom or ion where both levels of the reference transition have approximately the same BBR shift. Here the most promising candidate is $^{27}Al^{+}$ with a fractional BBR shift of the reference transition frequency of 10^{-17} at 300 K [3], followed by $^{115}In^{+}$ [37]. However, the latter approach limits the choice of candidates for tests of fundamental theories.

Our approach is based on the fact that, for the large majority of atomic transitions that are of interest as frequency standard reference transitions, the temperature dependence $\Delta(T)$ of the BBR shift is very well approximated by the law $\propto t^4$. Consider now two clock transitions with frequencies ω_1 and ω_2 exposed to the same thermal environment. The BBR shifts on each frequency can be represented as a temperature dependence given by

$$\omega_1(T) = \omega_1^{(0)} + a_1 t^4; \quad \omega_2(T) = \omega_2^{(0)} + a_2 t^4, \tag{17}$$

where $\omega_j^{(0)}$ is the unshifted frequency (at t = 0 K) and a_j is an individual characteristic of the j-th transition, which is proportional to the differential static polarizability of working levels. Introducing the coefficient $\varepsilon_{12} = \Delta_1(t)/\Delta_2(t) = a_1/a_2$, one can easily see, that the following superposition does not experience the BBR shift:

$$\omega_{syn} = \frac{\omega_1(t) - \varepsilon_{12}\omega_2(t)}{1-\varepsilon_{12}} = \frac{\omega_1^{(0)} - \varepsilon_{12}\omega_2^{(0)}}{1-\varepsilon_{12}} \tag{18}$$

Thus, one can use the frequency ω_{syn} as a new clock output frequency which is immune to the BBR shift and to fluctuations in the operating temperature, while the thermal shifts $\Delta_j(t)$ of the reference frequencies ω_j can be large.

As an example that permits the practical realization of the ideas presented above, we consider the ion $^{171}Yb^{+}$. As shown in figure 4, the level system of $^{171}Yb^{+}$ provides two narrow-linewidth transitions from the ground state in the visible spectral range, which can be used as reference transitions of an optical frequency standard: the quadrupole transition $^2S_{1/2}(F = 0) \rightarrow {}^2D_{3/2}(F = 2)$, $\lambda \approx 436$ nm and the octupole transition $^2S_{1/2}(F = 0) \rightarrow {}^2F_{7/2}(F = 3)$, $\lambda \approx 467$ nm. It should be noted that the case of $^{171}Yb^{+}$ is especially attractive because here both clock transitions lie in a technically convenient frequency range and experience exactly the same thermal environment if probed in the same ion.

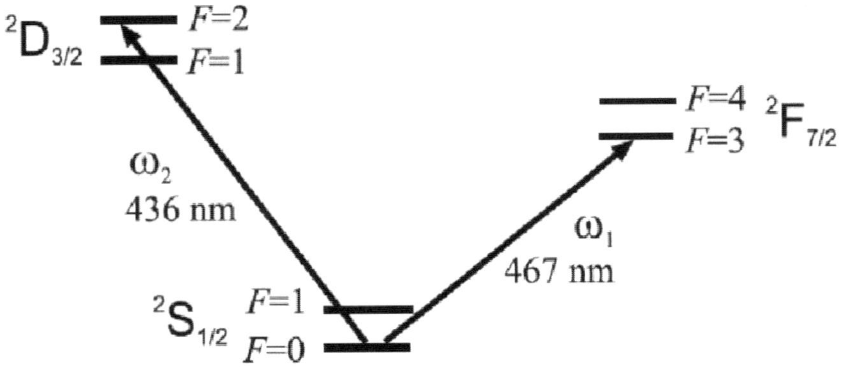

Figure 4. Section of the energy level scheme of $^{171}Yb^+$, showing the hyperfine levels of the $^2S_{1/2}$ ground state and the two lowest-lying excited states, which are metastable.

The BBR shifts of the quadrupole and octupole transitions of Yb^+ were calculated in [38]. The room-temperature (at 300 K) BBR shift of the quadrupole transition is calculated as $a_{quad} = -0.35$ Hz (fractional shift 5.1×10^{-16}) and that of the octupole transition as $a_{oct} = -0.15$ Hz (fractional shift 2.4×10^{-16}). Using the results of Ref. [21], for $^{171}Yb^+$ we find, that $\varepsilon_{12} = a_{oct} / a_{quad} = 0.43$, and the synthetic frequency equals to607 THz, corresponding to the synthetic wavelength 494 nm, which is sufficiently close to the initial reference transitions at 436 and 467 nm. The higher-order contributions ($\propto t^6$ and higher) to the BBR shift of the octupole reference transition are negligible compared to that of the quadrupole transition. As a result, we estimated that the BBR shift can be suppressed to the fractional level of 2.7×10^{-17} at 300 K with variations at the level of $\pm5\times10^{-18}$ in a broad interval of 300±15 K.

The concept of a synthetic atomic frequency standard can also be extended to the case that both reference frequencies $\omega_{1,2}$ lie in the microwave range. Atomic fountain clocks are based on transitions in the microwave range between the ground-state hyperfine sublevels of alkali atoms. For a synthetic atomic fountain frequency standard, for instance, the combination ^{87}Rb (ω_1 = 6.8 GHz) and ^{133}Cs (ω_1 = 9.2 GHz) can be considered. Here, at the synthetic frequency 1.9 GHz it is possible to suppress the fractional BBR shift of the individual standards by two orders of magnitude. It is interesting to note that nearly optimal conditions for the efficient suppression of the BBR shift are realized in the dual Rb/Cs fountain clock described in Ref. [39], because here both reference transitions are exposed to the same thermal environment.

4.3. Applications to the Ramsey and hyper-Ramsey schemes

Let us now show how to incorporate the synthetic frequency protocol in Ramsey spectroscopy for significant suppression of the probe-induced shifts in atomic clocks. Consider again an influence of the probe-induced shift Δ_{sh}, which arises only during the Ramsey pulses (see two-level scheme in figure 2), while this shift is absent during the dark time T. As a result, the stabilized frequency ω_T also becomes differing from unperturbed frequency ω_0: $\omega_T = \omega_0 + \delta_T$, where the index T denotes the fixed time of the free evolution interval under frequency stabilization. It can be shown that the dependence δ_T on the value T can be expressed as the following decreasing series in terms of powers of $1/T$:

$$\delta_T = \sum_{n=1}^{\infty} \frac{A_n}{T^n}, \qquad (19)$$

where the coefficients A_n depend on the pulse parameters (durations, amplitudes, phases, and the value Δ_{sh}).

Since the dark time interval T is precisely controlled in experiments, we can set a goal to eliminate the main contribution $\propto A_1/T$ in equation (19) using the synthetic frequency protocol. To solve this task we will apply two different dark intervals $T_1=T$ and $T_2=T/2$ (but with the same Ramsey pulses), which will give us the corresponding stabilized frequencies ω_T and $\omega_{T/2}$. Using Eqs. (16) and (19), we easily find the synthetic frequency $\omega_{syn}^{(1)}$ and its residual shift $\delta_{syn}^{(1)}$:

$$\omega_{syn}^{(1)} = 2\omega_T - \omega_{T/2},$$
$$\delta_{syn}^{(1)} = \omega_{syn}^{(1)} - \omega_0 = 2\delta_T - \delta_{T/2} = \sum_{n=2}^{\infty} \frac{B_n}{T^n}. \quad (20)$$

Moreover, we can go further to define a synthetic frequency of the second order $\omega_{syn}^{(2)}$, for which both contributions A_1/T and A_2/T^2 will be simultaneously canceled. Here we need to use three different time intervals ($T_1 = T$, $T_2 = T/2$, and $T_3 = T/3$) with the corresponding stabilized frequencies (ω_T , $\omega_{T/2}$, and $\omega_{T/3}$). The required superposition takes the form,

$$\omega_{syn}^{(2)} = 3\omega_T - 3\omega_{T/2} + \omega_{T/3},$$
$$\delta_{syn}^{(2)} = \omega_{syn}^{(2)} - \omega_0 = 3\delta_T - 3\delta_{T/2} + 3\delta_{T/3} = \sum_{n=3}^{\infty} \frac{C_n}{T^n}. \quad (21)$$

The main metrological advantage of the proposed synthetic frequency protocol in the Ramsey spectroscopy is an improvement of accuracy and long-term stability of the atomic clock, first of all. At the same time, we expect small deterioration in the short-term stability, because the averaging time to reach the same short-term instability in the measurement scheme based on Eqs. (20) or (21) is approximately two to three times longer than for a direct locking to the clock transition.

At first, let us consider the standard Ramsey spectroscopy, where the dominant contributions have the following linear dependencies on the small value $|\Delta/\Omega_0| << 1$:

$$\delta_T \approx \frac{2}{T}\frac{\Delta}{\Omega_0}; \quad \delta_{syn}^{(1)} \approx \frac{8}{\pi T}\frac{2\tau}{T}\frac{\Delta}{\Omega_0}; \quad \delta_{syn}^{(2)} \approx \frac{48}{\pi^2 T}\left(\frac{2\tau}{T}\right)^2\frac{\Delta}{\Omega_0}. \quad (22)$$

Here due to the smallness of the ratio $(2\tau/T) << 1$ (i.e., for short Ramsey pulses) we have the chain of inequalities: $\delta_{syn}^{(2)} << \delta_{syn}^{(1)} << \delta_T$. Thus, the synthetic frequency protocol can significantly suppress the shifts even for the standard Ramsey spectroscopy.

However, the most interesting and important results are obtained by the use of the hyper-Ramsey schemes. For the pulse sequence HR1 the dominant contributions take the form:

$$\delta_T \approx \frac{4}{T}\left(\frac{\Delta}{\Omega_0}\right)^3; \quad \delta_{syn}^{(1)} \approx \frac{48}{\pi T}\frac{4\tau}{T}\left(\frac{\Delta}{\Omega_0}\right)^5; \quad \delta_{syn}^{(2)} \approx \frac{865}{\pi^2 T}\left(\frac{4\tau}{T}\right)^2\left(\frac{\Delta}{\Omega_0}\right)^7, \quad (23)$$

and very similar results for the pulse sequence HR2:

$$\delta_T \approx -\frac{4}{T}\left(\frac{\Delta}{\Omega_0}\right)^3; \quad \delta_{syn}^{(1)} \approx \frac{48}{\pi T}\frac{4\tau}{T}\left(\frac{\Delta}{\Omega_0}\right)^5; \quad \delta_{syn}^{(2)} \approx -\frac{865}{\pi^2 T}\left(\frac{4\tau}{T}\right)^2\left(\frac{\Delta}{\Omega_0}\right)^7, \quad (24)$$

These formulas (23) and (24) demonstrate that the chain of inequalities $\delta_{syn}^{(2)} << \delta_{syn}^{(1)} << \delta_T$, can be realized due to the controlled smallness $|\Delta/\Omega_0| << 1$, first of all. Moreover, the condition $(4\tau/T) << 1$ leads to an additional suppression of the shifts. Thus, the relatively small initial shift δ_T and its fluctuations can be suppressed by several orders of magnitude to metrologically negligible values.

To summarize, the synthetic frequency protocol in Ramsey spectroscopy is a novel technique that offers a spectroscopic signal that is virtually free from probe-induced frequency shifts and their fluctuations. Our method has broad applications for any types of clocks, especially those based on ultranarrow transitions, two-photon transitions, lattice clocks based on bosonic isotopes with controlled collision shifts [5, 40], CPT-Ramsey [31-44], and POP-Ramsey clocks [45]. Moreover, our approach opens a prospect for the high-precision optical clocks based on direct frequency comb spectroscopy. High resolution matter-wave sensors [46] are also expected to benefit from the suppression of phase shifts in the interference patterns due to the excitation pulses.

5. Conclusion

We have presented three methods for ultra-precise spectroscopy of ultracold atoms and ions, which were proposed, developed, and investigated at the Institute of Laser Physics SB RAS. It should be noted that the experimental research on magnesium atoms and single ytterbium ions, which is aimed at the practical realization of these methods, has been actively carried out in the Institute since 2012. We plan to reach a level of relative instability of our frequency standards better than 10^{-16} by 2016-2017.

To conclude, we would like to make a forecast for the development of the research in the field of optical frequency standards:

(1) further improvement of metrological characteristics for single-ion (Yb^+, Al^+, In^+, Sr^+, Hg^+) and lattice-based atomic (Sr, Yb, Hg, Mg, Tm) optical frequency standards by using special methods of control over the frequency field shifts of different natures (thermal radiation, trap and lattice fields, probe field); reaching the uncertainty level of 10^{-18} in the nearest future; redefinition of frequency and time units based on optical frequency standards; precise tests of fundamental physical laws by using optical frequency standards;

(2) development of fundamentally new optical frequency standards (for example, based on nuclear transitions, magnetic dipole transitions in highly charged ions, etc.) that would allow reaching the frequency relative instability of the order of 10^{-19} and lower;

(3) development of compact mobile optical frequency standards based on ultracold atoms and ions, including those working in space and used for high-precision navigation and geodesy systems;

(4) improvement of metrological characteristics of optical frequency standards by realizing quantum networks of atomic clocks and using fiber and space communication lines.

Acknowledgments

The studies presented in this paper were supported by the Russian Scientific Fund grants 16-12-00052 (sections 3 and 4) and 16-12-00054 (section 2).

References

[1] Bagayev S N *et al.* 2000 *Appl. Phys.* B **70** 375
[2] Diddams S A *et al.* 2004 *Science* **306** 1318
[3] Rosenband T *et al.* 2008 *Science* **319** 1808
Chou C W *et al.* 2010 *Phys. Rev. Lett.* **104** 070802
[4] Huntemann N *et al.* 2016 *Phys. Rev. Lett.* **116** 063001
[5] Akatsuka T, Takamoto M and Katori H 2008 *Nature Physics* **4** 954
[6] Hinkley N *et al.* 2013 *Science* **341** 1215
Bloom B J *et al.* 2014 *Nature* **506** 71
[7] Schmidt P O *et al.* 2005 *Science* **309** 749
[8] Taichenachev A V *et al.* 2006 *Phys. Rev. Lett.* **96** 083001
Barber Z *et al.* 2006 *Phys. Rev. Lett.* **96** 083002
[9] Yudin V I *et al.* 2010 *Phys. Rev.* A **82** 011804(R)
Huntemann N *et al.* (2012) *Phys. Rev. Lett.* **109** 213002
[10] Yudin V I *et al.* 2011 *Phys. Rev. Lett.* **107** 030801

[11] Yudin V I *et al.* 2014 *Phys. Rev. Lett.* **113** 233003
[12] Katori H 2002 *Proc. of the Sixth Symposium on Frequency Standards and Metrology* (World Scientific, Singapore, 2002) pp 323–330
[13] Takamoto M *et al.* 2005 *Nature* **435** 321
[14] Ludlow A D *et al.* 2006 *Phys. Rev. Lett.* **96** 033003
[15] Hoyt C W *et al.* 2005 *Phys. Rev. Lett.* **95** 083003
[16] Hong T *et al.* 2005 *Opt. Lett.* **30** 2644
[17] Baillard X *et al.* 2007 *Opt. Lett.* **32** 1812
[18] Derevianko A, Katori H 2011 *Rev. Mod. Phys.* **83** 331
[19] Hobson R *et al.* 2016 *Phys. Rev.* A **93** 010501(R)
[20] Kulosa A P *et al.* 2015 *Phys. Rev. Lett.* **115** 240801
[21] Hosaka K *et al.* 2009 *Phys. Rev.* A **79** 033403
[22] Fischer M *et al.* 2004 *Phys. Rev. Lett.* **92** 230802
[23] Badr T *et al.* 2006 *Phys. Rev.* A **74** 062509
[24] Fortier T M *et al.* 2006 *Phys. Rev. Lett.* **97** 163905
[25] Stowe M C *et al.* 2008 *Adv. At. Mol. Opt. Phys.* **55** 1
[26] Ramsey N F 1950 *Phys. Rev.* **78** 695
[27] Zanon-Willette T *et al.* 2006 *Phys. Rev. Lett.* **97** 233001
[28] Morinaga A *et al.* 1989 *Appl. Phys.* B **48** 165
[29] Yudin V I *et al.* 2016 *Phys. Rev.* A **94** 052505
[30] Tabatchikova K S, Taichenachev A V and Yudin V I 2013 *JETP Letters* **97** 311
[31] Tabatchikova K S *et al.* 2015 *JETP* **120** 203
[32] Itano W *et al.* 1982 *Phys. Rev.* A **25** 1233(R)
[33] Oskay W H *et al.* 2006 *Phys. Rev. Lett.* **97** 020801
[34] Levi F *et al.* 2010 *IEEE Trans. UFFC* **57** 600
[35] Angstmann E J, Dzuba V A, and Flambaum V V 2006 *Phys. Rev.* A **74** 023405
[36] Safronova M S *et al.* 2010 *IEEE Trans. UFFC* **57** 94
[37] Becker Th *et al.* 2001 *Phys. Rev.* A **63** 051802(R)
[38] Lea S N, Webster S A and Barwood G P 2006 *Proc. of the 20th European Frequency and Time Forum* (PTB, Braunschweig, Germany) p 302
[39] Guena J *et al.* 2010 *IEEE Trans. Ultrason. Ferroelectr. Freq. Control* **57** 647
[40] Lisdat Ch *et al.* 2009 *Phys. Rev. Lett.* **103** 090801
[41] Hemmer P R *et al.* 1989 *J. Opt. Soc. Am.* B **6** 1519
[42] Zanon T *et al. Phys. Rev. Lett.* **94** 193002
[43] Chen X *et al.* 2010 *Chin. Phys. Lett.* **27** 113201
[44] Blanshan E *et al.* 2015 *Phys. Rev.* A **91** 041401(R)
[45] Micalizio S *et al.* 2012 *Metrologia* **49** 425
[46] Yver-Leduc F *et al.* 2003 *J. Opt.* B **5** S136

Laser-driven plasma wakefield electron acceleration and coherent femtosecond pulse generation in X-ray and gamma ranges

V I Trunov[1], K V Lotov[2,3], K V Gubin[1], E V Pestryakov[1,2], S N Bagayev[1,2] and P V Logachev[3]

[1]Institute of Laser Physics SB RAS, 13/3 Acad. Lavrentyev ave., Novosibirsk 630090, Russia
[2]Novosibirsk State National Research University, 2 Pirogova str., Novosibirsk 630090, Russia
[3]Budker Institute of Nuclear Physics SB RAS, 11 Acad. Lavrentyev ave., Novosibirsk 630090, Russia

E-mail: trunov@laser.nsc.ru

Abstract. The laser wakefield acceleration (LWFA) of electrons in capillaries and gas jets followed by inverse Compton scattering of high intensity femtosecond laser pulses is discussed. The drive and scattered pulses will be produced by the two-channel multi-terawatt laser system developed in ILP SB RAS.

At the present time, the impressive progress in laser wakefield acceleration (LWFA) of charged particles gives grounds to consider LWFA as a perspective method of electron beam production in the GeV energy range [1, 2]. Electrons of this energy can be used in advanced light sources for future application. Combining the LWFA of electrons and Compton backscattering of a probe light beam on them opens the possibility to create a tabletop source of femtosecond light beams in X-ray and gamma ranges [3, 4]. This kind of source can have high coherence and polarization, quasi-monochromatism, and tunability of radiation parameters.

LWFA experiments are currently prepared at the Institute of Laser Physics (ILP) in collaboration with Budker INP. The experiments are based on the two-channel multi-terawatt femtosecond high contrast, high angle stability laser system with the pulse repetition rate 10 Hz, which is developed at ILP [5].

Traditionally, there are two basic schemes of LWFA (figure 1). In one scheme, the acceleration occurs in plasma filled dielectric capillaries. Highest electron energies are obtained this way in wide capillaries with preformed plasma that has the density minimum on the axis. In another scheme, a supersonic gas jet is used.

The capillaries (or narrow waveguides) can prevent laser beam diffraction directly by reflection of the pulse from the walls and extend acceleration length up to many Rayleigh lengths [6, 7]. We study metallic capillaries. This scheme can be considered as an extreme case of the plasma-walled capillary and can lead to high energies of accelerated electrons at moderate laser beam powers [7]. In this case, metal walls behave like a cold dense plasma with a sharp boundary. We have experimentally studied propagation of high-intense (up to 10^{17} W/cm^2) high-contrast ($<10^{-8}$) 50 fs laser pulses through the 20 mm long 50 mkm wide triangular copper channel (figure 2a) in vacuum and in helium (~ 1÷1000 mbar) [8].

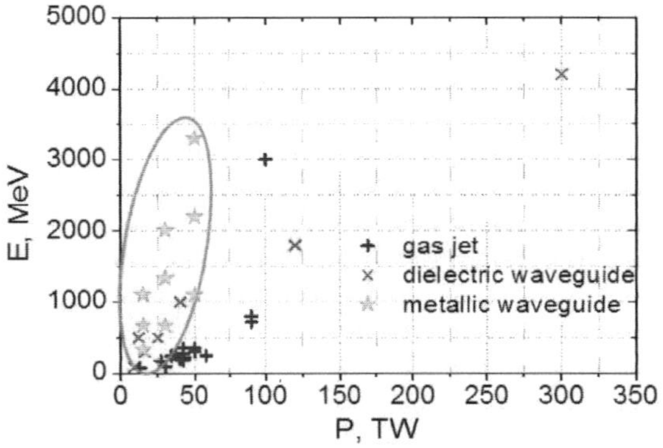

Figure 1. The energy of LWFA-accelerated electron versus the driver pulse power. The energy gain in metallic waveguides is estimated similarly to Ref.[7], other points are experimental ones taken from literature.

The main results are:

- Relative transmission 70% through the channel (15 Rayleigh lengths) is measured;
- Single mode regime of pulse propagation is demonstrated;
- No difference between vacuum and helium inside the capillary is observed;
- Decrease of the transmission to zero after hundreds or thousands of pulses is observed (figure 2b), which is caused by local plug formation [9].

As a result, we can conclude that plasma filled metal capillaries have perspectives as accelerating structures and deserve further studies, but high repetition rate sources of electrons need another LWFA scheme in its basis.

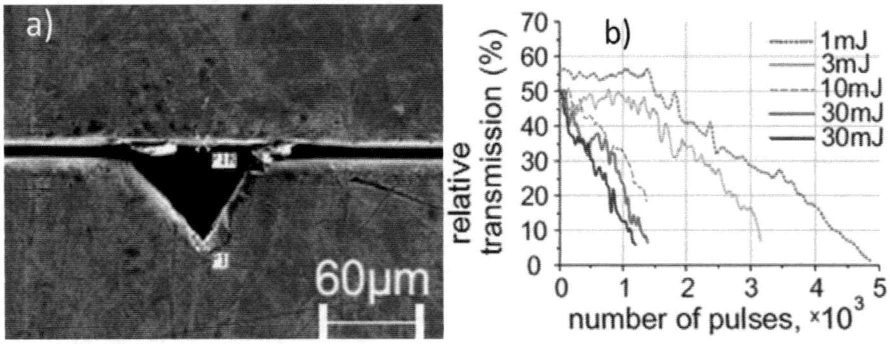

Figure 2. Capillary front view (a) and capillary degradation diagram (b).

To pursue further studies of laser-based acceleration techniques, a specialized experimental facility was designed. Main parts of the facility (Laval nozzle with pulse valve, laser beam compressor and parabolic mirrors, electron beam diagnostics) will placed into three vacuum chambers: the chamber for the laser pulse compressor, the chamber for a supersonic jet and focusing mirrors, and the chamber for electron beam diagnostics and beam dump. A sketch of the latter two chambers is shown in figure 3. In the default scenario, the same sub-PW high-contrast femtosecond laser pulse will be responsible for gas ionization and formation the plasma channel, wakefield excitation, and trapping of plasma electrons by the wave. This scenario and design are traditional for LWFA devices.

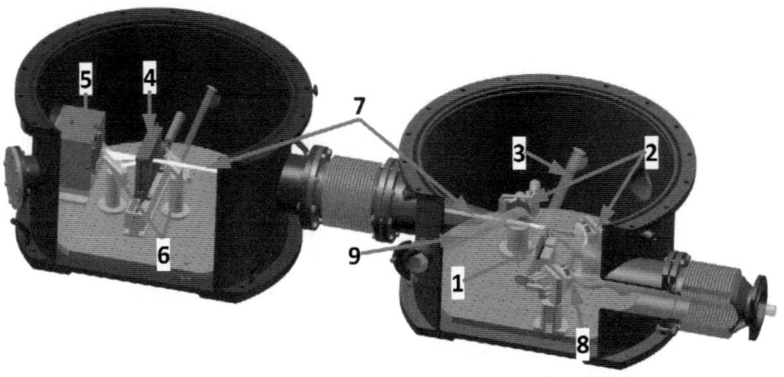

Figure 3. Two experimental chambers (without the compressor chamber): 1 – supersonic gas jet, 2 – focusing mirrors, 3 – laser beam for diagnosing the jet density, 4 – electron spectrometer magnet, 5 – Faraday cup, 6 – luminophor screens, 7 – electron beam, 8 – driving laser beam, 9 – scattered laser beam.

The numerical analysis of electron trapping and acceleration shows the following:

- For effective electrons trapping, we require to focus the laser beam down to the diameter ~ 10 mkm. Rayleigh length and effective length of acceleration will be about 0.5 mm.
- The optimum plasma density is $3 \div 5 \cdot 10^{18}$ cm^{-3}.
- In the case of the laser pulse energy 0.3 J, the maximum electron energy can reach 100 MeV. For the laser pulse energy 0.1 J, we can obtain ~ 65 MeV electrons at best.
- The maximum energy of Compton γ-quanta can reach 240 keV and 100 keV, respectively (figure 4a), and the main part of the scattered radiation will be in the angle ~ $1/\gamma$, where the relativistic factor of electrons $\gamma \sim 150 \div 200$ in our conditions, as it is shown in figure 4b.
- The supersonic gas jet is necessary to produce a flat top density profile of the jet, as required for minimizing the energy spread of electrons.

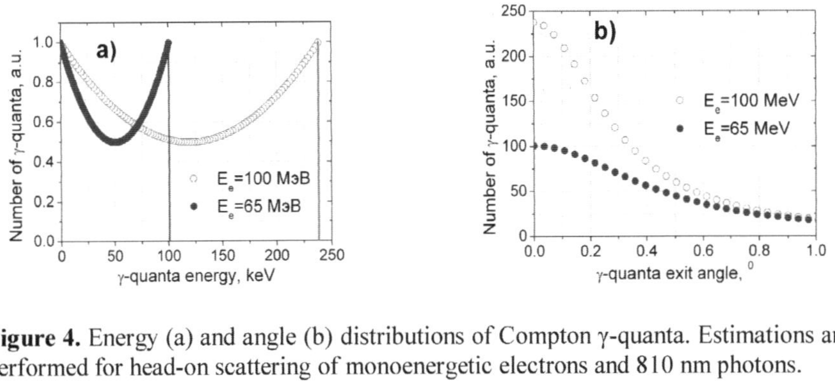

Figure 4. Energy (a) and angle (b) distributions of Compton γ-quanta. Estimations are performed for head-on scattering of monoenergetic electrons and 810 nm photons.

Numerical analysis of the gas flow shows that we can use a simple conical Laval nozzle with the waist diameter $0.3 \div 0.4$ mm, output diameter 1.5 mm, and cone angle 14°. In this case, we can obtain the jet density up to 10^{19} cm^{-3} and Mach number 4 with the input gas pressure below 10 atm. We use the fast pulse valve Festo that enables jet operation with the frequency up to 10 pulses per second.

Several variants of nozzles were produced and tested (figure 5), and now the gas system with the fast pulse valve is complete. Measured jet parameters are close to simulated ones.

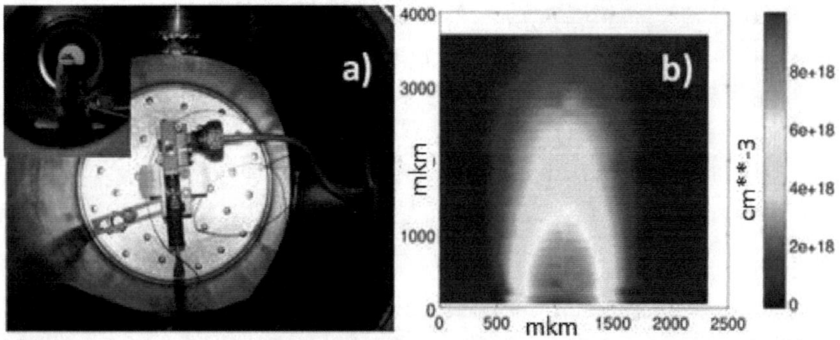

Figure 5. A Laval nozzle with the pulse valve installed at the testing stand (a), and the measured jet density profile (b).

Faraday cup (FC) and magnetic spectrometer are chosen as the basic electron diagnostic. The Faraday cup (figure 6) is designed for full stopping of 100 MeV electrons with minimized backscattering. FC is a cube-like capacitor of the size ~ 20 cm in each dimension. One plate of the capacitor is made of 6 cm thick tungsten with 1 cm thick aluminium cover plate. The capacity is as low as to 14 pF. This FC allows measuring charge of ultrashort electron bunch ($\tau \leq 1$ ps) with high precision (≤ 1 pC) without any additional complicated electronics and does not need special calibration procedures. The Faraday cup is fabricated and successfully tested under 120 MeV beam of VEPP-5 accelerator complex at Budker INP, Novosibirsk (figure 7).

Figure 6. Faraday cup in the experimental hall of VEPP-5 accelerator complex at Budker INP during tests under the electron beam.

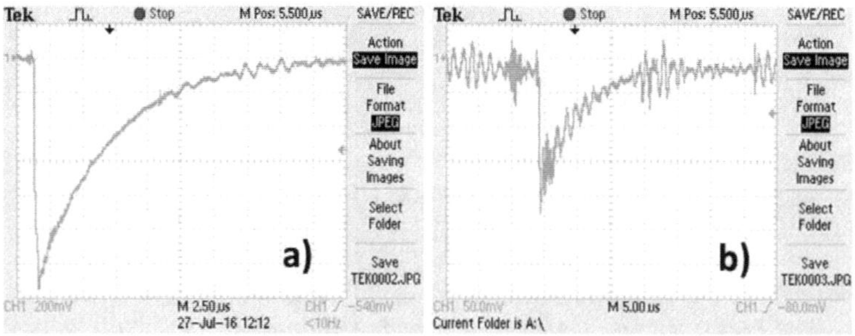

Figure 7. Signals from the Faraday cup measured during tests: single electron bunch of charges 72.3 pC (a) and 14 pC (b).

The electron beam spectrometer is based on the magnetic analyzer with the deflecting field 0.75 T. The analyzer is made from permanent magnets (figure 8). The deflected beam is detected by the Renex luminophor screen and registered by a special CCD camera. At present, both devices are fabricated and being tested under the real high-energy electron beam at Budker INP.

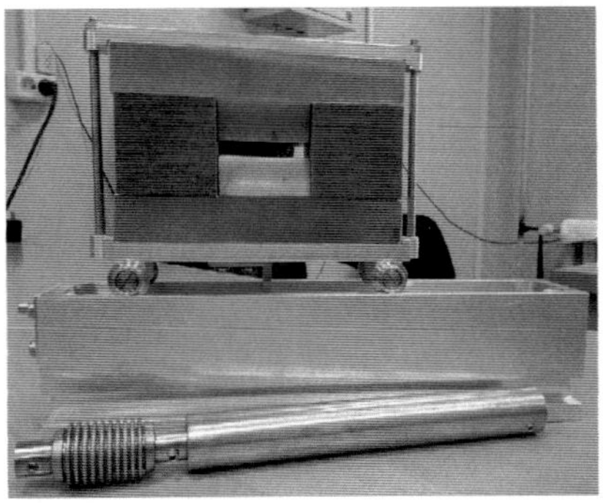

Figure 8. The deflecting magnet of the electron spectrometer. Permanent magnets are replaced by dielectric imitations for preliminary assembling.

The main immediate aim of the experimental work is producing LWFA-accelerated electrons with energies up to 100 MeV. The next stage is to produce high-energy Compton backscattered gamma-quanta with the maximum energy up to ~ 200 keV.

Acknowledgments
This work is supported in part by RAS Program "Extreme laser radiation: physics and fundamental applications", project registration number 115113010008.

References

[1] Leemans W P, Gonsalves A J, Mao H.-S., Nakamura K, Benedetti C, Schroeder C B, Tóth C, Daniels J, Mittelberger D E, Bulanov S S, Vay J.-L., Geddes C G R and Esarey E 2014 *Phys. Rev. Lett.* **113** 245002
[2] Kim H T, Pae K H, Cha H J, Kim I J, Yu T J, Sung J H, Lee S K, Jeong T M and Lee J 2013 *Phys. Rev. Lett.* **111** 165002
[3] Ta Phuoc K, Corde S, Thaury C, Malka V, Tafzi A, Goddet J P, Shah R C, Sebban S and Rousse A 2012 *Nature Photonics* **6** 308–11
[4] Chen S, Powers N D, Ghebregziabher I, Maharjan C M, Liu C, Golovin G, Banerjee S, Zhang J, Cunningham N, Moorti A, Clarke S, Pozzi S and Umstadter D P 2013 *Phys. Rev. Lett.* **110** 155003
[5] Leshchenko V E, Trunov V I, Frolov S A, Pestryakov E V, Vasiliev V A, Kvashnin N L and Bagayev S N 2014 *Laser Phys. Lett.* **11** 095301
[6] Lotov K V 1999 *Part. Accel.* **63** 139–46
[7] Lotov K V 2001 *Laser and Particle Beams* **19** 219–22
[8] Lotov K V, Gubin K V, Leshchenko V E, Trunov V I and Pestryakov E V 2015 *Phys. Plasmas* **22** 103111
[9] Gubin K V, Lotov K V, Trunov V I and Pestryakov E V 2016 *J. Appl. Phys.* **120** 113103

Detection of phase randomization in a chain of Bose condensates

A V Turlapov[1,2]

[1] Institute of Applied Physics, Russian Academy of Sciences, ul. Ulyanova 46, Nizhniy Novgorod, Russia
[2] N. I. Lobachevsky State University of Nizhni Novgorod, Nizhniy Novgorod, 603950, Russia

E-mail: turlapov@appl.sci-nnov.ru

Abstract. Interference in a long chain of Bose condensates is observed. Spatially quasi-periodic interference pattern appears even when the phases of the condensates are uncorrelated. However, the spatial fringe period depends qualitatively on whether the adjacent condensates are in phase or not. This is used for measuring the degree of phase coherence.

Chain of interfering elements is a paradigmatic model in several areas of physics. The relative phases of the field in each element determine the stationary state and the dynamics of the chain. The evolution of a chain with equal phases was studied in optics back in 1836 [1]: in the free space, the electromagnetic field from a chain of identical sources reassumes its initial form after certain propagation distance, which is presently referred to as the Talbot effect. Similar effects are observed in vacuum electronics, acoustics, plasmonics, and matter-wave optics. For matter fields the effect may be seen in the time domain, without propagation.

Solid state physics offers a variety of situations, where the phase of each element fluctuates either due to thermal effects or purely quantum reasons. In the Josephson junction chains such fluctuations drive phase transitions between superconducting and isolating states [2, 3]. Phase slips and the resulting negative interference may prevent the electric current from flowing through the chains. Another example of a chain is high-temperature superconductors, which are composed of layered structures. In systems composed of thin layers, the phase may also fluctuate in the layer plane. The amount of in-plane fluctuations signals transitions between Bose-condensed state, non-condensed Berezinskii-Kosterlitz-Thouless superfluid [4, 5], and the normal state.

The most direct information about the phases in the chain elements is obtained in interference experiments. In solids, where most interesting problems are encountered, the ability to observe interference is far below that in optics. In experiments with ultracold atoms and molecules, along with other interesting physics [6, 7], one may observe the effects known from solid state physics [8, 9, 10]. This is combined with direct observation of matter-wave interference [11].

Here, in experiment with ultracold bosonic molecules, we show that for a chain of randomly phased fields, the spatial quasi-order appears shortly after the onset of the free evolution. Within a simple model, we show that the spatial periodicity is the direct consequence of disordered phases. The spatial period differs from that of the original Talbot effect. Therefore, from the spatial period of the interference one may distinguish whether the initial state of the chain was ordered or not. Moreover, for partially disordered phases two effects combine: One may see the

IOP Conf. Series: Journal of Physics: Conf. Series **793** (2017) 012029 doi:10.1088/1742-6596/793/1/012029

Talbot effect on top of the disordered-phase interference. From the relative strength of the two effects one may judge the degree of the phase disorder.

In experiments we use a chain of molecular Bose-Einstein condensates (BECs) trapped in a one-dimensional optical lattice as shown in figure 1. Each cloud is a kinematically two-dimensional (2D) system, where most of the molecules occupy the lowest state of motion along the lattice and many states of motion along the layers. In uniform 2D systems the Bose condensation is prohibited at $T > 0$. However, in a harmonic potential a noninteracting Bose gas may condense at finite temperatures. In a repulsive BEC, a molecule locally sees a flat potential, which is the sum of the harmonic trap and the repulsive mean field. From the straight interference fringes we judge that the gas condenses at relatively high temperatures.

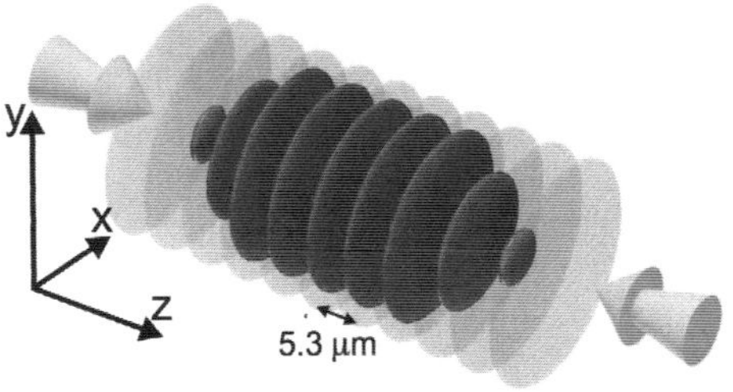

Figure 1. Trapping ultracold atoms in antinodes of a standing optical wave. The isolated clouds of atoms shown in dark red, the standing-wave intensity shown in light purple.

The experimental setup is similar to that of Ref. [12] and references therein. The bosons are weakly-bound Li_2 molecules in a long-lived excited vibrational state, each composed of two fermionic ^{6}Li atoms. A series of BECs is prepared in an optical lattice potential formed by two counter-propagating laser beams of wavelength 10.6 μm. The maxima of the standing-wave intensity are the minima of the potential. Weak transverse confinement appears due to the Gaussian shape of the mode. In the z direction, the period of the potential is $d = 5.3$ μm. A sequence of about 30 wells is populated by condensates containing about $N = 1200$–1300 molecules each.

To observe the interference, the trapping potential is turned off nearly instantaneously at $t = 0$. The condensates start to expand and interfere in free space. Dynamics in the z direction is most notable, while the expansion in the orthogonal directions is slow and unimportant here.

If the initial state were a BEC with identical phases φ_j, the evolution would show the Talbot effect: The initial periodic density distribution would reappear at the integer multiples of the Talbot time $T_d = Md^2/\pi\hbar = 1.693$ ms, where M is the molecular boson mass.

The observed interference, as shown in Figs. 2(b–e), differs from the Talbot effect dramatically. At $t = T_d/2$ the period of the interference fringes equals to the initial one in agreement with the temporal Talbot effect. At $t = T_d$, however, the prime spatial period is $2d$, which is twice larger than the biggest period possible within the Talbot effect. Snapshots taken after larger evolutions time also show that the density modulation is periodic with the period growing linearly with time. In particular, the period is $4d$ at $t = 2T_d$ and $8d$ at $t = 4T_d$ as seen in figure 2(c) and figure 2(d) respectively. The presence of spatial order and increase in periodicity are also seen in the Fourier transforms of the density distribution along z. We display the respective Fourier transforms in Figs. 2(a′)-(d′). The increase of the spatial periodicity is seen as the peaks at the fractional spatial frequencies.

We interpret the observation as the interference of molecular BECs whose phases φ_j are random relative to each other. The relative phases of the BECs establish due to the competition between the tunneling, which tends to lock the phases and dephasing, which may happen either due to quantum fluctuations or the temperature.

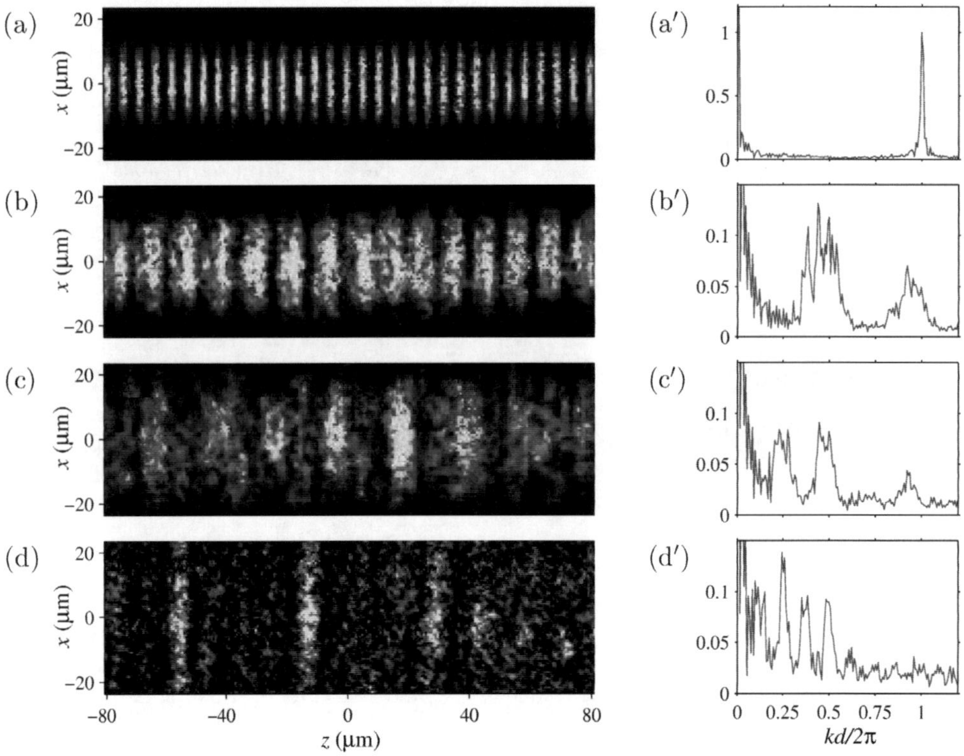

Figure 2. The interference of BECs prepared at $t = 0$ with random phases relative to each other. (a)–(d): Images taken at $t = 0$, $t = T_d$, $t = 2T_d$, and $t = 4T_d$ respectively. In each image one may see periodic density distribution. (a′)–(d′) are Fourier transforms of density distribution along z for the respective times. Increase of the spatial period is seen as peaks at fractional spatial frequencies.

The author acknowledges the financial support by the Russian Foundation for Basic Research (grant nos. 14-22-02080-ofi-m, 15-02-08464, 15-42-02638).

References

[1] Talbot H F 1836 *Philos. Mag.* **6** 401
[2] Bradley R M and Doniach S 1984 *Phys. Rev.* B **30** 1138
[3] Glazman L I and Larkin A I 1997 *Phys. Rev. Lett.* **79** 3736
[4] Berezinskii V L 1972 *Sov. Phys. JETP* **34** 610
[5] Kosterlitz J M and Thouless D J 1973 *J. Phys. C: Solid State Phys.* **6** 1181
[6] Taichenachev A V, Yudin V I and Bagayev S N 2016 *Phys. Usp.* **59** 184
[7] Ryabtsev I I, Beterov I I, Tretyakov D B, Entin V M and Yakshina E A 2016 *Phys. Usp.* **59** 196
[8] Bloch I, Dalibard J and Zwerger W 2008 *Rev. Mod. Phys.* **80** 885
[9] Giorgini S, Pitaevskii L P and Stringari S 2008 *Rev. Mod. Phys.* **80** 1215
[10] Vishnyakova G A, Golovizin A A, Kalganova E S, Sorokin V N, Sukachev D D, Tregubov D O, Khabarova K Y and Kolachevsky N N 2016 *Phys. Usp.* **59** 168
[11] Cronin A D, Schmiedmayer J and Pritchard D E 2009 *Rev. Mod. Phys.* **81** 1051
[12] Makhalov V, Martiyanov K and Turlapov A 2014 *Phys. Rev. Lett.* **112** 045301

High-efficiency lasing and spectroscopy of domestic 1%Nd:YAG and 1%Ho:YAG ceramics

S M Vatnik[1], I A Vedin[1], P F Kurbatov[1], V V Osipov[2], K E Luk'yashin[2], R N Maksimov[2] and V I Solomonov[2]

[1] Institute of Laser Physics SB RAS, 13/3 Acad. Lavrentyev ave., Novosibirsk 630090, Russia
[2] Institute of Electrophysics, Ural Branch of Russian Academy of Sciences, 106 Amundsen str., Ekaterinburg 620016, Russia

E-mail: vatnik@laser.nsc.ru

Abstract. We report on spectroscopy and high-efficiency lasing of YAG ceramics synthesized at the Institute of Electrophysics, Ural Branch, Russian Academy of Sciences. The best slope efficiency is to be 36% for 1%Nd:YAG ceramics and 40% for 1%Ho:YAG ceramics, in the latter case the emission was centred at 2090 nm. Internal losses in the samples of domestic ceramics were estimated.

1. Introduction

The lasing and spectroscopic properties of oxide laser ceramics have been studied since 1990s. Lasing in 1%Nd:YAG ceramics was obtained for the first time in 1995 [1], and already in 2001 the lasing efficiency of Nd:YAG ceramics was almost the same of single crystals [2]. Significant progress in the synthesis of oxide laser ceramics achieved in recent years allowed one, in particular, to obtain large size (up to 50 mm in diameter) samples of Nd:YAG ceramics with comparatively low intrinsic losses and residual porosity [3, 4] by domestic pressing and sintering technologies [5, 6]. Laser ceramics has some important advantages compared to single crystals, among which are the relatively simple synthesis of large-size samples, the possibility of producing multilayer and composite ceramic structures, and a comparatively low production cost [7]. In common, highly transparent ceramics are very promising for high-volume applications such as large gain media for multi-kilowatt diode pumped lasers and amplifiers. At the same time, domestic ceramics are inferior to the best foreign samples in microstructural homogeneity, which is reflected on lasing parameters, including slope efficiency and threshold pump power. In this connection, it is very interesting to study the lasing characteristics of samples synthesized by different methods. In this work, we report on optical properties and laser performance of 1%Nd:YAG and 1%Ho:YAG ceramics synthesized at the Institute of Electrophysics (IEP), Ural Branch, Russian Academy of Sciences. Detailed comparative analysis of the methods of producing highly transparent YAG ceramics from nanopowders is given in [8 – 10]. At the same time, despite a substantial progress in production methods and technology, the optimization of synthesis parameters in order to improve the micro structural homogeneity and optical characteristics of ceramics still remains a topical scientific practical problem in the field of creation of highly efficient laser media.

2. Experiment

In our previous works [11, 12] we reported on optical properties and laser performance of the samples Nd:YAG optical ceramics had the form of disks with dimensions of Ø16×3.1 mm (IREE, samples no. 8 and 10), as well as Ø11×1.1 mm and Ø47×2.1 mm (IEP, samples nos. 408, 414, 418, and B2). In the present work, the results of experimental study of spectral and lasing characteristics of ceramic disks 1%Nd:YAG with dimensions ∅11×1.5 mm and 1%Ho:YAG ∅11×1 mm (IEP, samples nos. 1570 and 1083) are presented for the first time. We also compared the optical properties and laser performance of the studied samples and the sample 1%Nd:YAG synthesized at the Baikowski Ltd. (Konoshima Chem., no. R100097, ∅10×2.2 mm). The corresponding transmission spectrum is shown in figure 1 and figure 3 for 1%Nd:YAG and 1%Ho:YAG, respectively.

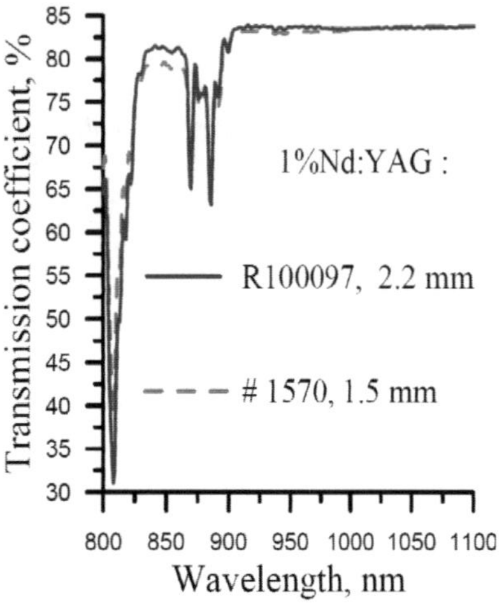

Figure 1. Transmission spectra of 1%Nd:YAG ceramic samples R10097 (solid curve) and No. 1570 (dashed curve).

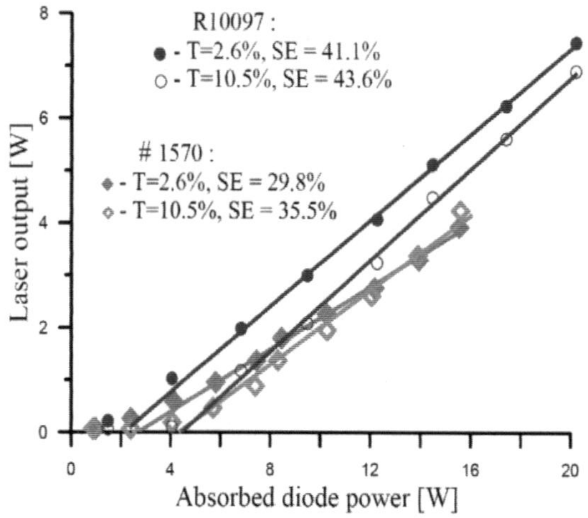

Figure 2. Dependences of the output laser power on the absorbed pump power for 1%Nd:YAG laser ceramic samples No. 1570 (rhombs) and R10097 (circles). SE – slope efficiency.

After measuring the transmission coefficients of the samples on a Shimadzu UV-3101PC spectrophotometer in the spectral range 800 – 1100 nm, the faces of Nd:YAG ceramics disks were coated by dielectric films: the first face was coated by a broadband antireflection film with the residual reflectance $R < 0.15\%$ at the pump (808 nm) and laser (1064 nm) wavelengths, and the second face was coated by a combined reflecting film consisting of a rear dielectric mirror ($R > 99.8$ % at $\lambda = 1064$ nm) and an additional metallization layer to achieve a high reflection coefficient for the pump beam incident at angles of 0–30°. For efficient heat removal, the ceramic disks were mounted on copper heat sinks with an intermediate indium foil layer 100 μm thick using different soldering and pressing methods. For pumping of ceramic samples nos. 1570 and R100097 we used two diode bars with the total power up to 50 W at a wavelength of 808 nm, whose radiation was focused on the active element using a two-mirror collimator and auxiliary optics [11], which formed an approximately round spot ~1.2 mm in diameter in the focus; the power loss in the entire optical channel did not exceed 10%. All laser experiments were performed in a short linear cavity with the physical length $L = 20$ mm, which was formed by an external concave mirror and a rear mirror deposited on the active element from the side of the heat sink as is shown in figure 1 in [11]. As an output mirror, we used a spherical concave mirror with the curvature radii $r = -40$ mm and the transmittance $T = 2.6$ % and 10.5% at the wavelength $\lambda = 1064$ nm.

The sample of 1%Ho:YAG ceramic was pumped by the intracavity scheme shown in the inset in figure 3. The cavity was formed by a highly reflecting mirror on a surface of a 250-μm-thick crystal of double potassium – lutetium tungstate 5%Tm:KLuW [13] and a partially transmitting mirror (T_{out} = 0.6% in the range 1.85 – 2.1 μm) deposited on the external surface of the ceramics. All optical surfaces inside the cavity had antireflection coatings in the spectral range of 1.85 – 2.1 μm (with the residual losses on each surface of at most 0.1% in the range 1.85 μm and ~0.5% in the range 2.1 μm). The lens made of fused KI silica with a focal length of 18 mm was placed at equal distances from the ceramics and disk element, and the physical cavity length was 50 mm. The active disk element (5%Tm:KLuW) was pumped by collimated radiation from two arrays of laser diodes (at a wavelength λ_p = 806 nm) as in [13]; the diameter of the pump beam was 0.95 mm. For reducing thermal fluxes all measurements were performed in a quasi-cw regime with the duty factor of 14 %, the duration of current pulses of diode pumping was 7 ms at a pulse repetition rate of 50 ms. The absorbed power of diode pumping was always determined as the difference between passed and incident powers, and the optical powers of light beams (pumping, generation) were detected by an Ophir L30A power meter. Lasing spectra were measured with an MDR-24 monochromator, FR-185 photoresistor and selective Unipan-233 nanovoltmeter employed as a pre-amplifier; the spectral resolution was ~0.5 nm. In general, the experimental scheme is rather close to the scheme described in [13] except for several technical improvements related to the conditions of thermal stabilization of diode bars and active elements.

3. Results and discussion

The transmission spectra of 1%Nd:YAG ceramic samples nos. 1570 and R100097 in the range of 800 – 1100 nm are shown in figure 1. As was expected, the absorption maxima in the near IR spectral region lie at a wavelength of 808 nm, which is typical for yttrium aluminum garnets doped with trivalent neodymium. Comparing the transmission coefficients T_{exp} of the ceramic samples in the transparency range (1000 – 1100 nm) with the calculated transmission coefficient T_c of the optical material with the refractive index n = 1.80 (for the ideal homogeneous ceramics taking into account only Fresnel reflection losses), it is possible to estimate the absolute absorption and scattering losses $T_{loss} = T_c - T_{exp}$. In the course of preliminary study of the lasing characteristics of the samples, it was found that a change in the physical cavity length by 20 – 40 mm only slightly (within 5 %) affects the output laser power, while an increase in the cavity length by more than 40 mm causes a sharp decrease in the power because the cavity leaves the stability region. Therefore, all laser experiments were performed with the cavity length L = 20 mm, at which the output laser power was maximal. At these cavity parameters, the Gaussian beam diameter is ~0.23 mm. Since the pump beam spot diameter was considerably larger (~1.2 mm), the comparison of laser ceramics was performed with respect to the maximum output power and lasing efficiency, the mode composition being ignored. The results of investigation of the lasing characteristics of samples of 1%Nd:YAG ceramics are presented in figure 2 and in Table 1. In these experiments, the optical power of the pump unit was increased to 50 W (cw regime) due to the use of an additional diode array. Comparing the slope efficiencies of lasing in the ceramic samples Nos 1570 and R100097 at different transmittances of the output mirror (29.8 % and 35.5 % for T_{out} = 2.6% and 10.5%), we estimated the total losses in the laser cavity [14] to be 0.7% and 0.2%, respectively. Assuming that the absorption and scattering losses in the volume of the reference sample R10097 are considerably lower than the losses at optical surfaces (antireflection coatings and mirrors, including the output one), we find that the losses in the studied sample No. 1570 are ~0.5% at the laser wavelength (two passes), i.e., the specific losses are ~1.6×10^{-2} cm^{-1}. At these losses, the slope efficiency of ceramic sample No. 1570 (35.5%) reaches 80% of the corresponding parameter for the reference sample R10097 (43.6%).

Lasing characteristics were studied in sample No. 1083 of 1 %Ho:YAG ceramics with the size of Ø11×1 mm. The dependence of the average output power of the laser on the average absorbed pump power for the cavity geometry described above is shown in figure 4. Curve (*1*) was obtained for the optical 1%Ho:YAG ceramics with dielectric coatings and characterises the radiation power with wavelengths of 1.85 and 2.09 μm. Curve (*4*) corresponds to the radiation power with λ = 1.85 μm (obtained by using a plane dielectric mirror with the transmittance of 0.6 % instead of the ceramics).

Curves (*2*) and (*3*) correspond to the powers at these wavelengths obtained by recalculating the relative intensities of the lasing spectra according to data from figure 5. Starting with a certain threshold pump power (~1 W according to figures 4 and 5), there are two light fields excited simultaneously in the cavity, which correspond to the $^3F_4 \rightarrow {}^3H_6$ transition of Tm^{3+} ions in the KLuW matrix and the $^5I_7 \rightarrow {}^5I_8$ transition of Ho^{3+} ions in the YAG matrix. At a wavelength of 1.85 μm the absorption of 1%Ho:YAG ceramics (in two passes) is 3.0% in the regime of unsaturated absorption, and in the regime of lasing taking into account inversion ~20 % [15] (see Figs 4, 5) the absorption is T_{abs} = 2.4%.

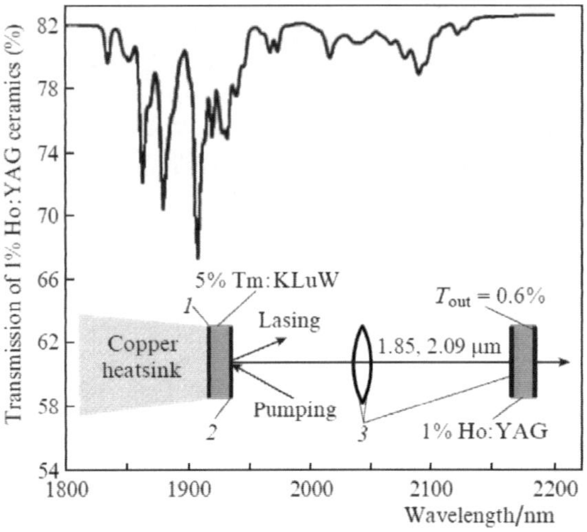

Figure 3. Transmission spectrum of the sample of optical 1%Ho:YAG ceramics with a thickness of 1.0 mm. The inset shows the schematic diagram of the experiment: (*1*) highly reflecting dichroic mirror with the reflection coefficient R > 99.9 % in the range 1.85 – 2.1 mm and at λ = 0.8 μm; (*2*, *3*) antireflection coatings in the region specified.

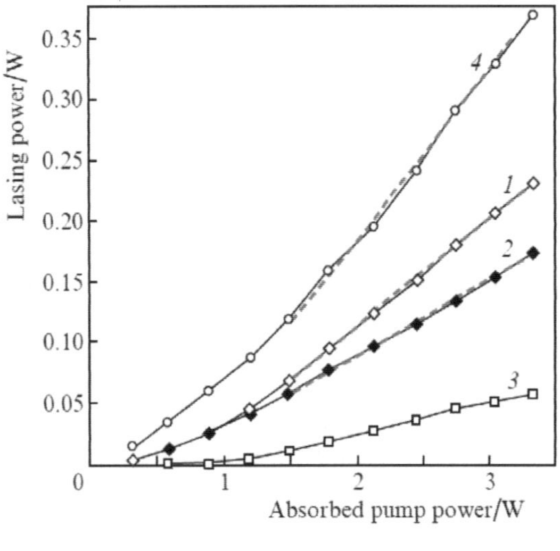

Figure 4. Output lasing power vs. absorbed pump power: (*1*) for the optical 1%Ho:YAG ceramics and radiation with λ = 1.85 and 2.09 μm; (*2*) and (*3*) for the same ceramics and radiation with λ=1.85 and 2.09 μm, respectively, recalculated from curve (*1*) taking into account relative spectral intensities (see figure 5); (*4*) for the dielectric plane mirror instead of the ceramics) with the transmission Tout = 0.6 % at 1.85 μm. Dashed curves in (*1*), (*2*) and (*4*) are linear approximations of the corresponding dependences for calculating the slope efficiency η_{exp}.

Thus, the total losses in the cavity with 1%Ho:YAG ceramics are $T_{tot} = T_{abs} + T_{out} + T_{loss}$, where T_{loss} are parasitic losses inside the cavity (per single pass) to absorption and scattering reduced to the wavelength 1.85 μm. If the ceramics is substituted for the outcoupling mirror with the transmission

T_{out}, then absorption is absent, i.e., $T_{abs} = 0$. According to general relations [16], the slope efficiency η_{exp} determined experimentally can be written in the form

$$\eta_{exp} = \eta T_{out} /(T_{abs} + T_{out} + T_{loss}), \quad (1)$$

where η is the 'limiting' lasing slope efficiency for the ideal lossless cavity. By applying (1) to dependences (*4*) and (*2*) in figure 4 one obtains the relation

$$(0.6\% + T_{loss})/(3.0\ \% + T_{loss}) = 6.2\%/14.0\% = 0.44 \quad (2)$$

(here, 14.0% and 6.2% correspond to the lasing slope efficiencies for these dependences, respectively).

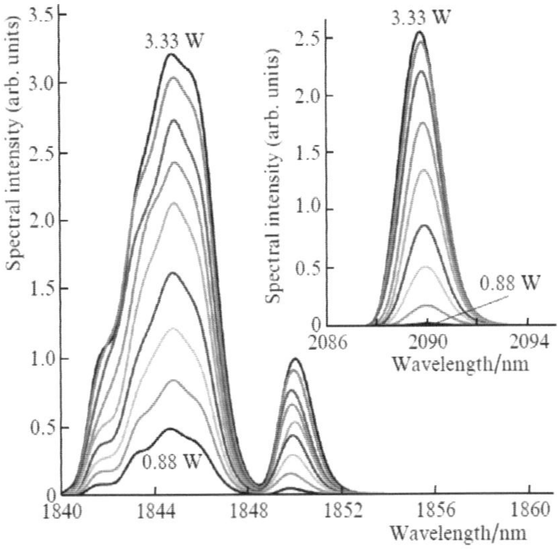

Figure 5. Spectral intensities of lasing at the absorbed pump power of 0.88, 1.19, 1.49, 1.79, 2.13, 2.45, 2.75, 3.05 and 3.33 W.

Hence, relations (1) and (2) give a possibility to estimate the parameter T_{loss} which is ~1.2% at η = 45%; this is in good agreement with results from [17, 18]. The estimate T_{loss} ~ 1.2 % also seems reasonable taking into account a large number of optical surfaces inside the cavity. According to these estimates, the light power absorbed by ceramics is five times greater than the output lasing power corresponding to curve (*2*) in figure 4. On the other hand, the total losses at a wavelength of 2.09 μm on antireflection coatings are approximately 3 % [from our estimates $T_{loss}(2.09) = 3.0\% \pm 1\%$], i.e., the intracavity losses are almost five times greater than the transmission coefficient of the outcoupling mirror $T_{out}(2.09) = 0.6\%$. Thus, by comparing curves (*2*) and (*3*) with (*1*) taken into account one can estimate the 'limiting' lasing slope efficiency for 1%Ho:YAG ceramics on the $^5I_7 \rightarrow {}^5I_8$ transition under intracavity pumping with respect to the absorbed power at a wavelength of 1.85 μm: $\eta = 40\% \pm 10\%$. This value is approximately twice the similar parameter for Ho:YAG crystals [15, 19], which may be related with possible structural microscopic defects in ceramics. Note that in some papers, the lasing parameters of holmium crystals were studied under intracavity pumping by radiation of thulium lasers; in that case the lasing slope efficiency (with respect to the absorbed power of diode pumping) was from 20 % [20] to ~ 40 % [21]. In the present work for the 1%Ho:YAG ceramic sample under investigation this parameter was 2.6 % [curve (*3*) in figure 4], which is explained, first of all, by the fact that intracavity losses are substantially higher than the transmission of the outcoupling mirror. Hence, for further enhancing the efficiency and output power of lasing, the cavity parameters should still be optimized.

Table 1. Lasing characteristics of samples YAG optical ceramics.

Sample number	Specific losses T_{loss}/H	Slope efficiency
Nd:YAG (IREE, IEP) [11]	10…30 % cm^{-1}	5…10%
1% Nd:YAG (IEP) [12]	6 % cm^{-1}	19%
0.8% Nd:YAG (IREE) [12]	5 % cm^{-1}	21%
1% Nd:YAG (#1570 IEP)	**1.6 % cm^{-1}**	**36%**
1% Nd:YAG (R10097 Kon.Chem.)	**> 0.2 % cm^{-1}**	**44%**
1% Ho:YAG (#1083 IEP), λ = 2.1 μm	**5 % cm^{-1}**	**~40%**

4. Conclusions

Laser oscillation has been demonstrated for all samples under study, and the threshold and slope efficiency prove to be in a good qualitative agreement with the estimated ceramic losses. In particular, the best slope efficiency of 1%Nd:YAG (36%) and 1%Ho:YAG (40%) ceramics synthesized at IEP quite correlates with its low losses, 1.6×10^{-2} cm^{-1} and 5×10^{-2} cm^{-1}, respectively. We have shown that the value of losses ~ 1% cm^{-1} seems to be an important milestone for "bad" or "good" ceramics in concern to their laser performance. Further improvement of the optical homogeneity of domestic ceramics remains a topical scientific and technical problem, which is of high priority for the development of high-efficiency multi-kilowatt solid-state laser systems.

Acknowledgments

This work was supported by the Research Programs of Presidium of Russian Academy of Sciences "Extreme Laser Light: Physics and Fundamental Applications" (project no. 4.4-2016) and. R@D program "Fundamentals of Breakthrough Double Technologies for National Security" (project no. 32).

References

[1] Ikesue A, Kinoshita T, Kamata K *et al.* 1995 *J. Am. Ceram. Soc.* **78** 1033
[2] Lu J, Murai T, Takaichi K, Uematsu T *et al.* 2001 *Appl. Phys. Lett.* **78** 3586
[3] Osipov V V, Khasanov O L, Shitov V A *et al.* 2008 *Ros. Nanotekhn.* **3** 98
[4] Bagaev S N, Osipov V V, Solomonov V I *et al.* 2012 *Opt. Mater.* **34** 1482
[5] Bagaev S N, Osipov V V, Shitov V A *et al.* 2012 *Atm. and Ocean. Opt.* **25** 292
[6] Bagaev S N, Osipov V V, Solomonov V I *et al.* 2012 *Perspekt. Mater.* **4** 18
[7] Sanghera J, Kim W, Villalobos G *et al.* 2012 *Materials* **5** 258
[8] Kotov Yu A, Osipov V V, Ivanov M G *et al.* 2005 *Proc. SPIE Int. Soc. Opt. Eng.* **6344** 1
[9] Osipov V V, Kotov Yu A, Ivanov M G *et al.* 2006 *Laser Phys.* **16** 116
[10] Osipov V V, Khasanov O L, Shitov V A *et al.* 2008 *Ros. Nanotekhnol.* **3** 98
[11] Vatnik S M, Osipov V V, Vedin I A, Kurbatov P F 2013 *Quantum Electron.* **43** 288
[12] Vatnik S M, Osipov V V, Luk'yashin K E *et al.* 2014 *Quantum Electron.* **44** 585
[13] Vatnik S M, Vedin I A, Kurbatov P F, Pavlyuk A A 2014 *Quantum Electron.* **44** 989
[14] Caird J A, Payne S A, Staber P R *et al.* 1988 *IEEE J. Quantum Electron.* **24** 1077
[15] Lippert E, Fonnum H *et al.* 2010 *Opt. Express* **18** 26475
[16] Svelto O 2010 *Principles of Lasers* (Heidelberg: Springer)
[17] Vatnik S M, Vedin I A *et al.* 2010 *Laser Phys. Lett.* **7** 435
[18] Vatnik S M, Vedin I A, Segura M *et al.* 2012 *Opt. Lett.* **37** 356
[19] Haakestad M W, Fonnum H, Arisholm G *et al.* 2010 *Opt. Express* **18** 25379
[20] Stoneman R C, Esterowitz L 1992 *Opt. Lett.* **17** 736
[21] Schellhorn M, Hirth A, Kieleck C 2003 *Opt. Lett.* **28** 1933

Quantum speed meter based on dissipative coupling

Sergey P Vyatchanin
Faculty of Physics, M.V. Lomonosov Moscow State University, Leninskie Gory, Moscow 119991, Russia

Andrey B Matsko
OEwaves Inc., 465 North Halstead Street, Suite 140, Pasadena, CA 91107

Abstract. We consider dissipative coupling Fabry-Perot cavity, i.e. its input mirror transmittance depends on position of probe mass. We show that dissipative coupling provide possibility to realize quantum speed meter by natural way, without additional setup for subtraction of position $x(t)$ and delayed position $x(t-\tau)$. Quantum speed meter is a quantum non demolition (QND) meter which allow to overcome Standatd Quantum Limit — we show it for speed meter based on dissipative coupling.

The sensitivity of the mechanical position measurement in an opto-mechanical system usually is limited due to quantum back action by so called standard quantum limit (SQL) [1, 2]. The SQL was studied in various configurations ranging from macroscopic kilometer-sized gravitational wave detectors [3] to microcavities [4, 5]. Detection of a classical force acting on a mechanical degree of freedom of an opto-mechanical system is an example of such a measurement. SQL of force measurement is not a fundamentally unavoidable limit. It can be overcome with variational measurement [3, 6, 7], opto-mechanical speed measurement [8, 9], and measurements using opto-mechanical rigidity [10, 11].

In this report we show that usage of dissipative coupling for detection of small signal force acting on a free mechanical test mass (figure 1) provides a direct possibility to realize speed meter and to beat the SQL [12].

The idea of speed meter enabling the force detection is that output optical field contains information about difference $\sim [x(t)-x(t-\tau)] \simeq \tau\dot{x}(t)$ (where x is the position, τ is a delay time). In frequency domain it corresponds to $\sim x(\Omega)\left(1-e^{-i\Omega\tau}\right) \simeq -i\Omega\tau\, x(\Omega)$. In proposed earlier speed meters such subtraction is realized by interaction of a mechanical degree of freedom with two coupled optical modes — it increases complexity of experimental realization [8, 9]. We show that the dissipative coupling without the dispersive one [13] provides this kind of subtraction automatically, enabling a simple way of measurement of speed of a free test mass, which is a quantum nondemolition (QND) variable, suitable for a sensitive detection of a classical force.

Our analysis gives formula for test mass position $\hat{x}$ and output field a_{out} in frequency domain:

$$\hat{x}_\Omega = -\frac{\hbar\eta A\sqrt{\kappa_0}}{m\Omega(\kappa_0-2i\Omega)}(\hat{b}_+-\hat{b}_-^\dagger)-\frac{F_\Omega}{m\Omega^2}\,,\qquad a_{out+}=\frac{\kappa_0+2i\Omega}{\kappa_0-2i\Omega}\hat{b}_+ +\eta A\sqrt{\kappa_0}\frac{-i\Omega x_\Omega}{\kappa_0-2i\Omega}\,, \quad (1)$$

where $\hat{x}_\Omega$, a_{out+}, F_Ω, and $\hat{b}_\pm \equiv \hat{b}(\omega_0 \pm \Omega)$ are Fourier amplitudes of corresponding operators, ω_0

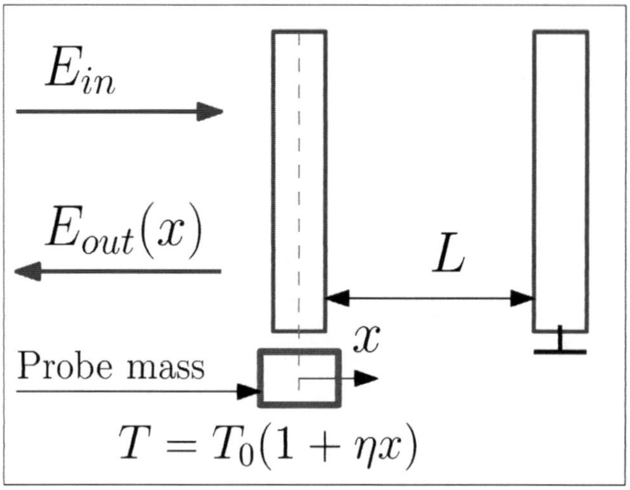

Figure 1. Speed meter based on dissipative coupling. Resonantly pumped mode of an optical Fabry-Perot cavity is dissipatively coupled with a mechanical degree of freedom represented by a free mass m a force of interest, F_s, acts upon. The coupling changes the power transmission coefficient of the cavity front mirror $T(x)$. In liner approximation this change can be presented as $T(x) = T_0(1 + \eta x)$, where x is the mirror displacement, η is a coupling coefficient.

is resonant frequency of cavity. We see that the output field provides information on speed of the probe mass $-i\Omega x$ if $\kappa_0 \gg \Omega$, but not the displacement.

In order to find sensitivity of detection of the classical force F_Ω acting on the test mass we find phase and amplitude quadratures q_a, q_p of output field through similar quadratures d_p, d_a of input one:

$$q_p = \frac{\kappa_0 + 2i\Omega}{\kappa_0 - 2i\Omega}\, d_p\,, \qquad q_a = \frac{\kappa_0 + 2i\Omega}{\kappa_0 - 2i\Omega}\left\{d_a - \mathcal{Q}\cdot d_p + \sqrt{2\mathcal{Q}}\cdot f_s\right\}, \tag{2}$$

where

$$\mathcal{Q} \equiv \frac{P\kappa_0^2}{\kappa_0^2 + 4\Omega^2}\,, \quad P \equiv \frac{2|A|^2\eta^2}{m\kappa_0}\,, \quad f_s \equiv e^{i\beta}\,\frac{F_s(\Omega)}{F_{SQL}}\,, \quad F_{SQL} \equiv \sqrt{2m\Omega^2}\,.$$

Here f_s is the signal force normalized by Standard Quantum Limit (SQL). We see that q_a contains shot noise term ($\sim d_a$), back action term ($\sim \mathcal{Q}d_p$) and signal term ($\sim \sqrt{2\mathcal{Q}}f_s$). We assume that output signal is registered by balanced homodyne detector, i.e. we are able to measure arbitrary quadrature amplitudes q_θ of the output light: $q_\theta = q_a\cos\theta + q_p\sin\theta$. One can find minimal f_s choosing homodyne angle θ by optimal way:

$$f \equiv \frac{e^{-i\beta}}{\sqrt{2}}\left\{\frac{d_a}{\sqrt{\mathcal{Q}}} + \left(-\sqrt{\mathcal{Q}} + \frac{\tan\theta}{\sqrt{\mathcal{Q}}}\right)d_p\right\}, \qquad \tan\theta_{opt} = \frac{P\kappa_0^2}{\kappa_0^2 + 4\Omega_0^2}\,. \tag{3}$$

Single-sided spectral density $S_f(\Omega)$ of noise f may be expressed through spectral densities S_a and S_p of corresponding quadratures d_a and d_p. We assume that input light is in coherent state (c.f.[3]), i.e. $S_a = S_d = 1$. The plots of S_f presented in figure 2, show that the sensitivity of the

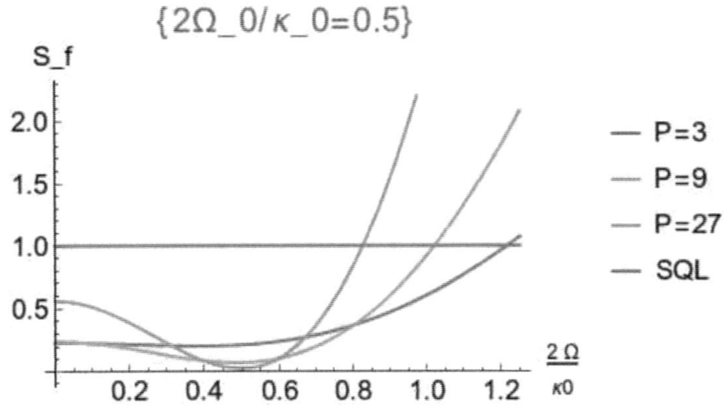

Figure 2. (color online) Normalized spectral density of the noise f for the optimal procedure of the force measurement using the speed meter technique evaluated for the optimal homodyne angle at frequencies $2\Omega/\kappa_0 = 0.5$ for different power parameters P QP. The horizontal line describes SQL.

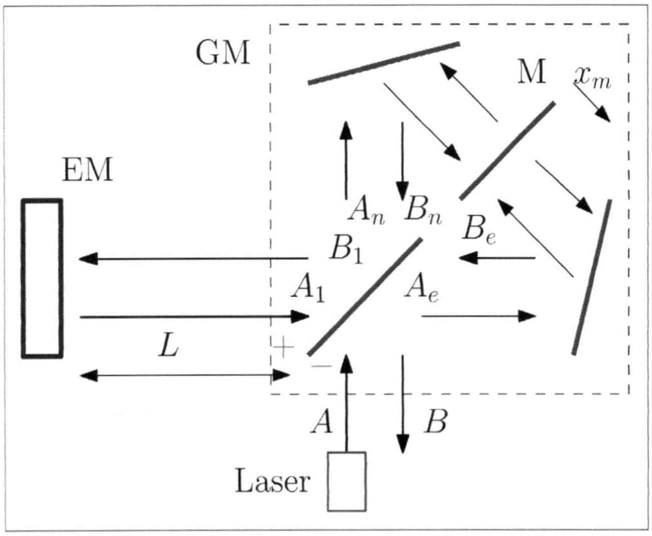

Figure 3. Michelson-Sagnac Interferometer as a input generalized mirror (GM) of a Fabry-Perot cavity. Perfectly reflecting mirror M represents a free test mass. Distance L between the generalized mirror and the end mirror (EM) is much larger than the size of interferometer.

force detection better than SQL can be achieved in a relatively large frequency band. Realization of pure dissipative coupling was suggested [13] on base of Michelson-Sagnac interferometer (see figure 3).

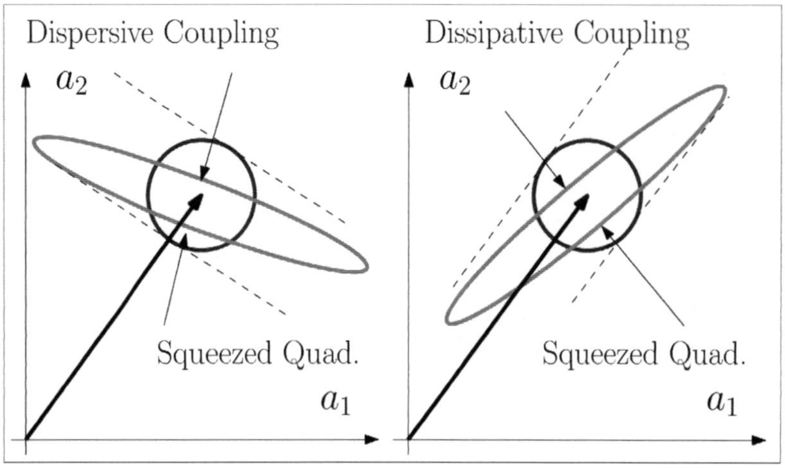

Figure 4. Illustration of the difference between types of ponderomotive squeezing achieved in the cases of dispersive and dissipative coupling. The blue circles on the both diagrams represent distribution of quantum fluctuations of the input wave prepared in the coherent state, red ellipses describe quantum squeezing of the fluctuations of the output waves.

Lets compare speed meter with displacement meter based on dispersive coupling (end mirror of resonantly pumped cavity is free mass) we rewrite output amplitudes from [3]:

$$q_a^{disp} = e^{-2i\alpha}\, d_a\,, \qquad q_p^{disp} = e^{-2i\alpha}\left\{d_p - \mathcal{K}\, d_a + \sqrt{2\mathcal{K}}\, f_s^{disp}\right\}, \tag{4}$$

with

$$\mathcal{K} \equiv \frac{8\kappa_0\omega_0^2 A^2}{mL^2\Omega^2\left(\kappa_0^2 + 4\Omega^2\right)}\,, \quad e^{-2i\alpha} \equiv \frac{\kappa_0 + 2i\Omega}{\kappa_0 - 2i\Omega}\,.$$

We see that amplitude quadrature amplitude q_a in (2) is replaced with phase quadrature amplitude q_p^{disp} in (4) as well as $q_p \Rightarrow q_a^{disp}$. Ponderomotive squeezing takes place for the both cases of coupling at large enough pump power ($|\mathcal{Q}|,\ |\mathcal{K}| \gg 1$). However, squeezed quadratures are different (figure 4).

Summing up, purely dissipative coupling naturally provides a realization of a quantum speed meter.

References

[1] Braginsky V B 1968 *Sov. Phys. JETP* **26** 831
[2] Braginsky V B and Khalili F Ya 1992 *Quantum Measurement* (Cambridge: Cambridge University Press)
[3] Kimble H J, Levin Yu, Matsko A B, Thorne K S and Vyatchanin S P 2001 *Phys. Rev.* D **65** 022002 (*Preprint* arXiv:gr-qc/0008026v2)
[4] Kippenberg T and Vahala K 2008 *Science* **321** 1172
[5] Dobrindt J M and Kippenberg T J 2010 *Phys. Rev. Lett.* **104** 033901
[6] Vyatchanin S P and Matsko A B 1993 *Sov. Phys. JETP* **77** 218–21
[7] Vyatchanin S and Zubova E 1995 *Phys. Lett.* A **201** 269–74
[8] Braginsky V B and Khalili F Ya 1990 *Phys. Lett.* A **147** 251–6
[9] Braginsky V B, Gorodetsky M L, Khalili F Y and Thorne K S 2000 *Phys. Rev.* D **61** 044002
[10] Braginsky V B and Khalili F Ya 1999 *Phys. Lett.* A **257** 241

[11] Khalili F Ya 2001 *Phys. Lett.* A **288** 251 (*Preprint* arXiv:gr-qc/0107084)
[12] Vyatchanin S and Matsko A 2016 *Phys. Rev.* A **93** 063817
[13] Xuereb A, Schnabel R and Hammerer K 2011 *Phys. Rev. Lett.* **107** 213604

Characterizing and imaging magnetic nanoparticles by optical magnetometry

A Weis, S Colombo, V Dolgovskiy, Z D Grujić, V Lebedev and J Zhang

Physics Department, University of Fribourg, Chemin du Musée 3, 1700 Fribourg, Switzerland

E-mail: antoine.weis@unifr.ch

Abstract. We review our ongoing work on deploying optical (atomic) magnetometry for measuring the magnetic response of magnetic nanoparticle (MNP) samples, yielding MNP size distributions, and other sample parameters like Néel relaxation time τ, saturation magnetisation M_s, anisotropy constant K and magnetic susceptibility χ. We address magnetorelaxation (MRX) signals, in which the decaying magnetisation $M(t)$ following a magnetising pulse is recorded by a single atomic magnetometer or by a novel magnetic source imaging camera (MSIC) allowing spatially resolved MRX studies of distributed MNP samples. We further show that optical magnetometers can be used for a direct measurement of the $M(H)$ and $\mathrm{d}M/\mathrm{d}H(H)$ dependencies of MNP samples, the latter forming the basis for an optical magnetometer implementation of the MPI (Magnetic Particle Imaging) method. All experiments are in view of developing biomedical imaging modalities.

1. Introduction

The deployment of magnetic nanoparticles (MNPs) for biomedical applications has steadily increased in the past decade [1]. The use of MNPs as MRI contrast agents and for hyperthermia is well investigated. More importantly, functionalised MNPs injected into the blood stream have a high potential for targeted drug delivery to biological entities, such as tumors or organs. In view of clinical applications, most MNP-based methods require quantitative imaging methods for monitoring the spatial MNP distribution in the biological tissue. Magnetorelaxation (MRX, [2]) imaging and Magnetic Particle Imaging (MPI, [3]) are two imaging modalities that are actively being developed towards this goal. On the other hand, there is also an evident need for measuring and optimising the MNP properties, in particular their size distribution. The average size and the size dispersion as well as specific magnetisation and magnetisation relaxation rates of MNPs in sample are important parameters in view of obtaining the highest possible MNP detection sensitivity with the lowest possible quantity of injected MNPs.

The imaging and characterisation of MNP samples both rely on the particles superparamagnetic character, related to the fact that MNPs of sufficiently small size are single ferromagnetic domains. As such, the magnetisation versus magnetic field dependence, $M(H)$, of an MNP ensemble is described by the Langevin function

$$L\left(x=\frac{H}{H_k}\right) = \frac{M(x)}{M_s} = \coth x - \frac{1}{x}\,, \tag{1}$$

where the saturation field

$$H_k = \frac{k_B T}{\mu_0 M_s V}, \tag{2}$$

is a function of the (particle size and material dependent) saturation magnetisation, M_s, and the MNPs core volume, V. The characteristic features of this magnetisation behaviour are (a) the absence of hysteresis and (b) the relatively modest value (few mT) of the saturation field $B_k = \mu_0 H_k$.

All applications mentioned above rely on magnetising the MNP located, say at $\vec{r}$=0, by a magnetic excitation field $\vec{H}(\vec{r}=0)$, and detecting the magnetic flux density $\vec{B}(\vec{r}) \propto \vec{M}(0)$ produced by the sample magnetisation $\vec{M}(0)$ with a magnetometer located at $\vec{r}$. In most conventional magnetic MNP experiments one deploys excitation fields $\vec{H}(\vec{r}=0,t)$ that harmonically oscillate at $\omega_{\rm exc}$, in combination with detection of $\vec{B}(\vec{r},t)$, or rather $d\vec{B}(\vec{r},t)/dt$ by a pick-up loop based on Faraday's induction law. Because of the time-derivative, the pick-up signal grows like $\omega_{\rm exc}$, thus favouring high-frequency excitation.

We strongly believe that atomic magnetometers (AM) present an interesting alternative to the pick-up loop detection of MNP signals, since their performance does not degrade with decreasing frequency, thus allowing efficient low-frequency, and even DC detection. Here we review several proof-of-principle experiments that we have performed along these lines.

2. Magneto-relaxation (MRX)

Magnetised MNPs, when immobilised (e.g., when attached to a cell surface, or embedded in a matrix/tissue) have Néel relaxation times in the range of seconds to minutes, depending on the particles' size. This property is used in MRX to discriminate signals from blocked MNPs against signals produced by MNPs in body fluids, whose magnetisation relaxes on the sub-ms time-scale. MRX consists in magnetising the particles by a static field $\vec{H}_{\rm exc}$ of several mT for a given amount of time, and then monitoring the decaying magnetic flux density's magnitude $B(t)$, after $\vec{H}_{\rm exc}$ has been switched off. In [4] we have shown that two AMs (operated as a gradiometer) can be used to monitor $B(t)$ decaying from a few nT down to a few pT over two minutes. In that paper we have also revised the functional time-dependence of the MRX-decay, and discussed under which simplifying assumptions the signal reduces to

$$B(t) \propto \ln\left(1 + \frac{\tau}{t}\right). \tag{3}$$

We have compared relaxation curves from various samples having average MNP particle radii ranging from 14 to 21 nm. Fits of the results have further allowed us extracting the saturation magnetisation M_s and the so-called magnetic anisotropy constant K, defining the Néel relaxation time

$$\tau = \tau_0 \exp\frac{KV}{k_B T} \quad \text{with} \quad \tau_0 \approx 10^{-9}\text{s}, \tag{4}$$

and to study the dependence of M_s and K on particle size.

3. Magnetic source imaging camera (MSIC)

The conventional MRX recording using a single magnetometer does not yield direct information about the MNPs' spatial distribution. The latter can be assessed by simultaneously recording MRX signals with an AM array. Based on the methods described by Fescenko and Weis [5], we have developed a magnetic source imaging camera (MSIC), i.e., a device yielding a direct visualisation of the two-dimensional spatial distribution of a specific component B_i of the flux density vector produced by a magnetised MNP sample. The MSIC principle is sketched in figure 1a. A sheet of circularly-polarised resonant laser light prepares a layer of spin-polarised

Cs atoms contained with a buffer gas in a cubic vapour cell. A CCD camera records fluorescence from that layer.

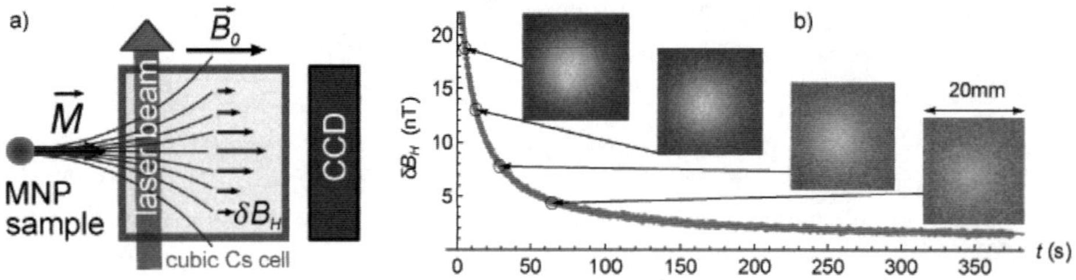

Figure 1. a): MSIC principle: The fluorescence emitted by each point $\vec{r}$ of the spin-polarised atomic layer depends on the projection δB_H of $\delta\vec{B}_{\rm MNP}(\vec{r})$ on $\vec{B}_0$; b) CCD frames representing fluorescence produced by dried MNP sample and the (time-logarithmic) MRX decay of the integrated fluorescence signal, fitted by equation 3.

A weak magnetic field $\vec{B}_0$ defines – by virtue of the Hanle effect [6] – the detected component of the flux density $\delta\vec{B}_{\rm MNP}(\vec{r})||\hat{B}_0$ produced by the object of interest, here a magnetised MNP sample. $\delta\vec{B}_{\rm MNP}(\vec{r})$ changes the total field, and hence the atomic spin polarisation at each position $\vec{r}$ in the sheet. Since the fluorescence yield depends on the degree and orientation of the local spin polarisation, the fluorescence from each point thus changes in a specific manner as discussed in [5, 6]. Figure 1.b shows CCD frames that represent the fluorescence changes (proportional to δB_H) induced by magnetised MNPs [6] and the samples magnetisation decay (MRX signal).

4. AC susceptometry (ACS) and Magnetic Particle Imaging (MPI)

In contrast to blocked MNPs, the magnetisation of MNPs suspended in a fluid undergoes a very fast (sub-ms) Brown relaxation through rotational diffusion. This fact prevents their detection by MRX, but allows the particles to react quasi-instantaneously to oscillating drive fields $\vec{H}(t)$ producing a related response $\vec{B}_{\rm MNP}(t)$. Because of the nonlinear $B_{\rm MNP}(H)\propto M_{\rm MNP}(H)$ relation given by equation 2, a monochromatic drive oscillating at $\omega_{\rm drive}$ produces odd overtones at $(2n+1)\omega_{\rm drive}$ in the Fourier spectrum of the $B_{\rm MNP}(t)$ response. We have developed AM-based methods [7] allowing the direct recording (figure 2) of the $B_{\rm MNP}(H)$ dependence and its derivative $dB_{\rm MNP}/dH(H)$. The method can be applied to MNP samples containing down to 1 μg of iron and can be used to extract MNP size distribution [8].

The fact that the $dB_{\rm MNP}/dH$ signal peaks at H=0 can be used for measuring spatial MNP distributions. Suppose that a bulk sample containing an inhomogeneous MNP distribution is exposed to an inhomogeneous drive field, such that H=0 at one specific point (zero field point, ZFP) in the sample. Suppose further that the H field in other parts of the sample is sufficiently strong to saturate the MNP magnetisation at those points. The magnetometer will then detect $dB_{\rm MNP}/dH$ signals only from MNPs located at or near the ZFP. A record of the magnetometer signal while scanning the ZFP position through the bulk will thus yield information of the spatial MNP distribution.

The method of Magnetic Particle Imaging (MPI) was invented a decade ago by researchers at Philips [3]. It relies, in its originally proposed an realised variant, on the analysis of the MNP response in the frequency-domain. The method deployed by us is related to a variant of MPI known as X-space MPI that was demonstrated so far only with pick-up coil detection [9].

Currently, both frequency-space and X-space implementations of MPI deploy drive fields oscillating at 25 kHz, together with pick-up coil detection. The high frequency at which the high

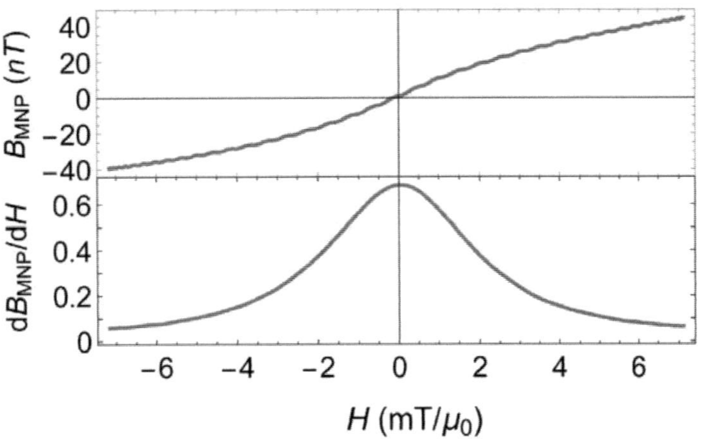

Figure 2. Magnetisation curve and its derivative of a water-suspended MNP (Ferrotec EMG707) sample containing 3 mg of iron. Details of the experimental methods are given in Ref. [7].

power drive field oscillates leads to concerns regarding heating and peripheral nerve stimulation of patients. We believe that AM-detection offers a promising perspective for developing a low-frequency MPI-scanner that may circumvent those issues.

Acknowledgments

Work funded by grants 200021/160128 and 200021/149542 of the Swiss National Science Foundation.

References

[1] Huang S H and Juang R S 2011 *J. Nanopart. Res.* **13** 4411–4430
[2] Liebl M, Wiekhorst F, Eberbeck D, Radon P, Gutkelch D, Baumgarten D, Steinhoff U and Trahms L 2015 *Biomed. Engin.-Biomed. Tech.* **60** 427–443
[3] Gleich B and Weizenecker J 2005 *Nature* **435** 1214–1217 ISSN 0028-0836
[4] Dolgovskiy V, Lebedev V, Colombo S, Weis A, Michen B, Ackermann-Hirschi L and Petri-Fink A 2015 *J. Magn. Magn. Mater.* **379** 137–150 ISSN 0304-8853
[5] Fescenko I and Weis A 2014 *J. Phys. D: Appl. Phys.* **47** 235001
[6] Dolgovskiy V, Fescenko I, Sekiguchi N, Colombo S, Lebedev V, Zhang J and Weis A 2016 *Applied Physics Letters* **109** 023505
[7] Colombo S, Lebedev V, Grujic Z, Dolgovskiy V and Weis A 2016 *International Journal on Magnetic Particle Imaging* **2** 1606002
[8] Colombo S, Lebedev V, Grujic Z, Dolgovskiy V and Weis A 2016 *International Journal on Magnetic Particle Imaging* **2** 1604001
[9] Goodwill P W, Saritas E U, Croft L R, Kim T N, Krishnan K M, Schaffer D V and Conolly S M 2012 *Adv. Mater.* **24** 3870–3877

Atomic spectroscopy in periodic fields

V I Yudin[1-3], **M Yu Basalaev**[1,2] **and A V Taichenachev**[1,2]

[1] Novosibirsk State University, 2 Pirogova str., Novosibirsk 630090, Russia
[2] Institute of Laser Physics SB RAS, 13/3 Acad. Lavrentyev ave., Novosibirsk 630090, Russia
[3] Novosibirsk State Technical University, 20 K. Marx ave., Novosibirsk 630073, Russia

E-mail: mbasalaev@gmail.com

Abstract. Using the density matrix formalism, we prove the existence of the periodic steady-state for an arbitrary periodically driven system described by linear dynamic equations. The presented derivation simultaneously contains a simple and effective computational algorithm, which automatically guarantees a full account of all frequency components.

For the last several decades rapid scientific and technological progress has been substantially connected with an expansion of the lasers and laser technologies at different areas of the science, engineering, and industry. Many impressive successes in these directions are due to the theoretical support, motivation, and interpretation of experimental researches. In this context, of paramount importance is the formulation of mathematical models (equations) and finding of their solutions, which adequately describe the physical picture of investigated problems. During long time, steady states (which arises under the interaction of a quantum system with stationary external fields) play a key role in the theoretical description of the basic problems in laser physics and spectroscopy (for example, see [1–3]). However, in the last few years the devices in which different parameters of electromagnetic fields are periodically modulated have gained a greater importance. First of all, the so-called frequency comb generators use the periodic pulse modulation of a laser field. Such sources of pulse radiation are actively used now in modern atomic clocks for frequency measurements, and they have promising perspectives for direct frequency comb spectroscopy. Also, the phase (frequency) and/or amplitude periodic modulation of the laser field is now widely used for different tasks and applications (including atomic clocks and magnetometers). In all these examples the standard concept of steady state based on the time-independent equation is inapplicable.

We generalize the steady-state concept for an arbitrary quantum system under arbitrary periodic external influence. In this way we prove the following existence theorem: if the coefficients of density matrix dynamic equation

$$\frac{\partial}{\partial t}\hat{\rho}(t) = -\frac{i}{\hbar}[\hat{H},\hat{\rho}(t)] + \hat{\Gamma}\{\hat{\rho}(t)\}, \qquad \mathrm{Tr}\{\hat{\rho}(t)\} = 1 \tag{1}$$

have the period T , then the periodic solution with the same period T exists always. Acompletely unexpected result is that so universal and fundamental a statement is based only on the normalization condition for the density matrix. Due to the relaxation processes this solution is realized as an asymptotics ($t \to +\infty$) and, therefore, can be characterized as a periodic steady state. The developed simple algorithm allows us to directly construct this solution independently of initial conditions and without the use of either Floquet or Fourier formalisms. Our approach considerably simplifies the analysis regardless of the periodic modulation character: from smoothly harmonic type to ultrashort pulses.

Some authors supposed (without proof) the existence of the periodic steady-state solution for some certain problems. In this case they usually used the Fourier analysis for numerical calculations). However, an intuitive assumption about the periodic steady state is now rigorously substantiated [4]. At the same time, of special interest is the direct and simple method, which allows us to construct the periodic solution without Fourier expansion.

Let us describe one possible numerical algorithm allowing to construct the periodic solution of equation (1). First, rewrite the differential equation (1) for the density matrix in the vector form:

$$\frac{\partial}{\partial t}\vec{\rho}(t)=\hat{L}(t)\vec{\rho}(t), \quad \mathrm{Tr}\{\hat{\rho}(t)\}=1, \tag{2}$$

where the column vector $\vec{\rho}(t)$ is formed by the matrix elements $\rho_{ab}(t)$ using some definite rule, the linear operator $\hat{L}(t)$ corresponds to the right-hand member of equation (1). In accordance with equation (2), for other instant of time t_2 we can write

$$\vec{\rho}(t_2)=\hat{A}(t_2,t_1)\vec{\rho}(t_1) \tag{3}$$

where the two-time evolution operator $\hat{A}(t_2,t_1)$ is determined by the matrix $\hat{L}(t)$. Consider $\vec{\rho}(t)$ at arbitrary instant of time t. In conformity with equation (3), the vector $\vec{\rho}(t+T)$ is determined as

$$\vec{\rho}(t+T)=\hat{A}(t+T,t)\vec{\rho}(t), \tag{4}$$

where$T$ is time period in the operator $\hat{L}(t)$, i.e. $\hat{L}(t+T)=\hat{L}(t)$. Supposing the existence of the periodic solution $\vec{\rho}(t+T)=\vec{\rho}(t)$, it follows from equation (4) that this solution satisfies the equation

$$\vec{\rho}(t)=\hat{A}(t+T,t)\vec{\rho}(t), \quad \mathrm{Tr}\{\hat{\rho}(t)\}=1, \tag{5}$$

which always has a nonzero solution as it has been proven in our work [4]. We consider an arbitrary periodic dependence of the operator $\hat{L}(t)$. For instance, under an atom-field interaction such a dependence can be produced by the modulation of the field parameters (amplitude, phase, polarization, etc.).

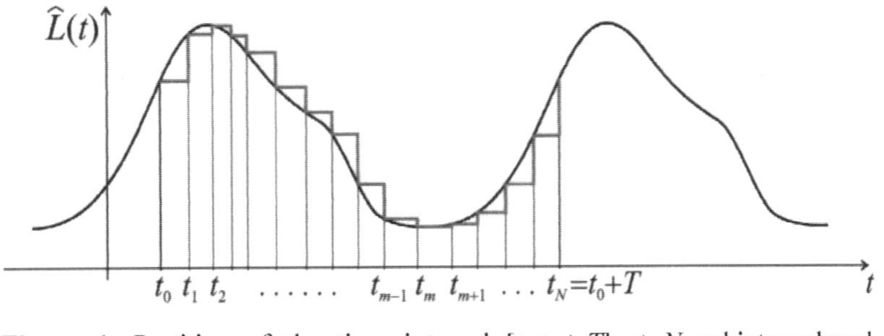

Figure 1. Partition of the time interval $[t_0,t_0+T]$ at N subintervalsand symbolic approximation of the dependence $\hat{L}(t)$ by the step function (red step line).

The selected time interval $[t_0, t_0+T]$ is divided into N small subintervals, where $t_N=t_0+T$. The character of partition (uniform or nonuniform discrete mesh) and number of subintervals are determined in conformity with the studied problem. The dependence $\hat{L}(t)$ we will approximate by step function where the matrix $\hat{L}(t)$ has the constant value $\hat{L}(t_{m-1})$ inside of subinterval $(t_{m-1}, t_m]$. In this case the vector $\vec{\rho}(t_0)$ in initial point $t_0$is determined by equation (5), where the evolution operator $\hat{A}(t_0+T,t_0)$ has the form of a chronologically ordered product of the matrix exponents:

$$\hat{A}(t_0+T,t_0) \approx \prod_{m=1}^{m=N} \mathrm{e}^{(t_m-t_{m-1})\hat{L}(t_{m-1})} = \mathrm{e}^{(t_N-t_{N-1})\hat{L}(t_{N-1})} \times \ldots \times \mathrm{e}^{(t_1-t_0)\hat{L}(t_0)} \quad (6)$$

The vectors $\vec{\rho}(t_m)$ in other points of the interval $[t_0, t_0 + T]$ are determined by the recurrence relation

$$\vec{\rho}(t_m) = \mathrm{e}^{(t_m-t_{m-1})\hat{L}(t_{m-1})}\, \vec{\rho}(t_{m-1}) \quad (7)$$

In summary, in the framework of density matrix formalism we have rigorously proven the existence theorem of the periodic steady state for an arbitrary periodically driven system. Due to the relaxation processes this state is realized as an asymptotic ($t \to +\infty$) independently of initial conditions, i.e., periodicity is the main attribute of steady state. The proof simultaneously contains a computational algorithm, which uses neither Floquet nor Fourier theories. Our method radically simplifies the calculations for arbitrary types of periodic modulation (including the ultrashort pulses) and opens up great possibilities for analysis and development of new methods in laser physics, nonlinear optics, and spectroscopy.As animportant case, the developed method can be applied for calculation of the line shape and field-induced shift of the coherent population trapping resonances, which are widely used in atomic clocks.

Acknowledgments
The work was supported by the Russian Science Foundation (project no. 16-12-10147). The calculations for direct frequency comb spectroscopy of two-level atom were supported by the Russian Foundation for Basic Research (project no. 16-32-60050 mol_a_dk).

References

[1] Demtroder W 2003 *Laser Spectroscopy* (Berlin: Springer-Verlag) p 986
[2] Letokhov V S and Chebotaev V P 1977 *Nonlinear Laser Spectroscopy* (Berlin: Springer-Verlag) p 466
[3] Rautian S G and Shalagin A M 1991 *Kinetic Problems of Nonlinear Spectroscopy* (Amsterdam: North-Holland) p 454
[4] Yudin V I, Taichenachev A V and Basalaev M Yu 2016 *Phys. Rev.* A **93** 013820

IOP Conf. Series: Journal of Physics: Conf. Series **793** (2017) 012034 doi:10.1088/1742-6596/793/1/012034

Synthetic frequency protocol in the Ramsey spectroscopy

V I Yudin[1-3], M Yu Basalaev[1,2] and A V Taichenachev[1,2]

[1] Novosibirsk State University, 2 Pirogova str., Novosibirsk 630090, Russia
[2] Institute of Laser Physics SB RAS, 13/3 Acad. Lavrentyev ave., Novosibirsk 630090, Russia
[3] Novosibirsk State Technical University, 20 K. Marx ave., Novosibirsk 630073, Russia

E-mail: viyudin@mail.ru

Abstract. We develop an universal method to significantly suppress probe-induced shifts in any types of atomic clocks using the Ramsey spectroscopy. The frequency shifts can be suppressed considerably below a fractional level of 10^{-18} practically for any optical atomic clocks.

At the present time, huge progress occurs for high-precision optical atomic clocks based on both neutral atoms in optical lattices and trapped ions. Exceptional accuracy and stability at the $10^{-17}\div10^{-18}$ level are achieved. Potential possibilities to achieve the level of 10^{-19} become clearer for nuclear clocks and for highly charged ions On the way to these remarkable achievements, different barriers arise, which require the development of new unconventional approaches. As an example, for some of the promising clock systems, one of the key problems is the frequency shift of the clock transition due to the excitation pulses themselves. For the case of magnetically induced spectroscopy these shifts (quadratic Zeeman and ac-Stark shifts) could ultimately limit the achievable performance. Moreover, for ultra-narrow transitions (e.g., electric octupole and two-photon transitions) the ac-Stark shift can be so large in some cases to rule out high accuracy clock performance at all.

Unconventional solution to this important problem was proposed in the paper [1], in which so-called hyper-Ramsey method has been developed. Soon this approach was successfully realized in [2], where the huge suppression (by four orders of magnitude) of probe-induced shifts was experimentally demonstrated. We develop new universal method to dramatically suppress probe-induced shifts and their fluctuations in any type of atomic clocks. Our approach is based on adaptation of so-called synthetic frequency concept [3] to the Ramsey spectroscopy with the use of interrogations for different durations of free evaluation intervals. We show that this protocol in combination with the original hyper-Ramsey spectroscopy [1] makes most extremal results and is quite stable with respect to the decoherentization.

The essence of our approach consists in the following. Previously in the paper [3] the so-called synthetic frequency method, allowing to radically suppress thermal (BBR) shift in atomic clock, was proposed. However, an ideology of this method can be easy extended on cancelling of an arbitrary systematic shift. Indeed, let us consider two clock frequencies $\omega_1^{(0)}$ and $\omega_2^{(0)}$ (different in the general case). Assume that due to a certain physical cause we have the stabilized frequencies ω_1 and ω_2, which are shifted relative to the unperturbed frequencies at the values Δ_1 and Δ_2 :

$$\omega_1=\omega_1^{(0)}+\Delta_1\,,\omega_2=\omega_2^{(0)}+\Delta_2\,. \tag{1}$$

Also assume that the ratio $\varepsilon_{12}=\Delta_1/\Delta_2=const$ does not fluctuate, while the shifts $\Delta_{1,2}$ can be varied during experiment (i.e., $\Delta_{1,2}\neq const$). In this case, we can construct the following superposition:

$$\omega_{syn}=\frac{\omega_1-\varepsilon_{12}\omega_2}{1-\varepsilon_{12}}=\frac{\omega_1^{(0)}-\varepsilon_{12}\omega_2^{(0)}}{1-\varepsilon_{12}}, \quad (2)$$

which is insensitive to the perturbations $\Delta_{1,2}$ and their fluctuations. This frequency we will call as "synthetic frequency". A key advantage of this concept consists in the following: to construct the shift-free frequency ω_{syn} we do not need to know the real values of shifts $\Delta_{1,2}$, because we need to know only their ratio ε_{12}, which can be exactly calculated (or measured) for many cases.

Consider an influence of probe-induced shift Δ_{sh}, which arises only during the Ramsey pulses, while this shift is absent during the dark time T. As a result, the stabilized frequency ω_T also becomes differing from unperturbed frequency ω_0:

$$\omega_T=\omega_0+\overline{\delta}_T, \quad (3)$$

where the index T denotes the fixed time of the free evolution interval under frequency stabilization, and the resulting shift $\overline{\delta}_T$ exists due to the $\Delta_{sh}\neq 0$. On the basis of general principles, it can be shown that the dependence $\overline{\delta}_T$ on the value T can be expressed as the following decreasing series in terms of powers of $1/T$:

$$\overline{\delta}_T=\frac{A_1}{T}+\frac{A_2}{T^2}+\ldots+\frac{A_n}{T^n}+\ldots, \quad (4)$$

where the coefficients A_n depend on the pulse parameters (durations, amplitudes, phases, and the value Δ_{sh}).

Because the time T is precisely controlled in experiments, then we can set a goal to eliminate the main contribution $\propto A_1/T$ in equation (4) using the synthetic frequency protocol. To solve this task we will apply two different dark intervals T_1 and T_2 (but with the same Ramsey pulses), which will give us the corresponding stabilized frequencies ω_{T_1} and ω_{T_2}. Using equations (2)–(4) we easy find the synthetic frequency $\omega_{syn}^{(1)}$ and its residual shift $\overline{\delta}_{syn}^{(1)}$:

$$\omega_{syn}^{(1)}=\frac{\omega_{T_1}-(T_2/T_1)\omega_{T_2}}{1-(T_2/T_1)},\ \overline{\delta}_{syn}^{(1)}=\omega_{syn}^{(1)}-\omega_0=\frac{\overline{\delta}_{T_1}-(T_2/T_1)\overline{\delta}_{T_2}}{1-(T_2/T_1)}, \quad (5)$$

where the expression for $\overline{\delta}_{syn}^{(1)}$ does not contain the term $\propto A_1$. In the particular case of T_1 = T and $T_2 = T/2$:

$$\omega_{syn}^{(1)}=2\omega_T-\omega_{T/2},\ \overline{\delta}_{syn}^{(1)}=2\overline{\delta}_T-\overline{\delta}_{T/2}. \quad (6)$$

The value $\overline{\delta}_{syn}^{(1)}$ can be less than $\overline{\delta}_T$ [see equation (4)] by several orders of magnitude. Moreover, we can go further to define other synthetic frequency $\omega_{syn}^{(2)}$, for which both contributions A_1/T and A_2/T^2 will be simultaneously canceled. Here we need to use three different time intervals (T_1, T_2, T_3) with the corresponding stabilized frequencies (ω_{T1}, ω_{T2}, ω_{T3}). In particular, we consider the case of $T_1 = T$, $T_2 = T/2$ and T_3 = T/3, for which the required superposition takes the form:

$$\omega_{syn}^{(2)}=3\omega_T-3\omega_{T/2}+\omega_{T/3},\ \overline{\delta}_{syn}^{(2)}=3\overline{\delta}_T-3\overline{\delta}_{T/2}+\overline{\delta}_{T/3} \quad (7)$$

where the value $\overline{\delta}_{syn}^{(2)}$ can be less than $\overline{\delta}_{syn}^{(1)}$ by several orders of magnitude.

In conclusion, the synthetic frequency protocol in the Ramsey spectroscopy is a novel technique that offers a spectroscopic signal that is virtually free from probe-induced frequency shifts and their fluctuations. Most extremal results are obtained in combination with so-called hyper-Ramsey spectroscopy [1]. In the latter case, the probe-induced frequency shifts can be suppressed considerably below a fractional level of 10^{-18} practically for any optical atomic clocks, where this shift previously was metrologically significant. We have appliedthe synthetic frequency protocol to the Ramsey spectroscopyof coherent population trapping (CPT) resonances. Note that CPT clocks are one of the prospective variants of compact rf clocks with relatively valuable metrologicalcharacteristics. Because the probe-induced shift forCPT-Ramsey resonance satisfies the general dependence (4) onfree evolution interval T, we can expect good efficiency of our approach. Indeed, our calculations show significantsuppression of the light shift for CPT-Ramsey clocks with theuse of the synthetic frequency protocol.

Acknowledgments

The work was supported by the Russian Science Foundation (Project No. 16-12-10147). The calculations for two-level atom were supported by the Russian Foundation for Basic Research (Project No. 16-32-60050 mol_a_dk).

References

[1] Yudin V I, Taichenachev A V, Oates C W *et al.* 2010 *Phys. Rev.* A **82** 011804(R)
[2] Huntemann N, Lipphardt B, Okhapkin M *et al.* 2012 *Phys. Rev. Lett.* **109** 213002
[3] Yudin V I, Taichenachev A V, Okhapkin M V *et al.* 2011 *Phys. Rev. Lett.* **107** 030801

IOP Conf. Series: Journal of Physics: Conf. Series **793** (2017) 012035 doi:10.1088/1742-6596/793/1/012035

An experimental toolbox for the generation of cold and ultracold polar molecules

Martin Zeppenfeld, Thomas Gantner, Rosa Glöckner, Martin Ibrügger, Manuel Koller, Alexander Prehn, Xing Wu, Sotir Chervenkov and Gerhard Rempe

MPI for Quantum Optics, Hans Kopfermann Str. 1, 85748 Garching, Germany

E-mail: martin.zeppenfeld@mpq.mpg.de

Abstract. Cold and ultracold molecules enable fascinating applications in quantum science. We present our toolbox of techniques to generate the required molecule ensembles, including buffergas cooling, centrifuge deceleration and optoelectrical Sisyphus cooling. We obtain excellent control over both the motional and internal molecular degrees of freedom, allowing us to aim at various applications.

Molecules cooled to very low temperatures offer an exciting platform to investigate quantum phenomena. Thus, precision measurements on molecules allow the investigation of fundamental symmetries of nature [1]. Investigation of chemical reactions at low temperatures provides insight into chemical processes in interstellar space [2]. Moreover, the long-range dipole-dipole interactions between polar molecules and the variety of internal molecular states would be ideal for quantum information processing [3], and for investigation of many body physics in quantum degenerate gases [4].

In this summary, we present our multistep approach for generating internal state controlled molecular ensembles at cold ($\lesssim 1\,\mathrm{K}$) and ultracold ($\lesssim 1\,\mathrm{mK}$) temperatures. First, an initial sample of cold molecules is generated via velocity filtering [5] or buffergas cooling [6]. Second, molecules are decelerated via a centrifuge decelerator [7]. Third, molecules are trapped in a microstructured electric trap [8]. Fourth, cooling to sub-millikelvin temperatures is achieved via optoelectrical Sisyphus cooling [9, 10, 11]. Control of the internal molecular state is achieved either by buffergas cooling [6, 12], or by optical pumping via a vibrational transition [13].

For all our experiments, we make use of the strong interaction between polar molecules and static electric fields. For laboratory fields of up to $100\,\mathrm{kV/cm}$, interaction energies on the order of $1\,\mathrm{K}\times k_B$ are possible, allowing for guiding and trapping of low-field-seeking states. As a first application, filtering of the low-velocity tail of the Maxwell-Boltzmann velocity distribution emerging from an effusive molecule source using a quadrupole electric guide provides a robust high-flux source of cold polar molecules [5]. Alternatively, pre-cooling via a Helium buffergas at $\sim 5\,\mathrm{K}$ in a buffergas cell is possible [6].

For velocity filtering or buffergas cooling, the slowest molecules are reduced or eliminated, respectively, due to collisions with faster molecules and/or Helium atoms at the source [14]. To provide a large flux of slow molecules, we have developed a centrifuge decelerator for molecules [7], allowing for deceleration of continuous beams of molecules. Molecules in a lab-fixed quadrupole guide are injected into a quadrupole guide at the periphery of a rotating disk. A

Figure 1. Centrifuge decelerator for polar molecules. Molecules enter the centrifuge in the lower left and exit along the axis in the center.

quadrupole guide on the rotating disk guides the molecule to the center of the disk where they are transferred back to a lab-fixed guide along the axis of rotation. Due to the centrifugal potential in the rotating frame, molecules are thereby decelerated. Molecules with initial velocities $> 150\,\mathrm{m/s}$ can be decelerated almost to a standstill. A clever design of the transition from the lab-fixed quadrupole guide to the rotating quadrupole guide allows for deceleration of continuous rather than pulsed molecular beams [7].

Molecules with a kinetic energy below roughly $1\,\mathrm{K}\times\mathrm{k_B}$ can be loaded into an electrostatic trap. In our experiments, we make use of a unique trap design, with molecules trapped between a pair of microstructured capacitor plates, with an additional perimeter electrode for transverse confinement [8, 9]. In addition to providing record trap lifetimes of up to a minute [11], this trap design provides homogeneous electric fields in a large fraction of the trap volume. This is ideal for spectroscopic applications and to selectively address individual molecular rotational states via microwave and infrared radiation.

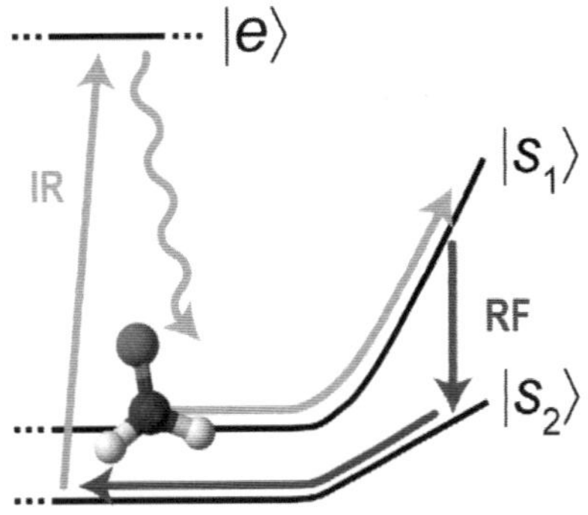

Figure 2. Operation principle of optoelectrical Sisyphus cooling, as explained in the text.

Most applications of cold molecules require temperatures substantially below the $\sim 1\,\mathrm{K}$ achievable with the previously mentioned techniques. For this purpose, we have developed optoelectrical Sisyphus cooling [9]. The operation principle of this generally applicable cooling scheme is shown in figure 2. Molecules move from weaker to stronger electric fields in a strongly trapped state $|s_1\rangle$ and are transferred to a more weakly trapped state $|s_2\rangle$ in high electric fields via an RF field. Moving back to weaker electric fields, the molecules regain less kinetic energy than they lost when moving to stronger electric fields. Optical pumping back to the strongly trapped state via excitation to an excited state $|e\rangle$ closes the cycle in a one-way process, thus providing the necessary entropy dissipation. We have demonstrated this cooling scheme in a proof of principle experiment by reducing the temperature of about a million methyl fluoride (CH_3F) molecules by more than an order of magnitude to 29 mK [10]. More recently, we have applied optoelectrical cooling to formaldehyde (H_2CO) producing an ensemble of 300, 000 molecules at 420 μK [11]. This represents the largest ensemble of ultracold molecules in any experiment worldwide.

A final requirement for many applications of cold molecules is gaining and maintaining control of the internal molecular state. Here, buffergas cooling plays a key role by providing internal state cooling already at the molecule source [6]. In this way, we obtain molecule beams with over 90 % of molecules in a single rotational state [12]. As an alternative, we have demonstrated internal state cooling in combination with optoelectrical Sisyphus cooling via optical pumping inside our electric trap [13]. In this way, we achieve ultracold molecules with over 80 % of molecules in a single rotational state [11].

The presented techniques enable a wide range of exciting measurements. The long lifetimes inside our electric trap, combined with internal state purity and sufficiently high densities enable state-resolved collision studies for molecules in a new temperature regime. Temperatures below 1 mK allow the realization of a molecular fountain, enabling greatly improved precision measurements on molecules. Finally, investigation of sympathetic or evaporative cooling offers a promising route to a quantum degenerate gas of polar molecules.

References

[1] Baron J *et al.* 2014 *Science* **343** 269
[2] Smith I W M and Rowe B R 2000 *Acc. Chem. Res.* **33** 261
[3] DeMille D 2002 *Phys. Rev. Lett.* **88** 067901
[4] Baranov M A *et al.* 2012 *Chem. Rev.* **112** 5012
[5] Rangwala S A *et al.* 2003 *Phys. Rev.* A **67** 043406
[6] Van Buuren L D *et al.* 2009 *Phys. Rev. Lett.* **102** 033001
[7] Chervenkov S *et al.* 2014 *Phys. Rev. Lett.* **112** 013001
[8] Englert B G U *et al.* 2011 *Phys. Rev. Lett.* **107** 263003
[9] Zeppenfeld M *et al.* 2009 *Phys. Rev.* A **80** 041401(R)
[10] Zeppenfeld M *et al.* 2012 *Nature* **491** 570
[11] Prehn A *et al.* 2016 *Phys. Rev. Lett.* **116** 063005
[12] Wu X *et al.* 2016 *ChemPhysChem* **17** 3631
[13] Glöckner R *et al.* 2015 *Phys. Rev. Lett.* **115** 233001
[14] Motsch M *et al.* 2009 *New J. Phys.* **11** 055030